AF248923

STRESS WAVE PROPAGATION IN SOLIDS

AN INTRODUCTION

MONOGRAPHS AND TEXTBOOKS IN MATERIAL SCIENCE

Edited by

R. F. Bunshah
UNIVERSITY OF CALIFORNIA
LOS ANGELES, CALIFORNIA

A. Sosin
UNIVERSITY OF UTAH
SALT LAKE CITY, UTAH

Additional volumes in preparation

Stress Wave Propagation in Solids

AN INTRODUCTION

Richard J. Wasley
Chemistry Department
University of California
Lawrence Livermore Laboratory
Livermore, California

MARCEL DEKKER, INC. New York   1973

MARCEL DEKKER, INC.

95 Madison Avenue, New York, New York 10016

LIBRARY OF CONGRESS CATALOG CARD NUMBER: 73-78561

ISBN: 0-8247-6039-5

PRINTED IN THE UNITED STATES OF AMERICA

The propagation of energy disturbances has fascinated scientists and engineers for many years. In fact, few subjects have attracted the interest of so many investigators for so long a time. However, because of the lack of adequate instrumentation, early work was principally theoretical in nature, and moreover, often speculative and philosophical. It occasionally gave rise to more than one explanation of the same phenomenon. The development of reliable equipment in the past two decades has made possible the conduct of definitive experimental work and enormous output of meaningful information in this general subject.

It is both presumptuous and impractical to attempt in one book the presentation and description of propagation of all kinds of disturbances in all types of media. Thus, the intention of this work is quite specific. It is to "set the stage" for the continued study of the propagation of relatively short-duration, high-intensity, principally nonelastic, mechanical stress disturbances in solids, with the application of certain relevant aspects of elasticity. More specifically, the first part of the book treats some of the dynamic analyses of elastic solid media which (sensibly) obey Hooke's law, *i.e.*, of solids that exhibit linear behavior with negligible internal friction. The last section treats, in large part, some of the theoretical and experimental aspects of two (interrelated) classes of nonelastic mechanical wave phenomena in solids that are of significant importance in contemporary technology, namely: (1) the study of high-strain-rate, elastic-plastic loading under conditions of macroscopic (continuum) one-dimensional stress; and (2) the study of shock-wave propagation under conditions of macroscopic one-dimensional strain.

The contents of the book are presented so that the discussions of the topics of nonelastic phenomena are associated with the discussions of the topics of elastic phenomena. In particular, elastic wave propagation in bounded media (Chapter I-4) is analogous and has application to uniaxial stress, high-strain-rate elastic-plastic loading (Chapter II-2); elastic wave propagation in extended media (Chapter I-2) is similarly associated with uniaxial strain shock-wave propagation (Chapter II-3); and reflection and refraction (Chapter I-3) have application to material interface considerations such as impedance mismatch effects, for loading conditions of both uniaxial stress and uniaxial strain. Moreover, a number of specific concepts exist among the topics considered which are similar, *e.g.*, the formation and treatment of elastic precursors in shock waves and weak shock/acoustic approximations. The scope, organization and detail of the subject matter are discussed in the Introduction.

It is believed by the author that a thorough understanding of fundamental ideas in dynamic Hookean elasticity, which in themselves can be reasonably well defined, forms a firm base for extending understanding and perspective to the particular topics in dynamic nonelastic phenomena with which we are concerned (as well as to other subjects within the field of disturbance propagation in general). Study of these topics without the benefit of such a grounding in dynamic elasticity is at the price of loss of much useful insight. This is true irrespective of whether one is studying analytical, numerical, or experimental non-Hookean problems. Thus, one of the main themes of the manuscript is the application of dynamic elastic concepts to certain dynamic nonelastic phenomena.

In the past couple of decades, work in the field of elastic and nonelastic stress waves has assumed a role of increasing importance in physical science. These subjects, and in particular the latter (which includes the treatment of shock phenomena), cannot be presented as a closed subject. Several reasons account for this upswing; prime in importance among them are: (1) the aforementioned rapid development of the sophisticated instrumentation required for appropriate experimental study of interesting and important new materials under novel and extreme conditions of loading; (2) the increasing use of computers to solve associated theoretical problems that had been analytically intractable or grossly tedious; and (3) the application of the results of study in these fields to practical engineering and scientific effort in general and to high-speed phenomena in physics, chemistry, and engineering in particular. Illustrative of (3) are studies of behavior of engineering materials under high transient loading for application in structural design, investigation to detect flaws in machine components, studies of certain fundamental mechanical properties and high-pressure equations of state of solids, and studies of seismic phenomena.

The book was written as an outgrowth of active work and lectures in these fields by the author at the University of California, Lawrence Livermore Laboratory. Although it is impossible not to reflect to some degree the writer's personal inclinations and specialized interests, considerable effort has been made to achieve a balanced analytical/experimental discussion, particularly in the treatment of nonelastic phenomena. The physical basis and general meaning of stress wave propagation are emphasized. Applications and explanatory examples are provided where appropriate. References are selected for their clarity and relative accessibility, but no attempt has been made to make the list of references exhaustive. For aid in additional study, the collateral reading also includes citations of works that are somewhat outside the scope of the subjects treated.

The book is intended to be of specific assistance to those scientists and engineers only partly familiar with the subject and desiring further knowledge and proficiency through self-study. Although principally directed to people in the nonacademic fields, *e.g.*, national laboratories and private research groups, it can also be used advantageously in upper division or early graduate level work. The word "Introduction" in the title is not intended to convey the meaning that the book is elementary. A grasp of the fundamentals of mathematics, mechanics and physical science is assumed, including some experience with the subject of elasticity. The reader is also presumed to have had a reasonable exposure to experimental work.

Acknowledgment is gratefully made to the Lawrence Livermore Laboratory for permitting the author to use some of its facilities in the preparation of this book. The author is presently affiliated with the Mechanical Engineering Department at the Laboratory, although the bulk of the work was completed while he was in the Chemistry Department. The author also extends his indebtedness to his colleagues, D. F. Abell and E. James, for their valuable criticism and encouragement; to his wife, Margery, for the typing of several drafts; to Vivian R. Mendenhall who edited the final draft and coordinated the preparation of the manuscript and final copy, and to Elizabeth Manchester and the Technical Information Department Composition Section who prepared the masters.

CONTENTS

INTRODUCTION

The intention in this chapter is to define more completely than in the Preface certain aspects of the scope and detail presented in the book since its content and arrangement are somewhat unique. Included are, first, a brief discussion of the general topic of disturbance propagation. Then, specific statements are given regarding the organization of the text together with some overall comments on restrictions that are placed on the material and applications to be studied. Finally, two associated topics are discussed, namely, the role of numerical techniques and high-speed computers and the notation used, the latter including some comments on elements of Cartesian tensor theory.

I. General: Propagation of Disturbances

To provide the reader perspective it is helpful first to preview, in a qualitative sense, some of the concepts and words associated with disturbance propagation which are encountered in the book. No attempt is made at this point to define these words or to include all the idealization and concepts that we will use.

Physical manifestations of waves are numerous. The transmission of the transverse shape of a wire, the distribution of density in a fluid, a perturbation of air in a pipe, a high-intensity shock in a solid, and the current in a conductor all represent examples of disturbance propagation. Probably the most easily visualized illustration of wave motion, as well as the oldest, is provided by the propagation of disturbances on the surface of water. In fact, through such observations, evidence had existed for many years that force could be communicated from one body to another without the transport of matter. The studies of these various motions are highly developed subjects in specific fields of physics and engineering. We select for investigation aspects and applications of certain kinds of these waves.

For our purposes in discussing elastic wave phenomena in Part I of the book, we interpret the phrase "propagating disturbance" to mean a process or an influence (as some form of energy) which moves through the medium with a finite velocity, but without causing bulk transfer of the medium as a whole. There can be localized motion of the medium, but only in an oscillatory sense. The motion can be linear or nonlinear, steady or unsteady. The last condition occurs, for example, in the case of transient disturbances or when boundaries of a region move and interact with the wave.

Although qualitative, the above interpretation is quite unrestrictive and can be applied to a large class of disturbances propagating in diverse kinds of media, *i.e.*, to both a wave in an elastic solid and to an electromagnetic field in a vacuum (wherein we take the word "medium" to mean "vacuum"). The word "finite" in the above statement does impose some restriction, but one that is not of importance to our considerations. For example, the theory of propagation in rigid solids and examination of certain phenomena associated with diffusion and quantum waves lead to propagation velocities that are infinite. In addition, distinction must be drawn between progressive waves and standing waves; in the main, concern is directed only to the former.

Specifically, we concern ourselves for the most part with wave propagation phenomena in deformable elastic solids which (sensibly) obey Hooke's law so that the response in linear and internal friction is negligible. Attention is focused on wave (primarily plane) propagation in isotropic, homogeneous, elastic solids, and analysis uses concepts of classical mechanics with the medium treated as a continuum.

Under such conditions, the shape, size, and velocity of the wave do not usually vary during propagation either through attenuation or material dispersion. Caution with respect to the generality of this statement, however, is given in that dispersion does arise in elastic waves due to geometry of the medium, and that certain nonplanar elastic disturbances do change both in size and shape during transmission. The former is of particular importance to us, and a chapter of the book is devoted to its investigation.

For our purposes in discussing nonelastic wave phenomena in Part II of the book, some of these above statements must be modified. First, the deformable medium through which the stress wave propagates does not obey Hooke's law. Also, the motion is not oscillatory in the sense of it being continuously periodic; in fact, the motion may not be oscillatory at all but rather can be a single pulse, *e.g.*, a shock disturbance. In most cases there is a bulk transfer of the medium. Superposition is no longer valid, at least for the nonelastic topics we treat, and the materials may behave both in a nonlinear and in a time-dependent manner. The medium is still considered to be a continuum, and many of the concepts of isotropy, homogeniety, and small strain theory carry over from elasticity. More specific statements are given in Chapter II-1 with regard to how transmission is affected by the dynamic physical response of the host material, *e.g.*, whether it behaves as a viscoelastic, plastic, or hydrodynamic material. The physical response of the host medium is governed to a large extent by the magnitude and rate of application of the initiating disturbance.

We will point out some of the interrelations that exist between elastic and nonelastic waves both by physical analogies and by mathematical similarities. A common mathematical structure exists for describing the phenomena. Indeed, Newton's second law is common and fundamental to the several mechanical wave motions treated, and all are related in the generic sense that they are represented by solutions of hyperbolic partial differential equations. However, the techniques of application of the second law are not identical in all the situations considered.

References that are noted in the text are listed at the end of each of the chapters. Other selected references are listed according to several categories at the end of the book.

II. Organization

In Part I, Chapter I-1 reviews the basic concepts of quasistatic and dynamic elasticity in a fashion sufficient to enable more detailed investigation and application of elastic and nonelastic wave phenomena. A few historical notes of some of the significant contributions to the development of elasticity is provided first. The concepts of stress and strain and their relationship are then discussed. Emphasis is placed on small deformation theory of isotropic elastic solid media. The last part of this chapter extends these quasistatic concepts to dynamic conditions. The equations of motion are developed, the wave equation relevant to elastic disturbance propagation is presented, and both vibration and wave solutions to this equation are studied. Again emphasis is placed on small deformation theory of isotropic elastic solid media. A conventional treatment of elasticity is not followed in this chapter; only those subjects pertinent to our study of mechanical wave phenomena are presented, and certain topics, such as compatability equations and uniqueness conditions, are intentionally omitted.

Then on the basis of geometrical considerations, each of three subsequent chapters is devoted to a separate division of elastic wave propagation, proceeding from the simpler to the more complex cases.

Chapter I-2 discusses the propagation of elastic waves in extended or unbounded media. There are comments, although brief, concerning the propagation of elastic waves in anisotropic media and the propagation of transient disturbances, *i.e.*, pulses, in elastic media.

Chapter I-3 treats propagation of elastic disturbances in the simplest bounded medium we can envision, namely a semi-infinite solid with a plane boundary. We investigate several kinds of plane elastic surface waves, with the emphasis placed upon Rayleigh waves moving in isotropic media. We also study reflections of elastic waves at the plane boundary under the conditions of both normal and oblique incidence. We briefly extend our considerations to the examination of reflection and refraction of such waves impinging upon uneven surfaces; those considerations lead to the phenomenon of scattering.

Chapter I-4 deals with transmission of elastic waves in bounded media. We concern ourselves primarily with transmission in circular cylinders of infinite length in which the Pochhammer-Chree treatment is appropriate. Because of the complexity of this subject, we also comment upon certain elementary and approximate methods of analysis. The geometry introduces dispersion wherein disturbances are transmitted at velocities which depend upon the mode and the harmonic of the wave, the frequency of excitation, the elastic constants, and the dimensions of the sample.

Chapter I-5 concerns itself with selected applications of the concepts of elasticity discussed above. We comment upon applications of quasistatic elasticity, but in keeping with the tenor of this book, greater emphasis is given to the dynamic loading of elastic materials. A section is devoted to some facets of nondestructive testing using resonance and pulse ultrasonic techniques. Other dynamic applications are also discussed, such as those involving earthquake seismology, geophysical prospecting, and lumped mass (vibration) analyses of structures. Finally, a section is included on the comparison of experimental observations with certain of the elastic theoretical models previously investigated, and in particular, the Pochhammer-Chree exact solution.

Part II of the book is primarily concerned with dynamic nonelastic phenomena and their interrelationship with elasticity, again proceeding from the (generally) simpler considerations to the more involved and complicated cases. The contents are considered from both an analytical and experimental point of view, although more emphasis is placed upon the latter. The last part of the manuscript is very relevant in today's applications.

Chapter II-1 gives some general and historical comments pertaining to the subject and its development. Various representations of nonelastic material behavior (equations of state) are then discussed. Four divisions of mechanical stress-wave propagation are defined for convenience; namely, elastic, viscoelastic, plastic, and shock waves. The chapter concludes with a discussion of the important distinction between one-dimensional stress and one-dimensional strain loading.

Chapter II-2 is directed to the study of mechanical wave propagation under the condition of one dimensional stress. We discuss specifically modifications of the Hopkinson pressure bar.

Chapter II-3 treats selected phases of the study of shock-wave propagation in condensed media, considered under conditions of one-dimensional strain, in conjunction with application of the Rankine-Hugoniot equations. The dynamic mechanical behavior of solids at the "lower" stress levels is emphasized where material strengths are important.

III. Numerical Techniques and Computers

Numerical methods used in conjunction with high-speed digital computers and applied to the solution of many mathematical problems in the fields of science and engineering are generally recognized for their valuable contributions to the understanding of these problems (see, for example, the books edited by Alder *et al* [1] and by Taub and Fernbach [2]). Although in this book emphasis is placed upon analytical and experimental considerations, it is important, nevertheless, to point out the several advantages and disadvantages of numerical formulation and computer solution. It is particularly relevant here since we examine complex, and often intractable, wave propagation phenomena.

Most scientific problems that can be formulated mathematically can be treated numerically. With the present availability of large digital computers, there now exists the realization that systems of differential equations never attempted before can be solved. Even the more simple one-dimensional, time-dependent systems being solved routinely on computers today would have required so much effort a decade or so ago that their solutions were not even attempted.

The central advantage of numerical/computer solutions results from the fact that parameters can very easily and quickly be varied in a problem to observe effects of such variation. Even if the problem can be correctly formulated only in part, several different methods of complete formulation can be assumed in order to determine which is the best or most sensible solution. The computer can draw upon its resources of speed and endurance for accomplishment of this effort. We call these techniques "computer experiments." Such an experiment may be more concisely defined as numerical analysis of a mathematical model of the process or phenomenon under investigation in which the computer plays the main role in evaluating the consequences of this model.

Several different types of computer experiments can be clearly distinguished. First, in a given problem the physics might be well known, but the mathematical equations are intractable. Then a numerical/computer approach can be used to evaluate the consequences of the equations. In a second class of problems the details of physics might be unknown, and computer experiments can be used to identify the most important physical factors and to test and select theoretical models. Often the results of such a computer experiment will suggest development of theoretical analysis, *e.g.*, what simplifying assumptions are possible, or provide a guide for definitive laboratory experiments. A third type of computer experiment consists of a parametric study of numerical solutions of a problem which leads to correlation formulas for results.

There can be no doubt that computers are a blessing. However, there are certain present limitations associated with the machines that must be realized.

The principal concern is that they can perform only those functions for which they are programmed. They can calculate a specific problem with specific geometry and definitely prescribed input conditions. Thus, in the present state of development, computers can only be considered creative in a very narrow sense of being able to manipulate the symbols and rules of an art or science and then only in that fashion and with those constraints devised and selected by the programmer. Work is underway in both random processes to enable a machine to improvise and in extrapolation of certain criteria of design by the computer itself in order to produce better engineered machines, but speculation of results in such areas of artificial intelligence for scientific application is premature.

Another limitation of significant importance develops if the scientist or engineer does not really understand the basic physics of the problem he is attempting to solve. There is a risk that too great a reliance will be placed upon results of numerical analysis that may satisfy the input conditions, but may nevertheless be physically senseless. Interesting details that sometimes aid in providing great insight into a problem can be masked by numerical/computer analysis without careful interpretation. As yet the use of computers can only *assist* scientists and engineers in their judgment to make intelligent decisions and to obtain answers.

Computer experiments cannot replace analytical solutions. In general, analytical solutions are more valuable than computer solutions, for the former contain extensive information about any given phenomenon in very condensed form and show exactly how various factors affect the phenomenon under study. Analytical solutions can also be applied in conjunction with a computer experiment to calibrate and/or normalize a particular numerical scheme.

Nor can computer experiments replace experimental effort. Since a computer experiment is only as good as the physical model on which it is based, it is apparent that the study of new physical phenomena will also require experimentation in a laboratory (although, as indicated, numerical/computer solutions can suggest a course of action in such experimentation and can result in a decrease in the number of experiments required). Optimum advancement of understanding in the field lies in the balanced use of numerical/computer technology with both theory and experiment.

Other disadvantages or limitations can be attached to the use of machine computation. High initial costs or large rental charges require that the larger and more versatile equipment be limited for the most part to the richer institutions and government-owned facilities. Considerable experience is necessary to operate the machines efficiently. Qualities of patience and cleverness are required to make effective and intelligent use of them. Numerical analysis is still essentially an art rather than a science, particularly in the instance of systems of partial differential equations. The stability and convergence of finite difference schemes can only be rigorously established in a few special cases.

IV. Notation

Because of the variety of symbols for parameters and material constants pertinent to mechanical wave transmission used in scientific and engineering literature, and because of the occasionally unfortunate situation in which identical symbols are assigned to parameters or constants that have totally different physical meanings, a list of the main notation used in this work is appropriate. To provide uniformity, this list is selected on the basis of those symbols most widely in use. A representative number of systems of notations describing the more common parameters in wave propagation (primarily elastic), together with the corresponding notation we use, are given in Table 1. Most of the difficulty from symbol duplication arises in elastic wave propagation. Comparatively little notational variation and misunderstanding exists in the treatment of nonelastic wave phenomena, although to conform with convention there is a partial shift from Table 1 in parts of Chapters I-4, II-2 and II-3. Additional or different notation that is introduced is defined where used.

In addition to simply specifying what letters or symbols are associated with particular parameters and constants, it is necessary in most problem formulation

Table 1.
Systems of Notations

Parameter	Goldsmith[a]	Kolsky	Love	Nye	Timoshenko and Goodier	This text
Cartesian coordinates						
Stress						
Normal	σ_x	σ_{xx}	X_x	σ_{11}	σ_x	S_{11}
Tangential (shear)	τ_{xy}	σ_{xy}	X_y	σ_{12}	τ_{xy}	S_{12}
Strain						
Normal	ϵ_x	ϵ_{xx}	e_{xx}	ϵ_{11}	ϵ_x	ϵ_{11}
Tangential (shear)	γ_{xy}	ϵ_{xy}	e_{xy}	ϵ_{12}	γ_{xy}	ϵ_{12}
Dilatation	Δ	Δ	Δ	Δ	e	Δ
Rotation	$\overline{\omega}_x$	$\overline{\omega}_x$	$\overline{\omega}_x$	$\overline{\omega}_{12}$	ω_x	ω_{12}
Elastic constants						
Young's modulus	E	E	E	E	E	E
Poisson's ratio	μ	ν	σ	ν	ν	σ
Bulk modulus	K	k	k	K	k	k
Rigidity modulus	G	μ	μ	μ or G	G	μ
Lamés (first) constant	λ	λ	λ	λ	λ	λ
Angular frequency	p^*	p	p	-	p	p
Wave length	Λ	Λ	-	-	ℓ	Λ

[a] see References.

and solution to have some formal method of writing, handling, and manipulating the relevant equations. One such method is with the use of *tensor* or *indicial* notation and operation. Tensor formalism plays a very important role in the description of physical properties in present-day continuum physics and engineering, and one encounters such formalism in much of the pertinent literature. To more easily illustrate tensor properties of certain quantities, for example, stress, strain, and rotation tensors in Chapter I-1, it is helpful to discuss some features of this formalism here. It is assumed the reader has some familiarity with the subject. Detailed accounts, each somewhat different in point of view and presentation, can be found in Chou and Pagano [3], Fung [4], Michal [5], and Wills [6].

A tensor is a mathematical representation of an actual physical quantity or phenomenon. Tensor (or indicial) notation is a shorthand and bookkeeping method to enable the condensed writing of algebraic, differential, and integral equations, and to make the operations desired easier to perform.

Tensor quantities form a kind of hierarchy as we proceed from the simplest to the more complicated. Each member of the hierarchy is characterized by a set of numbers, and the main distinguishing feature of the members is the number of numbers in their respective sets. The numbers in a set are called *components* and the number of components N_c for a member of *rank* (or *order*) r is

$$N_c = n^r \tag{1}$$

in which n is the dimension of the space in which these quantities are defined. The letter n can represent any finite positive number, although $n \leqslant 3$ are the ones of most need. What is meant by "rank" or "order" will become clear as we proceed.

The simplest member of the hierarchy is the *scalar*. In expression (1), $r = 0$. It is defined by the requirement that its value is independent of the coordinate system. That is, a scalar is a quantity with magnitude only, having only one number associated with it. Quantities having this property are said to be *invariant* under change of reference system. Physical realizations of scalars are density, temperature, hydrostatic pressure, and energy.

The next member of the hierarchy is the *vector*. In expression (1), $r = 1$. One may define a vector in several ways. Geometrically it is given as a quantity characterized by magnitude, direction, and sense, and which adds according to the parallelogram law. In actual computation with vectors, it is more convenient to use the algebraic definition. A vector is an *ordered triple* of numbers which are its components referred to a coordinate system. This definition refers to a vector in $n = 3$. It is important to note that it is not enough to say a vector is a quantity which has three components, for this statement does not enable one to conclude which is the x_1 component, which is the x_2 component, *etc.* Once a convention is established for the order of mention, then an ordered triple of numbers specifies the vector completely. A definition of a vector will also be given later in which its components are shown to obey certain transformation laws.

Examples of physical realizations of vectors are numerous, *e.g.*, displacement, velocity, acceleration, momentum, and force; however, they have meaning only in $n \leqslant 3$, and for $n > 3$ there are no such realizations.

There are also quantities which are neither vectors nor scalars, but are tensor quantities of higher rank. And next in our hierarchy is the tensor of rank two. It is more difficult to discuss precisely what is meant by such a tensor without proceeding somewhat formally; this is done in the brief presentation of the pertinent transformation laws below. We speak of a second-rank *tensor field* if the tensor depends on position and the field consists of the points of the domain together with the values of n^2 functions of these points.

We can continue, arriving at tensors of rank three, of rank four, *etc.*, and at their corresponding tensor fields. In each instance the tensor consists of n^r components (numbers), and the tensor field involves n^r functions. The geometrical visualization of tensors becomes impossible for $n > 3$, *e.g.*, in situations involving relativistic concepts, and one has to proceed formally.

Having described in general what tensors are, we may now specify a notational technique for their description, namely, with tensor or indicial notation. Frequently, mathematical expressions are given so that they differ from one another only in the subscripts that refer to the directions of the coordinate axes. Thus, knowing one such equation, it is possible to write the others by suitably changing the various subscripts, a procedure which corresponds to an orthogonal rotation of axes.

Indicial notation usually uses numerical subscripts to write equations in their component (scalar) form and alphabetic notation to condense them. For example, we will have need of an equation of the form

$$T_i = S_{ij} \cos \alpha_j, \quad i, j = 1,2,3 \tag{2}$$

where the terms T_i, S_{ij}, and α_j are quantities related to forces arising within a stressed elastic body, but are not further defined at this point. In Eq. (2), the repeated index in the same term j is called the *dummy* index, whereas the nonrepeated subscript i is called the *free* index. A limitation exists in that the same letter cannot appear more than two times in a term written in indicial notation since such an operation is not defined. In the expressions, the *summation* (or *Einstein*) convention is used, in which the dummy index requires summation over the values indicated for that index, *i.e.*, j equals 1,2, and 3. A type of indicial notation often seen in texts on strength of material utilizes the alphabetical subscripts x, y, and z for description of equations (we will have some need of this form in Chapter I-4).

The formal manner of considering tensor quantities involves examination of the representation of the quantities upon choosing a coordinate system and the change of this representation consequent to selection of a different coordinate system. This is a very important statement. One must know how the two coordinate systems are related; that is, one must know the equations of transformation from one coordinate system to another, in order to arrive at a satisfactory definition of tensors of rank two or higher.

Let us consider rectilinear Cartesian coordinates. A rectilinear Cartesian coordinate system is one in which the axes are linear and orthogonal and in which the units of measurement along the axes are the same (orthonormal). Cartesian tensor notation is not usually adequate to treat certain complicated equations and problems in curvilinear coordinate systems (a fortunate exception is discussed in Chapter I-4), and recourse must be made to general tensor notation in which distinction must be made between two kinds of tensor behavior under transformation. For our use, Cartesian tensor notation is sufficient.

In Cartesian coordinates, the simplest transformation equations may be written (in three-dimensional space) in tensor notation using the summation convention

$$x_i' = a_{ij}x_j + b_i. \tag{3}$$

Following usual convention, the statements giving the ranges for the indices have been dropped. It is understood that the indices are to be taken over the range from one to three unless otherwise specified.

The terms x_i' are frequently referred to as the "new" and the x_i as the "old" coordinates, but any preferential status which the terms "old" and "new" imply is limited to the arbitrary method of first writing a given physical problem in terms of the x_i. In the three equations of (3), the nine quantities a_{ij} and the three quantities b_i are not functions of coordinates; however, it is wise to bear in mind that they may be functions of time. Geometrically, a_{ij} is the cosine of the angle between the jth old coordinate axis and the ith new coordinate axis, $i.e.$, it is the *direction cosine*. The b_i give the location of the origin of the x system relative to the x' system. We can also assume a common origin if we desire so that $b_i = 0$. The components of a tensor are unchanged by a translation of the origin. Notice that the a_{ij} and the b_i "straddle" both coordinate systems and cannot be defined if only one system is available.

It can be seen by expansion that the equations interrelating the direction cosines can be written

$$a_{ij}a_{ik} = a_{ji}a_{ki} = \delta_{jk} \tag{4}$$

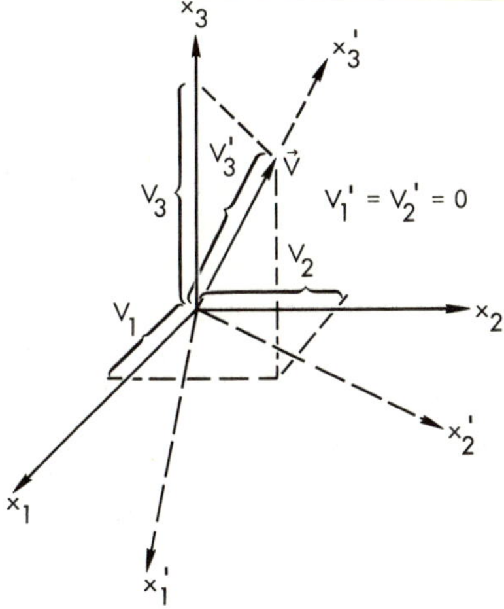

Fig. 1. Components of a vector in 3-space.

where δ_{jk} is the *Kronecker delta* or *unit tensor* and is defined as

$$\delta_{jk} = \begin{cases} 0 \text{ if } j \neq k \\[2mm] 1 \text{ if } j = k. \end{cases} \tag{5}$$

We now raise the question of the change of specification of a vector with change of coordinate system. That the question is not trivial can be seen in Fig. 1 which illustrates how a vector $\mathbf{V}$ (= $\vec{V}$) in 3-space which has all three components different from zero when resolved relative to the x system can have only one nonzero component when resolved relative to the x' system. Clearly, the numbers in the ordered triple which represents an actual physical quantity depend on how the coordinate system is chosen.

Transformation theory enables us to relate representations, and in Fig. 1 one can read that

$$V'_3 = a_{3j}\, V_j. \tag{6}$$

A similar sketch and some visualization of three-dimensional geometry will enable one to write

$$V'_i = a_{ij}\, V_j \tag{7}$$

for the transformation law in the more general situation where no coordinate axis of the new system lies along **V**. Note particularly that the vector **V** itself (or tensor quantity of any rank, for that matter) is invariant, for it corresponds to a physical reality independent of how one describes that reality.

Equations (7) may be taken as the mathematical definition of a vector, or tensor of the first rank. That is, a set of three quantities V_i referred to a coordinate system x_i and transformed to another system x_i' by expressions (7) is defined as a vector in three dimensions. Thus we see three definitions of a vector quantity are available to us, namely, the geometric definition, the algebraic definition, and the formal definition based upon the above transformation law.

We can proceed to the next member of our hierarchy, the tensor of rank two, and examine its transformation behavior. One would guess from the mathematical definition of a vector in 3-space, equations (7), that for a tensor of rank two $\widetilde{A}$, again written in 3-space, the transformation law should read

$$A_{ij}' = a_{ik}a_{j\ell}A_{k\ell}. \tag{8}$$

If desired, these relationships can be derived even more formally by examining the transformation properties of the products of the components of two vectors.

Thus, any group of nine scalar quantities referred to one three-dimensional coordinate system which transforms to a group of nine scalar quantities referred to another three-dimensional coordinate system by the rule (8) is called a tensor of the second rank. Note that two direction cosine factors appear in the transformation law for a tensor of rank two.[1] Roughly speaking, one factor is needed to take account of the changed orientation of the surface across which a component of the physical quantity acts, and the other appears because of the resolution of the component into new directions. The rule (8) is quite general and covers all possible transformations of second rank tensors in all dimensions merely by changing the range of the summation on the dummy subscript j and the range of the free index i, $i.e.$, we let i, j both run over the range $1, 2, \ldots, n$.

We may proceed to construct tensors of higher rank, each containing n^r components and each defined by a transformation law similar to equations (8). However, it must be emphasized that not every set of quantities can be called a tensor. The set of quantities must be shown to possess all the properties of a tensor. The existence of two sets of quantities in two different coordinate systems must first be established. If these sets are related by a transformation rule such as (8), the quantities are then components of a tensor of appropriate rank.

An important inference that can be drawn from the above discussion is that, in principle, one can completely solve any problem for any orientation of coordinate axes simply by solving the problem for one orientation of these axes. One tries, of course, to choose the orientation that will make the solution the easiest. The transformation equations then enable solution for any other desired coordinate system.

In addition to the use of a tensor of indicial notation, there are two other shorthand methods available to state mathematically the same quantity or parameter (and to facilitate operation thereon), namely, with *matrix* notation and *symbolic* notation. The matrix method, with the help of matrix algebra and matrix calculus, is generally more useful in problems involving a finite number of degrees of freedom, whereas the symbolic and tensor approaches are more suited to mechanics of continuous media. For example, the matrix operations are admirably appropriate for solving vibration problems using lumped-parameter analyses.

Nevertheless, a convenient and often-used method of writing the components of a tensor is with the use of matrix formulation. For example, we can write for a second rank tensor

$$\widetilde{A} = A_{ij} = \begin{pmatrix} A_{11} & A_{12} & A_{13} \\ A_{21} & A_{22} & A_{23} \\ A_{31} & A_{32} & A_{33} \end{pmatrix}. \tag{9}$$

Terms A_{11}, A_{22}, and A_{33} are often referred to as "diagonal" components. Certain properties of tensors, *e.g.*, invariant properties, and tensor components are more easily visualized in matrix notation. The fact that a set of elements appears in matrix form does not require that any particular operation need be performed on the elements or that these elements have any physical meaning. Nor does it imply that a matrix is a tensor. A tensor is a representation of an actual physical quantity or phenomenon, whereas a matrix is a mathematical tool often used in tensor manipulation.

In the symbolic representation, scalar, vector, and tensor quantities are given by different symbols; furthermore, each mathematical operation is assigned a symbol. For example, a scalar quantity might be denoted by ϕ, and a vector quantity by **V**. An example of vector addition is illustrated by expression (I-1.1) in Chapter I-1. This equation is simply a condensed version of the equations written in component form. Scalar and vector notation are not sufficient to treat second rank or higher rank tensor quantities, and other symbols must be introduced. A *dyadic* is used to indicate a second rank tensor; such a symbol might be written as $\mathscr{D}$. To consider third-rank tensors, *triadics* are needed; and for fourth rank tensors, *tetradics*.

A great advantage of the symbolic method over the indicial notational approach is that the symbols and mathematical operations defined by the symbols are independent of the coordinate system and thus invarient in any reference system. Therefore, if certain equations are expressed in this notational system, they are valid for any coordinate system, *e.g.*, cylindrical coordinates and polar coordinates. It is only necessary to express the required operations, such as divergence, curl, and gradient, in the particular coordinate system selected.

However, a serious disadvantage associated with this approach is that symbols for the various quantities and operations must be defined explicitly since the rules of scalar, vector, and tensor algebra and calculus, are in general, not interchangeable. Hence, as the rank of the tensor increases, more symbols are required for the numerous operations involved, and as a result, the utility of the technique diminishes. Specifically, in treating problems in elasticity, we will frequently use in our calculations a fourth rank tensor. Unlike symbolic notation, no radically new rules are required with the indicial method, and we can readily write and manipulate most of our expressions using this technique.

In treating mechanics of particles and rigid bodies, symbolic (vector) notation is sufficient and is usually used. Symbolic notation can also be used in the study of elementary mechanics of fluids and linear elasticity (with suitable definitions to account for fourth rank tensors), but in problems involving finite elastic deformations and plastic flow, the symbolic method becomes unmanageable and tensor notation must be used.

Notes

[1] It can be shown that the rank r is the number of factors a_{ij} in the transformation law.

References

[1] B. Alder, S. Fernbach, and M. Rotenberg (eds.), *Methods in Computational Physics*, Vol. 3, Academic, New York, 1964.

[2] A. Taub and S. Fernbach (eds.), *Computers and Their Role in the Physical Sciences*, Gordon and Breach, New York, 1970.

[3] P. C. Chou and N. J. Pagano, *Elasticity: Tensor, Dyadic, and Engineering Approaches*, Chaps. 8-12, D. van Nostrand, Princeton, N. J., 1967.

[4] Y. C. Fung, *A First Course in Continuum Mechanics*, Chap. 2, Prentice-Hall, Princeton, N. J., 1969.

[5] A. D. Michal, *Matrix and Tensor Calculus*, 7th printing, Wiley, New York, 1963.

[6] A. P. Wills, *Vector Analysis with an Introduction to Tensor Analysis*, Dover, New York, 1958.

Part I

CONCEPTS OF ELASTICITY

ELASTICITY: QUASISTATIC AND DYNAMIC RESPONSE

I. General: Historical Notes

Elasticity is defined here to mean the study and determination of the response of *continuous, perfectly elastic* solids subjected to applications of forces. A solid is considered a continuum when it can be divided into any number of parts, the smallest of which possesses the same continuous properties as the original solid. Clearly, there are conditions when this appellation becomes incorrect. It is apparent, for example, that if the solid is in fact essentially atomically subdivided, kinematical and dynamical variables cannot be assigned precisely as continuous functions over a finite volume of the solid. Based upon the considerable progress made in recent years with theories regarding the material structure, and with knowledge of the dependence of properties upon the mutual arrangement and interrelation of constitutive atoms, it appears likely a microscopic approach will become increasingly significant in physics and mechanics (for example, see the text by Cottrell [1]). However, the macroscopic continuum (or phenomenological) concept has been shown to be an excellent approximation for description of wave propagation in solids, as the wavelengths that can presently be generated experimentally are much greater than interatomic distances; moreover, it is both illustrative and mathematically convenient. Hence, it is the model adopted in this work.

Implicit in our treatment is that the solid is deformable. A deformable medium is one in which it is possible to change the separation between any two identifiable points by the application of appropriate forces. If the medium is perfectly elastic, the removal of forces results in the immediate and complete restoration of the original separation. In Chapters I-1 through I-5, conditions are further restricted, for the most part, to deformations below a certain limit so that linear relationship exists between these forces (more specifically, stresses) and deformations (more specifically, strains). Loss from the conversion of elastic mechanical energy into heat is not permissible within the scope of this definition. Obviously, no material completely satisfies these assumptions, but for many solids and conditions of loading they are a sufficient approximation.

Also implicit in our treatment is that the subject of elasticity is developed within the framework of classical mechanics. The classical approach means simply the mathematical consequences resulting from the Newtonian equations of state. Both the Newtonian equations and the concept of an equation of state are accorded significant discussion in the text.

Nothing in the definition of elasticity, *per se*, limits consideration to either a "slow" or a "fast" rate of loading. The division into categories of mechanical response as implied in the discussion above and in the chapter title is, in part, artificial rather than physical. Nevertheless, one can *define* various loading regimes, and they are often convenient and helpful for qualitative consideration. Figure I.1 presents one such convenient division in terms of strain rate. The

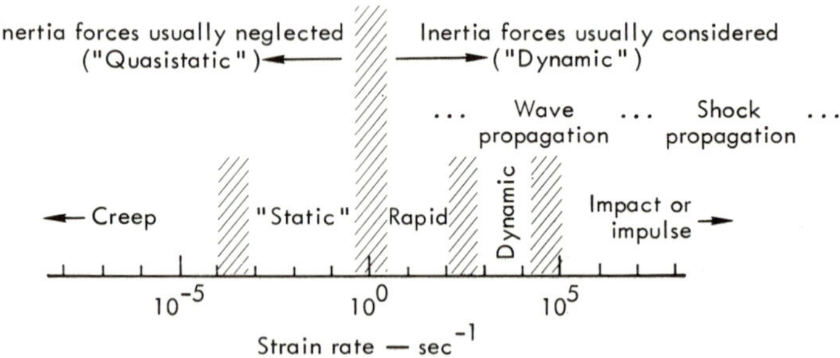

Fig. I.1. Regimes of mechanical loading response.

chart does not delineate discontinuities in mechanical material behavior. Although exceptions do occur, *e.g.*, in the formation of (idealized) shocks, change in behavior is usually at gradually changing strain rate. We consider the response to be quasistatic when the external forces are applied so slowly that equilibrium of the stress field throughout the entire body has been attained. We consider the response to be dynamic when inertial forces must be included in the treatment.

A few comments on the history of the subject are appropriate in order to place into proper context some of the more significant contributions in the development of elasticity relevant to our considerations. There are several texts available to which reference can be made for more detailed review, *e.g.*, Love [2], Rayleigh [3], and Todhunter and Pearson [4].

With some evidence to the contrary (see Truesdale [5]), the generally accepted notion is that Galileo, writing in 1638, indicated the thought processes and techniques necessary for rational scientific investigation and for the evolution of what, today, is called *classical mechanics*. He studied a variety of problems relating to the investigation of physical and mechanical phenomena, including the nature of free fall and uniformly accelerated motion of bodies, the simple pendulum, rigid body impact, rupture, and the general phenomenon of sound. Near the end of his life in 1642, he was in no doubt that the latter involved a wave process in a medium.

Before Galileo, the Pythagorean and Aristotlean viewpoints of rigid bodies dominated the teaching of mechanics. Although not physically reasonable in all aspects, *e.g.*, wave propagation velocities are infinite, the consequences that follow from rigid body dynamics still have application today. Although the theory is incapable of describing true wave phenomena and is limited to determination of the initial and final velocities of the bodies and the applied impulse, the approach is satisfactory where the duration of contact is large in comparison with the periods corresponding to the lowest natural frequencies of the bodies. For certain engineering applications, such as vehicle collision problem analysis, this extremely idealized technique can be very useful.

Robert Hooke, a contemporary of Galileo, also experimented with certain aspects of sound, such as pitch and frequency of vibration. More important to us however was his work in quasistatic elasticity. He published in 1678 the relation that extension is proportional to the force applied. This relation gave the necessary experimental foundation and formed the basis for the entire mathematical theory of elasticity, including elastic vibrations and the propagation of elastic waves.

Isaac Newton in 1686 established a firm mathematical basis for the (classical) dynamic mechanical behavior of materials, including wave propagation. As two of his contributions, for example, he obtained theoretically the velocities of propagation of surface waves on shallow water and of sound waves in the atmosphere. The latter was determined by treating the air as an ideal gas and assuming an isothermal process.[1]

In 1807, Thomas Young gave a definite physical meaning to elastic deformation by introducing a coefficient that represented an elastic material property independent of the size of the specimen (although apparently Euler had previously introduced a similar concept in some of his mathematical investigations).

In the late part of the 18th and in the early part of the 19th centuries, work in elasticity consisted of a composite of various solutions to special problems. Combination of these separate solutions into a unified mathematical theory and the development of analytical tools for the calculation of stresses and deformations in strained elastic bodies began with Navier's contribution in 1821. He determined, by means of a molecular approach and by variational methods, the general differential equations of equilibrium and motion (vibration) of elastic bodies. The forms of these equations were correct, but the derivations were based upon his oversimplified molecular concept. Thus, for small elastic deformations in isotropic media, he found only one constant necessary to specify such deformation. Cauchy, writing in 1822, developed the present formulation of the equations in terms of the displacement quantities, utilizing a generalization of Hooke's law to relate stress and strain components. He found, correctly, that two constants were necessary to define small deformation in isotropic elastic media. Later, in 1828, Cauchy generalized his equations to anisotropic media. However, he incorrectly determined that a total of only 15 coefficients were necessary to describe completely elastic deformation, the error resulting because of his improper assumptions regarding molecular interactions. George Green in 1827 repeated the development of the derivation for anisotropic media and correctly found 21 independent coefficients necessary. He based his work upon elastic energy considerations. In so doing, he also contributed significantly to a firmer understanding of conservation of energy and of minimum principles.

During much of this particular period of history, confusion existed as to how vibrations of different discrete frequencies could be formed (and in the case of waves, transmitted) independently in the same medium at the same time. One of the most significant and pertinent contributions that was eventually to be applied to the whole theory of small vibration (and wave) processes was the *principle of superposition* (or the *coexistence of small oscillations*) advanced by Young in 1801. The principle may be stated simply that the general motion of any vibration undergoing small movement can be given by the superposition of independent modes of vibration. D. Bernoulli had considered a somewhat limited version of this principle about 50 years before with regard to the vibration of strings under tension, but unfortunately was unable to convince his contemporaries of the validity or generality of his ideas.

Young's superposition principle was, in effect, generalized by Fourier in the early 19th century. Fourier did not direct his endeavors to the solution of problems in wave and vibration theory; rather, his studies were concerned with treating the conduction of heat flow in solids. However, the mathematical results were seen later to be of considerably wider relevance in that they provided a powerful, basic approach in the analysis of physical phenomena in which periodic behavior exists. Presented simply, Fourier's theorem enables a given function that is continuous and differentiable to be expressed as an infinite series of harmonic motion terms.

Lord Rayleigh wrote and experimented extensively in the field of acoustics in the 19th century (his 1877 publication, *Theory of Sound* is sometimes called the *Principia of Acoustics*). Among his vast range of specific interests within this and associated fields was elastic waves propagated along the surface of a solid. These surface waves, known as Rayleigh waves, are treated in Chapter I-3; comments on their relationship to seismic phenomena are given in Chapter I-5. Along with H. Lamb, he studied the propagation of wave in infinite elastic plates. He also introduced an approximate theory for treating the propagation of longitudinal waves in bounded media.

Energy methods, developed initially from investigations of Euler, Cauchy, and particularly Green, were becoming more useful in the 19th century to determine practically important stress distributions in elastically loaded structures in equilibrium. In addition, energy methods for treating problems in vibrations became prominent about the same time, a development in which Rayleigh was instrumental. One particularly valuable technique that he developed (later generalized by W. Ritz and known today as the Rayleigh-Ritz method) made use of the expressions for the maximum potential and kinetic energies for a complicated system of n degrees of freedom, in which the direct solution of the differential equation was impractical, to obtain the approximate determination of the lowest frequency of vibration.

Two of the workers whose contributions are of particular pertinence to our study were L. Pochhammer and C. Chree. Working independently in the late 19th century, they developed an exact analysis of harmonic waves propagating in elastic cylindrical bars by applying the mathematical theory of elasticity as postulated by Navier and Cauchy earlier in the century. One extremely important conclusion resulted from the boundary conditions imposed upon the problem: The velocity of propagation of longitudinal, flexural, and torsional (except the fundamental) harmonic waves are a function of wavelength. This conclusion is different from the simple longitudinal wave analysis developed by earlier workers. Discussion of waves in cylindrically bounded media is given in Chapter I-4.

As the 20th century was reached, most of the general theory of elasticity — both quasistatic and dynamic — had been formulated, and the treatments were able to assume their modern aspects. The development of the subject had been guided primarily by a relatively small group of theoreticians and an even smaller group of experimentalists. The historical order of development to this time appears to have followed a natural course in its evolution: first with special, usually empirical, solutions of engineering problems, then unification of these special solutions into a more fundamental (and mathematical) framework, and then particular application again, but guided strongly by these basic mathematical concepts.

The study of the dynamics of elastic materials had developed concurrently with the study of statics, but with less emphasis. Moreover, mathematicians had advanced significantly in the field of vibration theory, but somewhat less in the area of wave propagation analyses (although this is not to imply that no interrelation exists between the two topics). Because of the existence of rather limited test apparatus, much of the earlier dynamic experimental work was also more concerned with vibrations, and wave propagation studies were generally confined to large-scale tests. At the beginning of this century, no significant direct practical application of wave and vibration theory was actually required, *e.g.*, in building construction, and most dynamic work was only of academic concern. This was to experience a remarkable change after 1900, and particularly after 1940.

Recently, considerable advances have been made into such specialized analytical topics in elasticity as shell theory, complex variable theory for conformal mapping of two-dimensional problems, propagation of elastic waves in specific types of bounded media (for example, plates), transient or pulse disturbance transmission, and nonlinear elastic theory. Work in this last field of finite elastic strain representation was considered in the middle of the 19th century, but no serious effort was expended until several decades ago. Recent references in this area are books by Green and Zerna [6] and Gol'denblat [7].

In concluding this survey, it is important to leave the reader with the realization that Hookean elasticity is still of importance to physics and to present engineering design. These concepts have become integrated into all phases of mechanics that consider the behavior of deformable media. From this increasing interrelation, greater understanding is being achieved in such seemingly diverse subjects as electromagnetic theory, optics, hydrodynamics, acoustics, molecular theory, and, of principal importance to us, phenomena that are not elastic. Some of these points are examined further in Chapter I-5 and in Part II of the book.

II. Components of Stress and Strain

This phase of the subject is treated to some extent in nearly every book pertaining to elasticity. There is good reason for such repetition — understanding of the concepts of stress and strain is necessary before any further quantitative discussion can reasonably be pursued. To the extent that this repetition is considered relevant for our particular study, a review is given.

A. Stress

When a body is subjected to a load or system of loads, internal forces develop within the body. Two general types of internal forces must be considered when analyzing the response of the body to the loading — *body* forces and *surface* forces. The former are proportional to the mass of the body, are not a result of direct contact with other bodies, and are assumed uniformly distributed (in this work) throughout the volume of the body. Units of body forces are generally given in terms of force per unit volume. Examples of such forces are gravitational and magnetic forces. If the body is in accelerating or decelerating motion, inertia forces must also be so considered. Aside from inertia terms, for our purposes body forces become small in comparison with surface forces and may be reasonably neglected.

Surface forces are distributed over the surface of the body or across some arbitrary surface within the interior of the body. Examples of such forces are hydrostatic forces or the force resulting from contact of one body upon another.

Consider a tetrahedron of infinitesimal volume embedded in the interior of an elastic body at point P, and choose an x system of rectilinear Cartesian coordinates so that three faces of the tetrahedron lie in coordinate planes. The geometry is shown in Fig. I.2.

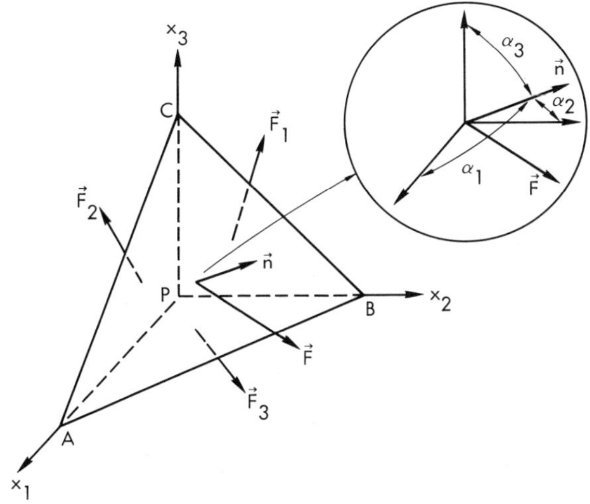

Fig. I.2. Forces on an infinitesimal tetrahedron embedded in an elastic body. $\vec{n}$ (=n) is a vector normal to the plane ABC.

We now examine forces $\mathbf{F}$, $\mathbf{F}_1$, $\mathbf{F}_2$, and $\mathbf{F}_3$ which act as vectors across the infinitesimal areas in Fig. I.2. The equilibrium of the tetrahedron under the forces exerted across its faces by the matter surrounding it requires

$$\mathbf{F} = \mathbf{F}_1 + \mathbf{F}_2 + \mathbf{F}_3, \tag{I-1.1}$$

written in vector form.

To treat these forces in a more amenable manner, they are considered in terms of intensity by introducing the notion of *traction*.[2] By definition,

$$\mathbf{T} = \lim_{dA \to 0} \frac{\mathbf{F}}{dA},$$
(I-1.2)

where $\mathbf{T}$ is the traction, $\mathbf{F}$ is the force, and dA is the differential area across which $\mathbf{F}$ is acting. Traction is a vector and has direction and sense which are the same as the direction and sense of the force. It is just force per unit area, but through the limit process it is defined at a point. To illustrate further what is meant by this notion of traction, let the area of plane ABC be dA. Then the area of the face perpendicular to the x_1 axis is $dA \cos \alpha_1$, where $\cos \alpha_1$ is the component of n resolved with respect to x_1. This is shown in Fig. I.2. Similarly, the areas perpendicular to the x_2 and x_3 axes are $dA \cos \alpha_2$ and $dA \cos \alpha_3$, respectively. These area components enable Eq. (I-1.1) to be written in terms of the tractions across the faces. Thus

$$\mathbf{T}\, dA = \mathbf{T}_1\, dA \cos \alpha_1 + \mathbf{T}_2\, dA \cos \alpha_2 + \mathbf{T}_3\, dA \cos \alpha_3$$

or

$$\mathbf{T} = \mathbf{T}_1 \cos \alpha_1 + \mathbf{T}_2 \cos \alpha_2 + \mathbf{T}_3 \cos \alpha_3$$
(I-1.3)

in which we have used the fundamental connection (I-1.1) between force and traction.

It is convenient now in our study of stress to resolve the traction $\mathbf{T}$ into components oriented in the directions of the coordinate axes chosen. We have for the components of $\mathbf{T}$,

$$T_1 = S_{11} \cos \alpha_1 + S_{12} \cos \alpha_2 + S_{13} \cos \alpha_3$$
(I-1.4)

and two other similar equations. Using tensor notation and making a slight change in specification to conform to the indicial representation, we can summarize these three equations:

$$T_i = S_{ij} \cos \alpha_j.$$
(2)

The quantities $S_{11}, S_{12}, S_{13} \ldots$ constitute a set of nine numbers; they are called the *stress components*. The first subscript denotes the coordinate axis direction along which the stress component is acting and the second the plane normal to the axis upon which the stress acts.

Figure I.3 shows our tetrahedron at point P in which both the tractions and stress components across the faces are drawn. One can see that the three stress components on a given face are simply the components of the traction on the face. The components S_{11}, S_{22}, and S_{33}, are called the *normal* components of

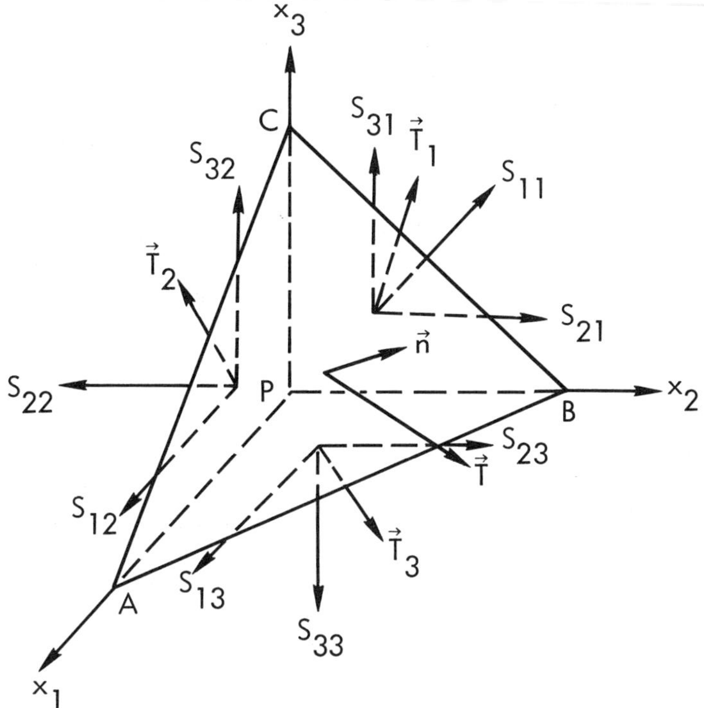

Fig. I.3. Tractions and stresses at point P.

stress; usual sign convention dictates that those normal stresses which are tensions are taken positive. The other components are called *shearing* or *tangential* components.

The specification of nine numbers is sufficient to characterize completely the state of stress at any point of the medium. That this is so follows from Eqs. (2) and (I-1.3) in which the quantities $S_{11}, S_{12}, S_{13}, \ldots$ at the point determine the traction **T** on any differential element of surface whose orientation is given by the unit vector **n**. In general, these numbers are different at different points so that these sets of numbers constitute nine functions of the points in the body. Such a set of functions, together with the points, is called a *stress field*.

The notion of stress is applicable to any continuum irrespective of the mechanical properties of the material, that is, regardless of whether the material behaves elastically, plastically, or as a fluid.

Using some of the formalism developed in the Introduction, we shall now prove that the stress $\widetilde{S}$ is a tensor of rank two by exhibiting that it transforms according to the rule Eq. (8). The argument is based on equilibrium of an infinitesimal mass distribution in a body under a system of forces, as we discussed with respect to Figs. I.2 and I.3. As implied by expressions (I-1.3) the tractions are added, with the result being zero according to the hypothesis of equilibrium against translation of the mass distribution embedded in the solid body that is under the set of applied forces.

We resolve the traction **T** into components relative to two coordinate systems, the x system as given previously by Eqs. (2) and (I-1.4), and an x' system in which Px_3 is chosen along the vector **n**. Using the definitions of the a_{ij} from the transformation equations (3) in the Introduction, relationships (2) may now be written

$$T_i = S_{ij}\, a_{3j}. \tag{I-1.5}$$

With respect to the x' system, let us include the x'_3 axis already defined and two axes so placed in face ABC that the three axes form the usual right-handed orthogonal triad; the direction cosines of these latter two axes are then denoted by a_{1j} and a_{2j}, respectively. We now examine the traction across face ABC by computing the normal stress and the tangential stress, the latter resolved into components along the axes of the x' system which lie in the face. By definition, this normal stress is S'_{33}, and the two components of the tangential stress are S'_{13} and S'_{23}. For the normal stress

$$S'_{33} = a_{3i}\, T_i = a_{3i}\, a_{3j}\, S_{ij}, \tag{I-1.6}$$

in which the first step holds because S'_{33} is just a traction component and transforms like a vector and the second step because of a substitution from Eqs. (I-1.5). Similarly,

$$S'_{13} = a_{1i}\, T_i = a_{1i}\, a_{3j}\, S_{ij} \tag{I-1.7}$$

and

$$S'_{23} = a_{2i}\, T_i = a_{2i}\, a_{3j}\, S_{ij}. \tag{I-1.8}$$

Expressions (I-1.6), (I-1.7), and (I-1.8) can be summarized by writing

$$S'_{k3} = a_{ki}\, a_{3j}\, S_{ij}. \tag{I-1.9}$$

For the full transformation law of $\widetilde{S}$, one needs the six additional expressions which would give $S'_{11}, S'_{21}, \ldots S'_{32}$ in terms of the S_{ij}. These expressions can be computed by considering the tractions across surfaces normal to x'_1 and x'_2. We avoid the laborious algebra involved, but it can be seen that such a transformation law can be generalized from Eq. (I-1.9). Thus

$$S_{k\ell} = a_{ki}\, a_{\ell j}\, S_{ij} \qquad\qquad\qquad (\text{I-1.10})$$

and we can regard $\widetilde{S}$ to be a tensor of rank two from rule (8), *i.e.*, the nine components of stress at a point constitute a second rank tensor. It is emphasized that the transformation equations (I-1.10) hold strictly only at some one point, but at that point it provides a complete description of the variation of stress with orientation of the surface across which the stress is transmitted. In the general instance of a stress field, the components of the stress tensor depend upon the position coordinates.

We can now proceed to illustrate two properties of the stress tensor of interest to us, namely, the symmetry and, to a lesser extent, the concept of principal stresses. These properties are closely interrelated; in face, symmetry of the stress tensor must be shown before the other can be proved to exist in general.

A second rank tensor is said to be symmetric if among its components the relation $S_{ij} = S_{ji}$ holds. This property of symmetry is extremely important because it allows complete specification of stress at a point in a body with at most six independent numbers. Symmetry will be shown with some generality by considering the body in motion,[3] allowing stress inhomogeneity, and including body forces. However, the hypothesis upon which the discussion rests is that there is no distribution of body torques. Unless this hypothesis holds, this tensor can never be shown to be symmetric. A distributed body torque occurs, for example, when an anisotropic material becomes polarized or magnetized in a field.

One approach is to develop the equations determining the changes of momentum using integral calculus. By allowing the volume of integration to pass to the limit zero, the symmetry of $\widetilde{S}$ immediately follows. This argument has the advantage of being correct to all orders of infinitesimals (because of the nature of an integral), but it has the disadvantage of complexity. Thus, we take an approach which uses differential calculus.

Consider a small rectangular parallelepiped situated within a stressed body with its center at point P, and with edges of lengths δx_i that are parallel to the axes x_i, respectively. The components of stress at the center P are taken as S_{ij}. Figure I.4 shows the average values of the stress components over the faces of the element viewed in the $x_2 x_3$ plane. We wish to find the equation of motion for rotation of the element about the Px_1 axis.

The couple due to shear components of stress on the two faces perpendicular

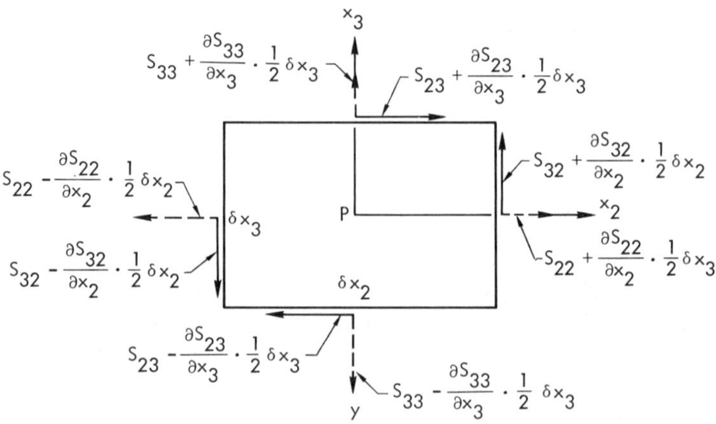

Fig. I.4. Inhomogeneous stresses in the $x_2 x_3$ plane; axis Px_1 is perpendicular to the plane.

to Px_2 is, to first order in an expansion of stress components in a Maclaurin's series,

$$\left(S_{32} - \frac{\partial S_{32}}{\partial x_2} \cdot \frac{1}{2}\delta x_2\right) \delta x_1 \, \delta x_3 \cdot \frac{1}{2}\delta x_2$$

$$+ \left(S_{32} + \frac{\partial S_{32}}{\partial x_2} \cdot \frac{1}{2}\delta x_2\right) \delta x_1 \, \delta x_3 \cdot \frac{1}{2}\delta x_2 = S_{32} \, \delta x_1 \, \delta x_2 \, \delta x_3.$$

The moment due to the shear component S_{23} is similarly

$$-S_{23} \, \delta x_1 \, \delta x_2 \, \delta x_3.$$

Thus, the equation of motion of rotation about Px_1 is

$$(S_{32} - S_{23}) \, \delta x_1 \, \delta x_2 \, \delta x_3 + \rho M_1 \, \delta x_1 \, \delta x_2 \, \delta x_3 = I_1 \, \ddot{\theta}_1 \qquad \text{(I-1.11)}$$

where ρ is the density, M_1 is the body torque per unit mass, I_1 is the moment of inertia, and $\ddot{\theta}_1$ is the angular acceleration, the last three with respect to axis Px_1. Because the resultant normal forces on the faces may not pass exactly through the midpoints of the faces, there may also be contributions to the above expression from the normal components S_{22} and S_{33}. However, the lever arm of each of these forces is considerably smaller than the lever arm for the shear forces, and so the contributions can be neglected. By similar reasoning, the small moment caused by the other shear components and body forces can be neglected in this relationship. From the definition of the moment of inertia, I_1 is proportional to $\rho \delta x^5$. Hence, as the element is allowed to become vanishingly small, the angular acceleration must increase rapidly to infinity as δx^{-2}. Such an increase is impossible in our continuum, and so

$$S_{32} - S_{23} + \rho M_1 = 0. \tag{I-1.12}$$

Two other similar equations may be derived. In the absence of electric and magnetic fields, a body torque does not (generally) exist, and we have

$$S_{ij} = S_{ji}. \tag{I-1.13}$$

For our purposes, symmetric tensors will suffice and relationships (I-1.13) hold. The symmetry properties of tensors of rank higher than 2 are a complicated topic, and since we shall not need these properties we avoid their discussion.

In recent years, a more general approach to linear elastic problems has been advanced in which the hypothesis is made that moments per unit area as well as forces per unit area act upon the surfaces of an infinitesimal element within an elastically stressed body. Such moments are called *couple stresses*, and it can be shown that the components of couple stress at a point are representable by a second rank tensor. The inclusion of couple stresses, as we have shown with body torques, will not allow a symmetric stress tensor. The use of this hypothesis requires the definition of a new elastic constant which relates couple stress and curvature. This constant has not yet been measured experimentally for any material.

A passing understanding of what is meant by the terms *principal stresses, principal axes*, and *principal planes* will prove helpful in later chapters, particularly in the sections involving the performance of experimental work. Such terms can be defined as follows: Three perpendicular planes can always be found at a point for which the shearing stresses vanish, *i.e.*, $S_{ij} = 0$ for $i \neq j$. Thus, for these conditions the resultant stresses are perpendicular to the three planes. These stresses are called the principal stresses (denoted by S_i) at the point, their directions the principal axes, and the three planes the principal planes. The magnitudes of the principal stresses at the point are invariants of the state of stress.

That these principal stresses, axes, and planes can always be constructed for the stress tensor (or for any symmetric tensor of rank two) in any finite number of dimensions can be shown quite generally by *diagonalizing* the tensor. What we mean is that a system of coordinates (the x' system) exists in which $S'_{ij} = 0$ for

$i \neq j$. Thus, all "off-diagonal" terms vanish [refer to the matrix in Eq. (9) in the Introduction] in the x' system, and the entire situation at the point under study consists of a tension (or compression, depending on the direction of the S_i and upon the sign convention used) across each coordinate plane. General proof of these statements is somewhat involved and of only peripheral interest to us in this book.

Considerably greater sophistication and completeness can be used in the presentation of the concepts of symmetry, couple stresses, and principal stresses. The reader may find such discussion in the texts by Eringen [8], and Malvern [9].

B. Strain

It is necessary to consider the effect that the stress field causes when imposed upon our elastic body. We examine two states of the body: an initial state, *i.e.*, before loading, and a terminal state, *i.e.*, after loading. The objective is to establish the kinematics of change of shape and size of volume elements within the medium; the kinetics of this change are not pertinent to the present discussion.

The movement of an arbitrary point in the body caused by the transition of the body from the first state to the second state is called the *displacement* of that point. If the displacement is such that the relative positions of any two points do not change with respect to each other, the movement is that of a *rigid body*. In this work, translations and rotations of the body as a whole are not of central interest to us. If the displacement is such that the relative positions of any two points do change from the first state to the second, the medium is considered to be *deformed*. The study of this deformation field constitutes the study of *strain*. Deformation is called *homogeneous*, or *uniform*, when the displacement gradients are constant throughout the body or, equivalently, when a change in configuration is expressible as a linear transformation of the coordinates. Thus, for example, any set of two planes parallel to each other in the initial state is transformed to another set of parallel planes in the terminal state when undergoing homogeneous strain, although the dimensions in the planes as well as the distance between them may change during transformation. If the displacement gradients are not constant throughout the body, *inhomogeneous* deformation results.

Consider the displacement of any material point P (see Fig. I.5) to be given symbolically by **u**. This vector may be resolved parallel to the Cartesian x_i axes into components u_i, respectively. Hence, if the coordinates of the material point P in the initial state are x_i, they become $(x_i + u_i)$ in the terminal state P'. We choose to represent our displacements in the Lagrangian sense in that we are directing our attention to what occurs to a given particle (a material point) using the coordinates of the particle in the initial state as the independent variables.

It is appropriate to mention here that there are two methods of displacement representation. These two methods are with the *Lagrangian* and *Eulerian* statements. The Lagrangian representation gives the mathematical equations for the trajectory of each particle, *i.e.*, the given particle is labelled and observed during the time of movement of interest. The Eulerian representation describes what happens to the material at a given point as observed from a certain position or origin. The chief difference between the Euler statement and the Lagrange statement is in point of view. Eulerian equations "snap-shot" the entire region of interest, whereas Lagrangian equations describe "traces" of a particle. Either representation allows use of a moving coordinate system. It is noted that in solid mechanics, concern is usually directed to displacements (or velocities) of material points rather than displacements at a position in space. Thus, the expression of displacement in Lagrangian coordinates is sometimes more expedient, unless the deformations are large. For relative displacements that need not be considered finite, we will see that either the Eulerian or Lagrangian statement is suitable.

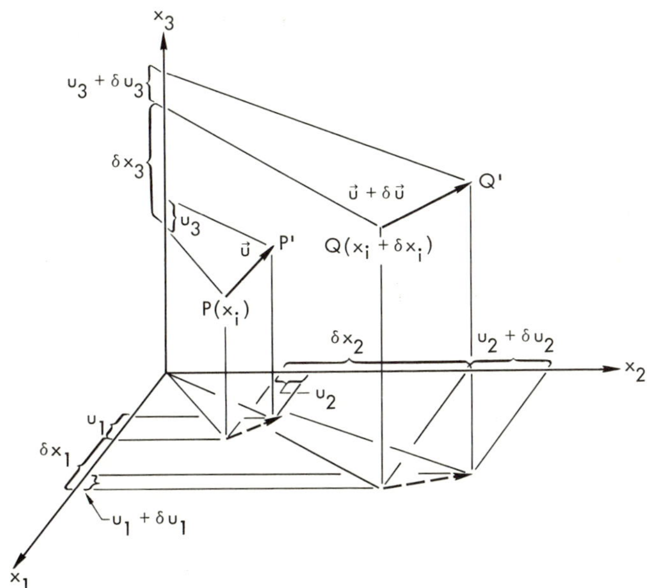

Fig. I.5. Relative displacement.

Now consider strain at the material point P to be given by how its position relative to adjacent material points has changed. Specifically, let the material point Q with coordinates $(x_i + \delta x_i)$ be a point near P and let the corresponding displacement which it has undergone have components $(u_i + \delta u_i)$ at material point Q'. Because we are treating the medium as a continuum and are assuming continuous single-valued displacements, we may write from Taylor's series

$$\delta u_i = \frac{\partial u_i}{\partial x_j}\, \delta x_j + \frac{1}{2!}\, \frac{\partial^2 u_i}{\partial x_j\, \partial x_k}\, \delta x_j\, \delta x_k$$

$$+ \ldots + \frac{1}{n!}\, \frac{\partial^n u_i}{(\partial x_j\, \partial x_k\, \partial x_\varrho \ldots)}\, (\delta x_j\, \delta x_k\, \delta x_\varrho \ldots) + \ldots . \qquad \text{(I-1.14)}$$

It is obvious that we have in Eqs. (I-1.14) nonlinear relationships. This expression can be greatly simplified without seriously restricting its usefulness. We can say that in a sufficiently small neighborhood of the point, the relative displacements are linear functions of the coordinates or, equivalently, the nonlinear relationships vary slowly so that the displacement gradients within a small region are constant with all higher derivatives being negligible. Thus, the deformation is sensibly homogeneous about that point, and the relative displacement may be written

$$\delta u_i = \frac{\partial u_i}{\partial x_j}\, \delta x_j. \qquad \text{(I-1.15)}$$

If the nine differential quantities in Eqs. (I-1.15) are known at P, then the displacement field may be calculated.

Expressions (I-1.15) may be rewritten in the form

$$\delta u_i = \frac{1}{2} \left(\frac{\partial u_i}{\partial x_j} + \frac{\partial u_j}{\partial x_i} \right) \delta x_j + \frac{1}{2} \left(\frac{\partial u_i}{\partial x_j} - \frac{\partial u_j}{\partial x_i} \right) \delta x_j \qquad \text{(I-1.16)}$$

and we may define the quantities

$$\epsilon_{ii} = \frac{\partial u_i}{\partial x_i}\, , \ i \text{ not summed}$$

$$2\epsilon_{ij} = \frac{\partial u_i}{\partial x_j} + \frac{\partial u_j}{\partial x_i} \qquad\qquad\qquad\text{(I-1.17)}$$

and

$$2\omega_{ij} = \frac{\partial u_i}{\partial x_j} - \frac{\partial u_j}{\partial x_i}.$$

The first six quantities, represented by the symbol ϵ, are called the *components of infinitesimal strain*. The strain components ϵ_{ii} correspond to tensile or compressive strains (depending upon the physical loading) and the ϵ_{ij} relate to shearing strains. The last three quantities, represented by the symbol ω, are called the *components of infinitesimal rotation*. If desired, the second set of expressions (I-1.17) can be reduced to the first set of (I-1.17) by putting $i=j$.

Expressions (I-1.17) are linear differential equations for the components of displacement. A system for which the relative displacement is sufficiently "small" so that these equations describe it adequately is a linear system and is one in which the principle of superposition holds. The above definitions are adequate for most of our purposes, even (somewhat surprisingly) for the topics in nonelasticity we treat.

In more sophisticated treatments (*e.g.*, see Eringen [8] or Novozhilov [10]), the strain components are first developed on the basis of finite deformation. The state of such deformation, or strain, is completely determined when we know the changes in lengths and angles between the initial and terminal states of corresponding linear elements, usually expressed in Eulerian coordinates relative to points in the deformed (terminal) condition (see Sokolnikoff [11]). These changes can be related to the square of the relative displacement (I-1.15). With considerable algebra, the results can be put into a form so that convenient definitions of strain in terms of the displacement gradient in (I-1.15) become evident, namely,

$$\epsilon_{ii} = \frac{\partial u_i}{\partial x_i} - \frac{1}{2}\frac{\partial u_k}{\partial x_i}\frac{\partial u_k}{\partial x_i}, i \text{ not summed}$$

$$\text{(I-1.18)}$$

$$2\epsilon_{ij} = \frac{\partial u_i}{\partial x_j} + \frac{\partial u_j}{\partial x_i} - \frac{\partial u_k}{\partial x_i}\frac{\partial u_k}{\partial x_j}$$

where the six quantities are called the *components of finite (homogeneous) strain*. Equations (I-1.18) are in Eulerian coordinates. The signs of the last terms in each of the expressions become positive in Lagrangian coordinates referenced to the undeformed (initial) state. It should be emphasized that it is customary to develop the theory of finite strain only on the basis of Eqs. (I-1.15), and not from (I-1.14).

There is considerable difficulty in the practical application of definitions (I-1.18). Thus, simplification to Eqs. (I-1.18) is often made by neglecting the products of the displacement gradients. This linearization constitutes the *small strain approximation*, and by this process we can arrive at the definitions given in Eqs. (I-1.17). The reason becomes apparent why, for this condition, there is no distinction between the Lagrangian and Eulerian representations.

It is important to note that strain has an arbitrary definition, depending upon the accuracy required, the constant factors to be included, and the desired mathematical features. However, displacement has not an arbitrary meaning. Displacement is fundamental in that it is a consequence of the nature of the body forces and boundary conditions imposed upon it; but strain is merely any convenient expression in terms of displacement and its derivatives. For example, the shearing strain components in Eqs. (I-1.17) are usually defined in the engineering literature without the constant factor 2; and, in fact historically, the relations were derived on the basis of this definition.

We now elaborate briefly on the physical meaning of strain and rotation quantities in (I-1.17). It is helpful to review their simplified geometrical aspects in plane strain for small displacements.[4]

The first set of strain quantities of (I-1.17), ϵ_{ii} (i not summed), correspond to fractional expansions and/or contractions of infinitesimal line elements passing through point P in Fig. I.5 and parallel to x_i, respectively. For example, referring to Fig. I.6, one can visualize what is meant by ϵ_{11} in terms of "finite" quantities. We can see

$$\epsilon_{11} = \lim_{x_1 \to 0} \frac{\Delta u_1}{\Delta x_1} = \frac{\partial u_1}{\partial x_1} \tag{I-1.19}$$

which is accurate to first order. The sign conventions often used for these quantities assign expansions, *i.e.*, tensile strains, to positive values and contractions, *i.e.*, compressive strains, to negative values; however, as long as consistency is maintained, it is unimportant which convention is selected.

The second three components of strain in expressions (I-1.17), ϵ_{ij}, are shearing strains. For example, in the $x_1 x_2$ plane, one can examine the distortion of the angle between the elements PA and PB shown in Fig. I.7.

If u_1 and u_2 are the displacements of point P in the x_1 and x_2 directions, respectively, the displacement of point A in the x_2 direction and point B in the

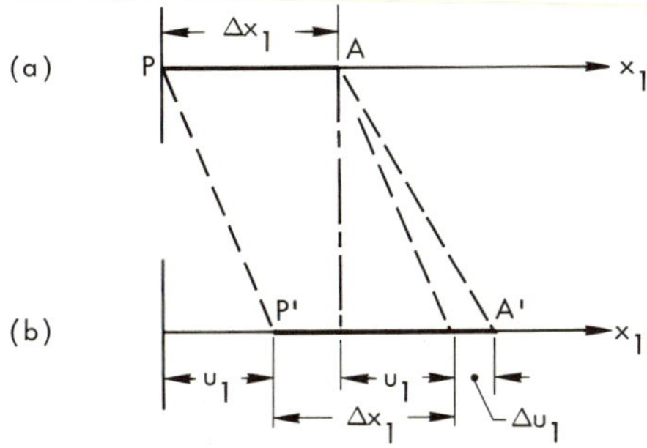

Fig. I.6. Deformation of a line element; (a) undeformed, (b) deformed.

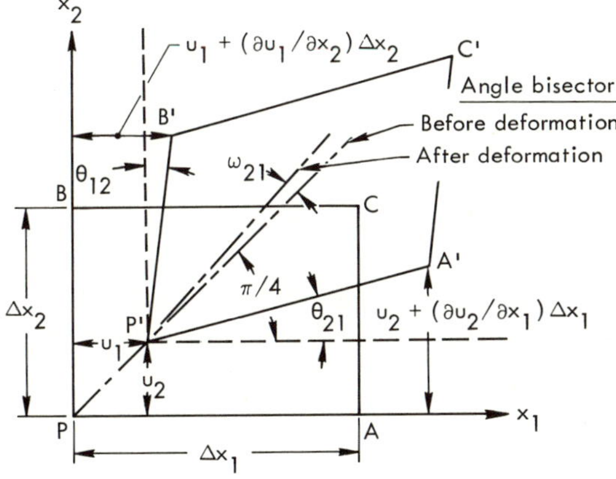

Fig. I.7. Shearing strain and rotation of an element.

x_1 direction are as shown. Because of these displacements, the new direction of the element $P'A'$ is inclined to the initial direction of PA by the small angle θ_{21} which to first order is equal to $\partial u_2/\partial x_1$. Similarly, $P'B'$ is inclined to the initial direction of PB by θ_{12}, equal to $\partial u_1/\partial x_2$ to first order. From these two statements, it will be seen that the right angle APB is diminished by the angle $(\partial u_1/\partial x_2 + \partial u_2/\partial x_1)$ and becomes $(\pi/2) - 2\epsilon_{12}$. The component ϵ_{12} represents one-half the change of angle between the line elements Δx_1 and Δx_2 and is a measure of the shear strain between the planes $x_1 x_3$ and $x_2 x_3$. Analogous discussion can be given for the other shear strain components.

Now consider the rotation components described by the last three equations of (I-1.17). In Fig. I.7, ω_{21} corresponds to the angle between the bisector of angle APB and the bisector of angle $A'P'B'$. An equivalent statement is that the quantity ω_{21} is seen to represent the algebraic average of the angular rotations of Δx_1 and of Δx_2, and can be related to infinitesimal rigid-body rotation. Similar statements may be given for the other rotation quantities.

In a fashion similar to that for stress in Section II-A, strain can also be shown to behave as a tensor of rank two. Because of the connection between the displacement vector and the strain tensor, it is easier to show that strain transforms as a tensor of rank two than it was to show that stress has this behavior. For the classical infinitesimal strain definition $(I-1.17)^5$ in the x' system, we have

$$2\epsilon'_{ij} = \frac{\partial u_i}{\partial x'_j} + \frac{\partial u_j}{\partial x'_i} \; . \tag{I-1.20}$$

The "chain rule" of partial differential calculus enables us to write

$$2\epsilon'_{ij} = \frac{\partial u'_i}{\partial x_k}\frac{\partial x_k}{\partial x'_j} + \frac{\partial u'_j}{\partial x_k}\frac{\partial x_k}{\partial x'_i} \; . \tag{I-1.21}$$

With the help of the equations of transformation from the x system to the x' system (3) in the Introduction, this result can be simplified to give

$$2\epsilon'_{ij} = a_{jk}\frac{\partial u'_i}{\partial x_k} + a_{ik}\frac{\partial u'_j}{\partial x_k} \; . \tag{I-1.22}$$

The simplification uses the property of the orthogonal transformation from Eqs. (3) and (4) that

$$a_{jk}\, x'_j = a_{jk}\, a_{ji}\, x_i = \delta_{ki}\, x_k \tag{I-1.23}$$

where δ_{ki} is defined in expressions (5). Now since the numbers u_i are the components of a vector, i.e., $u'_i = a_{im}\, u_m$ [see expressions (7)], (I-1.22) becomes

$$2\epsilon'_{ij} = a_{jk}\, a_{im}\frac{\partial u_m}{\partial x_k} + a_{ik}\, a_{jm}\frac{\partial u_m}{\partial x_k} \tag{I-1.24}$$

$$= a_{ik}\, a_{jm} \left(\frac{\partial u_k}{\partial x_m} + \frac{\partial u_m}{\partial x_k} \right) \tag{I-1.25}$$

in which (I-1.25) follows from (I-1.24) just by renaming two subscripts over which sums are carried out. By definition of ϵ_{ij}, Eqs. (I-1.25) have the form

$$\epsilon'_{ij} = a_{ik}\, a_{jm}\, \epsilon_{km} \tag{I-1.26}$$

which can be seen by comparison with the transformation law (8) to be what we set out to obtain.

We can prove that the components of infinitesimal body rotation in expressions (I-1.17) transform as tensors of rank two in exactly the same way as we showed the transformation behavior for strain. Indeed, one can see that the proof is word for word the same and is valid since that proof no where made use of the algebraic sign which in these definitions distinguishes ω_{ij} from the classical infinitesimal strain ϵ_{ij}.

By inspection of the definitions of the classical infinitesimal strain (I-1.17), and also of the more general strain (I-1.18), one can see that under interchange of the subscripts, i and j, the right members are invariant, *i.e.*, the same results are obtained. This invariance is sufficient to prove that the strain tensor by either of these two definitions is a symmetric tensor. Hence, as with stress, complete specification of strain at a point is obtained with at most six independent numbers.

In addition to encountering tensors which have the property of symmetry, one encounters similar quantities with the condition among their components such that

$$A_{ij} = -A_{ji}. \tag{I-1.27}$$

The relationship stated by (I-1.27) constitutes the condition of *antisymmetry* and refers here specifically to an antisymmetric tensor of second rank. It can be seen by inspection that the ω_{ij} in Eqs. (I-1.17) enjoy this property of antisymmetry.

As indicated previously, every symmetric second rank tensor can be put into diagonal form. The language used in connection with the diagonalized strain tensor is quite similar to the language used in connection with the diagonalized stress tensor. The axes in the coordinate system in which $\epsilon'_{ij} = 0, i \neq j$, are called the *principal axes* of strain and the three numbers ϵ'_{ii} (i not summed), are called the *principal strains*. In the special instance of an isotropic substance for which Hooke's law (discussed in the next section) is a valid connection between the stress and strain, the principal axes for the two tensors coincide. When the principal strain axes retain their direction during deformation, the deformation is said to be *irrotational*. Irrotational deformation occurs when there is no movement of any of the bisectors (on each of the three mutually perpendicular planes), *i.e.*, all the $\omega_{ij} = 0$.

In most conventional texts on the mechanics of deformable media, one encounters the term *compatibility equations*. Briefly, to make a problem statically or dynamically determinant, that is, to have as many equations as unknowns so as to completely specify the solution, it is necessary to develop certain specific relationships among the strain components (I-1.17). These relationships are the compatibility equations, and they are both necessary and sufficient to secure the existence of unique displacement quantities u_i connected with the strain components. The determination of the actual displacements from strains demands the introduction of these equations. However, the converse problem that we have been considering, in which we have developed the concept of strain from displacement, does not require them.

III. Stress-Strain Relationships

It is now necessary to connect in some fashion the concept of the stress field developed in Section I and the resulting strain field discussed in Section II. This interrelation is expressed in the form of some mathematical model. The treatments contained in the first two sections are applicable to all media that can be considered continuous, but now, in addition, we must be concerned with the characterization of the particular medium under analysis. We direct our primary attention in this section to: (1) solids that are (sensibly) perfectly elastic; (2) deformations that are "small"; and (3) loading conditions that are quasistatic.

The meaning of the first two conditions should be clear from previous discussion, but since we later extend our efforts to treat also the motion of the body as governed by the laws of Newtonian dynamics, it is desirable to be somewhat more explicit on the last point than the few general comments given in Section I. As indicated, the condition is imposed that the external forces are applied very slowly so that equilibrium of the stress field throughout the entire body has been attained. This means the equilibrium deformation position is achieved only after every point in the body has been notified of the applied forces and the resulting distortions signaled back to the points of their applications. Specifically, the requirement of equilibrium demands that (1) the time taken to apply force is long compared to the slowest mode of vibration of the body, *i.e.*, one for which the disturbance propagation velocity is least; or (2) any vibrations are damped out. These restrictions must obviously be relaxed later when we treat the propagation of stress waves.

A. Small Elastic Deformations – Generalized Hooke's Law

The theory of classical linear elasticity has its basis in a law formulated originally from experimental observation by Hooke in which he stated that the deformation resulting from a force applied to a body is linearly proportional to that force. Young gave this observation a definite physical meaning by defining the *modulus of elasticity* of the body in simple compression as the proportionality constant between stress and strain. Thus, as initially postulated, the law is applicable to investigation involving only quasistatic simple one-dimensional stress tensile or compressive loading. There is also the implied limitation here that we are dealing with the isothermal deformation of perfectly elastic materials.

One method of extending its usefulness is, by the process of induction, to generalize this law of simple proportionality between two quantities to a law of linear dependence among several quantities. Hence the original relationship can be restated: Each of the six components of stress at any point of a body is a linear function of the six components of strain at the point. We speak of this form of the equation as the *generalized Hooke's law*, and it represents the fundamental law of the classical theory of elasticity. While the law in this form cannot be proved directly by experiment, the consequences are found to be true for an elastic material whenever they have been tested under conditions of small strain. Hooke's law is an example of a mechanical equation of state. Equations of state are discussed in more detail in Chapter I-6.

Stated mathematically, Hooke's law in 3-space ($n = 3$) becomes

$$S_{11} = C_{1111}\epsilon_{11} + C_{1122}\epsilon_{22} + C_{1133}\epsilon_{33} + C_{1112}\epsilon_{12}$$

$$+ C_{1113}\epsilon_{13} + C_{1123}\epsilon_{23} \tag{I-1.28}$$

and five other similar equations. More concisely, we can write for the six equations,

$$S_{ij} = C_{ijk\ell}\epsilon_{k\ell}. \tag{I-1.29}$$

The numbers $C_{ijk\ell}$ are called the *elastic stiffness constants* of the material. The constants have the same dimensions as do the stress components. These numbers are facts of nature, determined, but not specified by man.

Note that it is basically illogical to suppose that we may have a directly proportional relationship between stress and strain as we have defined the terms, since stress refers to the condition in the material *after* deformation and strains are developed with reference to the *initial* state [but see Eq. (II-3.1)]. Provided we concern ourselves with small deformations so that the displacement derivatives are much less than unity, a satisfactory approximation results from the relation defined as in (I-1.29), and the internal inconsistency in the logic can be ignored. With finite relative displacements, the approximation is not valid [see discussion in connection with Eqs. (I-1.18)], and both tensor fields must be expressed using the same representation of coordinate systems.

The set of $C_{ijk\ell}$ in expressions (I-1.29) can be readily shown to represent a fourth rank tensor by the following reasoning: By virtue of being equated to stress, the right side of law (I-1.29) must be a second rank tensor. It has been contracted twice by having k and ℓ in both terms. The resultant of a sixth rank tensor so contracted is a tensor of second rank; and since the strain tensor is of second rank the $C_{ijk\ell}$ must be of fourth rank so that their product is sixth rank. This fact can also be proved rigorously using the appropriate fourth rank transformation law and suitable forms of Eqs. (I-1.10), (I-1.26), and (I-1.29).

It can be inferred from expression (I-1.28) that the $C_{ijk\ell}$ are a set of (at most) 36 independent numbers in three dimensions and may be written in a sixth-order, square-matrix array [see the formulation (9) in the Introduction]. Considering the point somewhat more formally, inasmuch as both the tensor components S_{ij} and $\epsilon_{k\ell}$ have been shown to be symmetric, the components $C_{ijk\ell}$ are symmetric with respect to the first two and to the last two indices, and are thus reduced from a set of 81 numbers (the number of components of a fourth rank tensor in 3-space) to a set of 36. Evidence of the economy of indicial notation is apparent when we examine expressions (I-1.28) and (I-1.29) and observe that the latter represents six equations, each with seven terms.

However, it can be shown that these 36 elastic stiffness constants are not all independent. The method of showing this involves the introduction of a *strain energy function W* having the form

$$W = \int S_{ij} d\epsilon_{ij}. \qquad \text{(I-1.30)}$$

As the name and the above expression imply, this function depends upon the state of deformation in a body, and it represents the potential energy contained in the strained body as referenced to a standard state of the body that is of uniform temperature and zero relative displacement.

It is necessary first to show that such a function exists for a given problem. From thermodynamic considerations it can be readily shown that a strain energy function does exist in general in two important physical situations; namely,

(1) where the deformation is considered to be produced slowly so that changes are essentially reversible and isothermal, as in quasistatic considerations; and (2) for reversible changes that are sufficiently rapid so that they can be satisfactorily represented as adiabatic processes, as in wave and vibration theory. Although the strain energy functions in the two situations are not usually the same, the difference is not significant for small deformations and distinction need not be made in practical studies.

From Eqs. (I-1.29) and (I-1.30), we have

$$dW = S_{ij}d\epsilon_{ij} = C_{ijk\ell}\epsilon_{k\ell}d\epsilon_{ij}$$

and

$$\frac{\partial W}{\partial \epsilon_{ij}} = C_{ijk\ell}\epsilon_{k\ell}.$$

We can differentiate both sides of this equation with respect to $\epsilon_{k\ell}$ and obtain

$$\frac{\partial^2 W}{\partial \epsilon_{ij}\partial \epsilon_{k\ell}} = C_{ijk\ell}.$$

Since the strain energy is a function only of the deformed configuration of the body, specified by the strain components, the order of differentiation makes no difference. Therefore,

$$C_{ijk\ell} = \frac{\partial^2 W}{\partial \epsilon_{ij}\partial \epsilon_{k\ell}} = \frac{\partial^2 W}{\partial \epsilon_{k\ell}\partial \epsilon_{ij}} = C_{k\ell ij}. \tag{I-1.31}$$

The sixth-order square matrix referred to above is thus symmetrical and the independent numbers of the $C_{ijk\ell}$ are reduced from 36 to 21 in the most general case.

A strain energy function such as expression (I-1.30) is also used as a means to treat problems involving large, perfectly elastic deformations. Such study has received some recent stimulation by the increasingly wide use of certain rubbers and plastics, many of which to a fair degree of approximation may be regarded as ideal elastic materials undergoing finite displacements when subjected to stresses of engineering interest.

The usual method of approach for problems of large elastic deformations involves use of expressions such as Eqs. (I-1.18) which retain the nonlinear terms in the strain definition. The exact solution then requires the combination of the stress-strain expression (Hooke's law, in which the proportionality constants become coefficients, *i.e.*, a function of strain) with the equations of equilibrium (or of motion) to yield a set of nonlinear partial differential relations for the determination of the parameters defining the deformation. Analytic treatment of the nonlinear system, is at best, difficult, although certain problems such as those involving simple tension and torsion of cylinders can be satisfactorily studied. Computers, using a suitable finite difference scheme, can also be employed to solve the equations directly.

The use of a strain energy function is an alternative method of solution that is useful in many problems. The significant advantage of this approach arises because stress and strain are both second rank tensors whereas energy is a scalar function, which fact generally renders the problem simpler mathematically. The book by Love [2] and the report by Wasley [12] contain additional discussion on the subject.

B. Anisotropy and Particularization to an Isotropic Medium

A body is called *isotropic* if its properties are identical in all directions. A volume element in an isotropic medium will retain the same properties regardless of its orientation. *Anisotropy* will be understood to describe those materials which are not isotropic. A body is called *homogeneous* if the properties of different volume elements in the body are the same; otherwise it is *inhomogeneous*. In an equivalent sense, a body can also be said to be homogeneous if that body has elastic properties describable by the generalized form of Hooke's law (I-1.29) and if the $C_{ijk\ell}$ do not have spatial variation. To describe inhomogeneous media, the $C_{ijk\ell}$ must be allowed to vary from point to point. It is helpful to reread the beginning of Section II-B and review the distinction between *body* homogeneity and *deformation* (or *strain*) homogeneity. Note that a body must be continuous to be considered either homogeneous or inhomogeneous. A body may possess different degrees of anisotropy and inhomogeneity with respect to a variety of properties, *e.g.*, dielectric and piezoelectric constants, and the refraction of light. We speak here, however, only of properties of the body referring to its mechanical behavior.

The condition of anisotropy is important in the study of a number of materials of physical interest. For example, when examining the mechanical behavior of certain crystals, there are found to be various degrees of geometrical symmetry of internal structure which can be introduced that allow elastic properties in particular directions to become identical (see the book by Nye [13]). If such symmetry exists, then the number of elastic stiffness constants $C_{ijk\ell}$ can be reduced to less than 21. Table I-1 gives examples of systems of crystal symmetry and the number of associated constants. The range of

Table I-1.
Crystalline symmetry versus independent elastic constants.

System	Number of $C_{ijk\ell}$	
Triclinic	21	
Monoclinic	13	
Orthorhombic	9	
Tetragonal	7 or 6	depending on class
Trigonal	7 or 6	
Hexagonal	5	
Cubic	3	
Isotropic	2	

symmetry varies from triclinic (completely anisotropic) with 21 independent constants required for description to isotropic with two such constants. A physical example of a monoclinic crystal is gypsum; an example of a hexagonal system is zinc.

Although there are no ideally isotropic bodies, many practically important structural materials approach this condition with reasonable approximation, providing consideration is directed to portions of the bodies that are large relative to any included anisotropic parts. In this sense, polycrystalline materials are isotropic. Thus, for the most part, it is appropriate to restrict our studies (both elastic and nonelastic) to the consideration of isotropic media.

The condition of isotropy reduces the number of independent elastic constants to two, denoted generally by λ and μ and called *Lamé's constants*. The quantity μ is also called the *shear modulus* or *modulus of rigidity*; in engineering literature it is usually designated by the symbol G. These constants are related to the general elastic tensor as follows:

$$C_{1122} = C_{1133} = C_{2233} = \lambda$$

$$C_{1212} = C_{1313} = C_{2323} = \mu$$

$$C_{1111} = C_{2222} = C_{3333} = \lambda + 2\mu$$

and the other 12 coefficients become zero. Inserting these values into Eqs. (I-1.29), Hooke's law becomes

$$S_{ij} = \lambda\Delta\delta_{ij} + 2\mu\epsilon_{ij} \tag{I-1.32}$$

where δ_{ij} is the Kronecker delta given by expressions (5), and where Δ is the *dilatation* defined by

$$\Delta = \epsilon_{ii}. \tag{I-1.33}$$

The dilatation has a simple geometrical meaning. It represents the first-order fractional change in volume of a small rectangular parallelepiped of the elastic material. Equation (I-1.33) is also called the *first invariant* of the strain tensor.

Engineering texts often treat the subject of Hooke's law from a less sophisticated approach (see, for example, the text by Timoshenko and Goodier [14]). It is helpful to consider briefly this approach so as (1) to examine the simplified physical aspects of Hooke's law (as in the discussion of strain in Section II-B), and (2) to gain some understanding of the significance and interrelation of the various elastic constants for isotropic media. Written in its simplest one-dimensional form, applicable for example to a thin rod, Hooke's law is

$$F = g\,(\delta x)$$

where F is the applied uniaxial force, x is the resulting displacement, and g is the proportionality constant. By dividing F by the original area A to obtain the nominal engineering stress, and by writing the displacement as the relative displacement $\delta x/x_0$, we can express Hooke's law in the form

$$F/A = E\,(\delta x/x_0)$$

or

$$S_{11} = E\epsilon_{11} \tag{I-1.34}$$

where E is the modulus of elasticity (Young's modulus) to which reference was made previously. The modulus equals the slope of the line in a plot of uniaxial stress versus strain in the elastic regime.

The reader can visualize in this simple situation what is meant by "small" relative displacements by considering that a mild steel rod with $E = 20.6 \times 10^{11}$ dynes/cm^2 (30×10^6 psi) and loaded to 20.6×10^8 dynes/cm^2 (30,000 psi) will produce a strain of only 0.001 cm/cm (or in./in.).

We examine further Eq. (I-1.34) in the form

$$\epsilon_{11} = S_{11}/E.$$

Accompanying this elongation in the 11-direction, there will be contractions (assuming such contractions are physically allowed) in the 22- and 33-directions, given by

$$\epsilon_{22} = \epsilon_{33} = -\sigma\epsilon_{11} = -\sigma(S_{11}/E) \tag{I-1.35}$$

where σ is a quantity that is constant within the elastic range and is called Poisson's ratio.[6] Thus, if a volume element is subjected to a traixial state of normal stress, the total strain in the 11-direction becomes to first order

$$\epsilon_{11} = \frac{1}{E}\left[S_{11} - \sigma(S_{22} + S_{33})\right] \tag{I-1.36}$$

with two other similar equations for ϵ_{22} and ϵ_{33}. The elastic stress-strain relation under a two-dimensional pure shear condition takes the form

$$\epsilon_{12} = \frac{1}{\mu}S_{12} \tag{I-1.37}$$

with two other similar equations for ϵ_{13} and ϵ_{23}. Equations (I-1.36) and (I-1.37) are found to be valid for any system of three-dimensional state of stress for an isotropic elastic material by simple superposition.

Inverting expressions (I-1.36) and (I-1.37), we have

$$S_{11} = \frac{\sigma E}{(1+\sigma)(1-2\sigma)}\Delta + \frac{E}{1+\sigma}\epsilon_{11} \tag{I-1.38}$$

with two similar equations for S_{22} and S_{33}, and

$$S_{12} = \mu\epsilon_{12} \tag{I-1.39}$$

with two similar equations for S_{13} and S_{23}. By defining

$$\lambda \equiv \frac{\sigma E}{(1+\sigma)(1-2\sigma)} \tag{I-1.40}$$

and

$$\mu \equiv \frac{1}{2}\left(\frac{E}{1+\sigma}\right), \tag{I-1.41}$$

we obtain relationships (I-1.32).

By inverting expressions (I-1.40) and (I-1.41), we obtain

$$E = \frac{3\lambda + 2\mu}{\lambda + \mu} \tag{I-1.42}$$

and

$$\sigma = \frac{\lambda}{2(\lambda + \mu)}.$$ (I-1.43)

Another quantity which is of use is the coefficient of bulk modulus k which is defined as the applied hydrostatic pressure p divided by the fractional volume change. With some manipulation of Hooke's law (I-1.32), we find

$$k = -\frac{p}{\Delta} = \lambda + \frac{2}{3}\mu = \frac{E}{3(1 - 2\sigma)}.$$ (I-1.44)

Although only two quantities are necessary to define the small strain elastic behavior of an isotropic solid, it is evident that five quantities are available.

IV. Equations Governing Deforming Elastic Solids

A. Hooke's Law Applied under Dynamic Conditions

In Section III we have written the generalized Hooke's law in 3-space in terms of six equations as applied to conditions of quasistatic response,

$$S_{ij} = C_{ijk\ell}\epsilon_{k\ell}.$$ (I-1.29)

In the isotropic form of law (I-1.29), the expression is then written

$$S_{ij} = \lambda\Delta\delta_{ij} + 2\mu\epsilon_{ij}.$$ (I-1.32)

The various terms have been defined previously.

By a very simple consideration, we may further extend and generalize Hooke's law in both the above forms so as to remove any remaining restrictions that the forces must be applied quasistatically to the entire body. In this generalization, the body is considered as being partitioned into a great number of small regions. The distortion of each region is then determined solely by the forces acting upon it by neighboring regions, irrespective of how rapidly these forces change.[7] In effect, one takes the size of the regions so small that all forces change only a negligible amount during a time interval equal to the period of the slowest mode of vibration of the region. As an indication of region "size," the wavelength of the impressed forcing function should be at least of the order of 10 times some representative dimension of the small region under consideration.

Thus, the only essential change in the application of Hooke's law in treating dynamic elastic problems is the removal of the word "entire" to describe the extent of the body under consideration. As indicated for quasistatic conditions in Section III, the application of this law under dynamic conditions cannot be proved directly by experiment, but experimental observations are found to agree with predictions made using the relationship.

It appears at first as if the procedure discussed above is rather artificial and arbitrary. That this is not so comes from the examination of Eqs. (I-1.29) and (I-1.32). It can be seen that Hooke's law will remain unaltered in dynamic problems since body forces do not appear anywhere. Moreover, the elastic members for small strains are constants and are neither rate-dependent nor functions of body motions.

B. *Equations of Motion*

In developing the concept of stress in Section II-A, we have discussed mechanical forces acting on a body that is in equilibrium. The subject of kinetics was not introduced there, except through the brief introduction of an expression for the equation of rotational motion to prove the symmetry of the stress tensor. We are prepared now to examine the mechanical behavior of an elastic body in terms of the propagation of stress disturbances; specifically, we want to develop the equations governing this propagation.

The equations of motion will be a special instance of Newton's second law, differing from the usual statement of the second law only in that we write them for unit mass. The second law will be applied in terms of linear motion. To obtain these equations for an elastic medium in Cartesian coordinates, we consider the variation in stress across our small (elemental) parallelepiped (refer to Fig. I.4 for a two-dimensional illustration). A total of six separate component stresses act parallel to each axis; in general, the components will vary from face to face. We take an approach for our derivation similar to that given in Section II-A. A more sophisticated mathematical approach in the development can be used if desired; see, for example, the text by Eringen [8].

The resultant force acting in the x_1 direction is treated first. We can write in a Maclaurin series expansion to first order about point P,

$$\left(S_{11} + \frac{\partial S_{11}}{\partial x_1}\frac{1}{2}\delta x_1\right)\delta x_2 \delta x_3 - \left(S_{11} - \frac{\partial S_{11}}{\partial x_1}\frac{1}{2}\delta x_1\right)\delta x_2 \delta x_3$$

$$+ \left(S_{12} + \frac{\partial S_{12}}{\partial x_2}\frac{1}{2}\delta x_2\right)\delta x_1 \delta x_3 - \left(S_{12} - \frac{\partial S_{12}}{\partial x_2}\frac{1}{2}\delta x_2\right)\delta x_1 \delta x_3$$

$$+ \left(S_{13} + \frac{\partial S_{13}}{\partial x_3} \frac{1}{2} \delta x_3 \right) \delta x_1 \delta x_2 - \left(S_{13} - \frac{\partial S_{13}}{\partial x_3} \frac{1}{2} \delta x_3 \right) \delta x_1 \delta x_2$$

$$+ \rho X_1 \delta x_1 \delta x_2 \delta x_3$$

where ρ is the density at the origin P and X_1 is the body force per unit mass. This expression simplifies to

$$\left(\frac{\partial S_{11}}{\partial x_1} + \frac{\partial S_{12}}{\partial x_2} + \frac{\partial S_{13}}{\partial x_2} + \rho X_1 \right) \delta x_1 \delta x_2 \delta x_3. \tag{I-1.45}$$

By the second law, we can equate the quantity (I-1.45) to

$$\rho \delta x_1 \delta x_2 \delta x_3 \frac{d^2}{dt^2} \left(u_1 + \frac{\partial u_1}{\partial x_1} \delta x_1 + \dots \right) \tag{I-1.46}$$

where the acceleration is written in terms of the displacement quantities, and the displacement is expanded in another Maclaurin series about P.

At this point, several simplifications can be made in the development. First, the higher-order terms in the acceleration expression are not to be retained. This is consistent with our restriction to small (or infinitesimal) disturbances so that the classical definitions of strain (I-1.17) are sufficiently accurate. In elastic wave propagation in solid media the body forces, such as gravity, can usually be neglected for most practical applications, and we make this approximation here (see Section II-A for occasional exceptions). A further simplification can be made which significantly decreases the complexity of the mathematics and yet does not seriously increase the limitations already imposed upon the results. Because we are restricting our attention to small displacements and velocities, the derivatives d/dt and d^2/dt^2 may be replaced by $\partial/\partial t$ and $\partial^2/\partial t^2$ respectively; specifically,

$$\frac{d}{dt} = \frac{\partial}{\partial t} + \frac{du_i}{dt} \frac{\partial}{\partial x_i} \approx \frac{\partial}{\partial t}. \tag{I-1.47}$$

This approximation is consistent with the earlier statement that restriction to small relative motion enables mathematical representations in either the Eulerian or Lagrangian form. The approximation would not be appropriate in problems where such restriction is not valid, *e.g.*, problems in hydrodynamics. As with the elastic stiffness constants, the mass density ρ of a body is generally treated as constant in elastic wave propagation, being a function of neither position nor time. We relax later the spatial restriction when treating reflection and refraction phenomena.

Upon division of the expression resulting from the combination of expressions (I-1.45) and (I-1.46) by the volume $\delta x_1 \delta x_2 \delta x_3$ and incorporating the above simplifications, we obtain the approximate *equations of small motion*,

$$\rho \frac{\partial^2 u_1}{\partial t^2} = \frac{\partial S_{11}}{\partial x_1} + \frac{\partial S_{12}}{\partial x_2} + \frac{\partial S_{13}}{\partial x_3} \tag{I-1.48}$$

and two other equations written in the x_2 and x_3 directions. We can summarize these equations by writing

$$\rho \frac{\partial^2 u_i}{\partial t^2} = \frac{\partial S_{ij}}{\partial x_j}. \tag{I-1.49}$$

Expressions (I-1.49) hold whatever the stress-strain behavior of the medium might be. To solve these expressions for an elastic material, we must introduce into them the stress-strain relations discussed in Section IV-A.

Substituting Hooke's law (I-1.29) into the equations of motion (I-1.49), we get the field relations

$$\frac{\partial^2 u_i}{\partial t^2} = \frac{\partial}{\partial x_j} (C_{ijk\ell} \epsilon_{k\ell}). \tag{I-1.50}$$

The definitions (I-1.17) for small strain can be used to write the strain components ϵ_{ij} in terms of the displacements u_i. Thus we obtain three equations in terms of the three displacement components,

$$\rho \frac{\partial^2 u_i}{\partial t^2} = \frac{1}{2} \frac{\partial}{\partial x_j} \left[C_{ijk\ell} \left(\frac{\partial u_k}{\partial x_\ell} + \frac{\partial u_\ell}{\partial x_k} \right) \right]$$

$$= \frac{1}{2} C_{ijk\ell} \left(\frac{\partial^2 u_k}{\partial x_\ell \partial x_j} + \frac{\partial^2 u_\ell}{\partial x_k \partial x_j} \right) + \frac{1}{2} \left(\frac{\partial u_k}{\partial x_\ell} + \frac{\partial u_\ell}{\partial x_k} \right) \frac{\partial}{\partial x_j} (C_{ijk\ell}). \tag{I-1.51}$$

Since we do not usually allow the $C_{ijk\ell}$ to have spatial variation (the body is homogeneous), expressions (I-1.51) can be written as

$$\rho \frac{\partial^2 u_i}{\partial t^2} = \frac{1}{2} C_{ijk\ell} \left(\frac{\partial^2 u_k}{\partial x_\ell \partial x_j} + \frac{\partial^2 u_\ell}{\partial x_k \partial x_j} \right). \qquad (I\text{-}1.52)$$

Equations (I-1.52) are the field equations of motion in terms of the three displacement quantities for an anisotropic, homogeneous, perfectly elastic solid under conditions of small (actually, infinitesimal) displacements and negligible body forces.

Now specializing to isotropic media, we substitute the constitutive relations (I-1.32) into the equations of motion (I-1.49) and obtain

$$\rho \frac{\partial^2 u_i}{\partial t^2} = \frac{\partial}{\partial x_j} (\lambda \Delta \delta_{ij} + 2\mu \epsilon_{ij}). \qquad (I\text{-}1.53)$$

As before, (I-1.53) can be written in terms of displacement quantities and in terms of the dilatation

$$\rho \frac{\partial^2 u_i}{\partial t^2} = \frac{\partial}{\partial x_j} \left[\lambda \Delta \delta_{ij} + \mu \left(\frac{\partial u_i}{\partial x_j} + \frac{\partial u_j}{\partial x_i} \right) \right]. \qquad (I\text{-}1.54)$$

Considering again the existence of a homogeneous condition of strain, (I-1.54) becomes

$$\rho \frac{\partial^2 u_i}{\partial t^2} = \lambda \delta_{ij} \frac{\partial \Delta}{\partial x_j} + \mu \left(\frac{\partial^2 u_j}{\partial x_i \partial x_j} + \frac{\partial^2 u_i}{\partial x_j \partial x_j} \right). \qquad (I\text{-}1.55)$$

Expressions (I-1.55) contain the quantity $\left(\partial^2 u_i / \partial x_j \partial x_j \right)$. It specifies that the displacements u_i are differentiated in a particular fashion by a specific operation. This operator is called the *Laplacian operator*. It is defined (with respect to a differentiable scalar function ϕ for example) as $\partial^2 \phi / \partial x_i \partial x_i = \phi_{ii}$ in Cartesian tensor notation and $\nabla^2 \phi = \nabla \cdot \nabla \phi = \text{div} \, \nabla \phi$ in symbolic notation. If the problem is one-dimensional (so that only one space coordinate is involved), then the Laplacian operator can be taken to refer to the process of ordinary differentiation.

By rearranging the terms in Eqs. (I-1.55), they may be given as

$$\rho \frac{\partial^2 u_i}{\partial t^2} = (\lambda + \mu)\frac{\partial \Delta}{\partial x_i} + \mu \frac{\partial^2 u_i}{\partial x_j \partial x_j}. \tag{I-1.56}$$

For illustration, the x_1 component of (I-1.56) can be expanded and written as

$$\rho \frac{\partial^2 u_1}{\partial t^2} = (\lambda + \mu)\frac{\partial}{\partial x_1}\left(\frac{\partial u_1}{\partial x_1} + \frac{\partial u_2}{\partial x_2} + \frac{\partial u_3}{\partial x_3}\right) + \mu\left(\frac{\partial^2 u_1}{\partial x_1^2} + \frac{\partial^2 u_1}{\partial x_2^2} + \frac{\partial^2 u_1}{\partial x_3^2}\right). \tag{I-1.57}$$

Equations (I-1.56) are the field equations of motion in terms of the three displacement quantities for an isotropic, homogeneous, perfectly elastic solid under conditions of small (infinitesimal) displacements and negligible body forces.

It is these equations of motion (I-1.52) and (I-1.56), which are basically associated with mechanical wave propagation and vibration effects in elastic media. Additional initial and boundary conditions are imposed that relate to the specific problem in question. We examine these statements further in the next section.

Relations (I-1.52) and (I-1.56) can be couched in coordinate systems other than that of a rectilinear Cartesian system. The choice depends upon several factors, one of the most important being the essential physical geometry of the given problem. In Eqs. (I-1.56), the Laplacian operator and the dilatation must also be expressed in the system selected. Chapter I-4 discusses these points.

For our purposes, the conservation of momentum (Newton's second law) is sufficient for deduction of a determinate set of expressions. This is not necessarily the situation when treating quasilinear, nonlinear, or nonconservative systems, and other conservation relations such as the equation of continuity and conservation of energy have to be included in the analysis (see Chapter II-3). A quasilinear equation is defined by the requirement that the dependent variable, but not its derivatives, may appear in the coefficients of the equation. A nonlinear equation has no such restrictions.

C. The Wave Equation: Vibration and Wave Solutions

In the course of establishing the governing dynamic equations for continuous elastic media we are usually led, somewhere in the process, to a linear, homogeneous, second order, partial differential equation with constant coefficients which is in the form of the *wave equation*. This important equation of physics is found to represent basically all types of wave motion (as well as vibratory effects) in all conservative media (or in a vacuum) where the velocity of propagation is constant. Although this expression is only an approximation of physical reality, it is an extremely satisfactory idealization appropriate to many problems of practical interest. The thorough awareness of its origin and treatment is essential in the study of dynamic elastic phenomena, and contributes significantly to the understanding of many facets of dynamic nonelastic phenomena.

In this section, we briefly define the wave equation and show how it can be obtained from a set of equations of motion and conditions of environment that are imposed to satisfy typical geometrical and kinematical constraints. We discuss some solutions to the wave equation and their physical meanings.

We can illustrate the manner in which the wave equation is generated using the equations of motion (I-1.56). The process of differentiating each equation by $\partial/\partial x_i$ in respective order and then adding gives

$$\rho \frac{\partial^2 \Delta}{\partial t^2} = (\lambda + 2\mu) \frac{\partial^2 \Delta}{\partial x_i \partial x_i}. \tag{I-1.58}$$

We recall that the parameters ρ, λ, and μ are treated as material constants for most of our work in elasticity, enabling Eq. (I-1.58) to be put into the form

$$\frac{\partial^2 \psi}{\partial t^2} = c^2 \frac{\partial^2 \psi}{\partial x_i \partial x_i} \tag{I-1.59}$$

where $\psi = \psi(x_i, t)$ is the dependent variable and is a measure of some property of the disturbance such as displacement or velocity and c is a physical constant, the meaning of which we investigate later.[8] Equation (I-1.59) is the classical *wave equation*, often called the *equation of wave motion* to avoid any confusion with modern wave (quantum) mechanics.

Partial differential equations can be classified into three types: elliptic, parabolic, and hyperbolic. The decision as to which type a particular equation belongs to is quite definite, for certain relations among the coefficients determine the type unequivocally. The wave equation (I-1.59) is an example of the hyperbolic type. It is essential to know the classification of an equation in question since the applicability of techniques of solution depends, in part, upon a knowledge of its type.

One of the chief difficulties of solving partial differential equations is that ordinarily there exists a vast collection of functions which can be taken as solutions. From this collection of functions, one must select the one (or few) solution(s) applicable to a given problem by imposing additional conditions. These additional conditions are usually concerned with the boundary geometry of the given problem and are hence called *boundary conditions*, or what we have previously termed conditions of environment. Boundary conditions which refer to the geometry at the initial instant of interest are termed *initial conditions*. The partial differential equation under consideration together with the appropriate boundary conditions is called in mathematics a *boundary value problem*.

There are two general alternate methods by which solutions to linear partial differential equations are obtained: (1) the method originated by Cauchy and Fourier in which the equation is separated into a set of ordinary differential equations, and in which in general each ordinary equation is treated by the method of power series; and (2) the method originated by Riemann and Hadamard in which the equation is transformed to a new set of independent variables by finding a set of "natural" coordinates that are related in a particular fashion to the original independent variables. The method of separation lends itself to a *standing wave* or *normal mode* solution of the wave equation appropriate to problems in elastic vibration in systems of finite physical dimensions, and the method of transformation lends itself to a *traveling wave* or *progressive wave* solution of that equation appropriate to problems in elastic wave propagation.

We summarize first the *separation of variables* technique used in conjunction with the standing wave solution. The method of separation is better adapted to the solution of linear elliptic differential equations than to the other two types, although it can be applied satisfactorily to the wave equation. We are not concerned in our work with solutions using this technique. For more detail than the comments given below, one can refer to Churchill [15]. For a special method of treating standing wave problems, see Chapter I-5, Section II-C.

The approach used involves an assumption that the solution (if one spatial dimension is involved) is of the form

$$\psi(x, t) = X(x)\, T(t) \tag{I-1.60}$$

where X is a function of only the spatial coordinate and T is a function of only the time. By substitution of the above relationship into the partial differential equation, we find the result can be expressed as one separate collection of terms involving functions of the spatial variable alone, equal to another separate collection of terms involving functions of the time coordinate alone. Thus, each collection of terms must equal a constant — the same constant — which is called the *separation parameter* or *separation constant*. We then have two linear ordinary differential equations, one describing the position variation and one describing the time variation.

The method of power series is the most general approach available for treating the resulting two ordinary differential equations. However, the equations may be of some known form, *e.g.*, exact; or some special techniques may be applicable, *e.g.*, variation of parameters or integrating factors, so that the use of power series is not necessary.

By introducing the boundary conditions into the general solution of the spatial equation a relationship is obtained, called the *frequency equation* (or the *characteristic equation*) to which nontrivial solutions result only for discrete vibrational frequencies, termed *eigenvalues* (also called *eigenfrequencies, characteristic frequencies*, or *proper frequencies*). Then for the nth eigenvalue, a nontrivial solution to the spatial differential equation can be found which is named the nth *eigenfunction* (or *wave function*).

In a similar fashion, there corresponds to the nth eigenvalue a nontrivial solution to the temporal differential equation which is again called the nth eigenfunction. Thus, a multiplication of these two functions constitute a solution to the original partial differential equation; since we are considering an equation of second order, we have two arbitrary constants in the combined expression. From the principle of superposition, any linear combination of these solutions will also satisfy the same partial differential equation and boundary conditions.

By invoking the initial conditions through the use of Fourier analysis, the two constants in the general solution can be uniquely found and the particular vibration which will occur can be determined.

In the uniform one-dimensional system we have discussed the eigenvalues are simple integral multiples or harmonics of a basic frequency called the *fundamental frequency*. The fundamental frequency (or fundamental eigenvalue) is the lowest frequency naturally occurring for the system in question and is assigned the value $n = 1$. The second harmonic is twice the fundamental and is labeled $n = 2$. The eigenfunctions are simple sine and cosine functions and are termed the *normal modes* of vibration, the lowest being the fundamental mode. The resulting collective motion that is composed of one or more modes is known as oscillatory *simple harmonic motion*, further discussion of which is given later in this section.

If higher space dimensions are treated, the same technique of separation can be used, but more separation constants result and the eigenvalues do not in general form a simple series. However, the eigenfunctions do enjoy orthogonal properties and represent normal modes of oscillation.

We summarize next with some greater elaboration the *transformation of variables* technique. This approach is often called the *method of characteristics*, and occasionally *D'Alembert's method of integration*. Although the two solutions to the wave equation are considered separately they are nevertheless interconnected; we shall illustrate this fact.

In the method of transformation, as applied to the wave equation in one spatial dimension, a set of natural coordinates is formed from two families of *characteristics* which define a grid or network of straight lines or curves in a plane; along each of these curves one of the solutions to the transformed partial differential equation is constant. The form of the original partial differential equation after transformation to the natural coordinates is called the *normal form*. To have physical meaning, the method of transformation is applicable to hyperbolic[9] equations, in which the two families of characteristics are real curves. An advantage of this approach compared with the separation method is that quasilinear equations may be treated — a point of considerable importance in subjects such as hydrodynamics and supersonic gas dynamics.

It is first necessary to determine just which transformation of independent variables in the wave equation is the most appropriate to use, that is, what new coordinate system is most useful for our purposes. In some cases, the transformation can be found formally. Often, however, experience plays an important role in obtaining a satisfactory transformation. One way to proceed is to examine further the unidimensional standing wave solutions to the wave equation. Solution (I-1.60) can often be written in the form

$$\psi = \psi \left(\sin \frac{n\pi x_1}{\ell} \, \cos \frac{n\pi ct}{\ell} \right) \tag{I-1.61}$$

where ℓ is some characteristic length (for example, the length of a wire between supports). We find from trigonometry and from periodicity of harmonic functions that

$$2 \sin \frac{n\pi x_1}{\ell} \, \cos \frac{n\pi ct}{\ell} = \sin \left[\frac{n\pi}{\ell} (x_1 - ct) \right] + \sin \left[\frac{n\pi}{\ell} (x_1 + ct) \right]. \tag{I-1.62}$$

It follows that the most general solution of the one-dimensional wave equation is

$$\psi = f(x_1 - ct) + g(x_1 + ct) \tag{I-1.63}$$

where f and g are arbitrary functions of the arguments $(x_1 - ct)$ and $(x_1 + ct)$, respectively. These quantities f and g must be consistent with the requirements of continuity, small amplitude, and the various imposed boundary conditions. Two functions are necessary because the wave equation is of second order. Examples of such arbitrary functions are $(x_1 - ct)^3$ and $\ell n(x_1 + ct)$. That (I-1.63) is the general solution of the uniaxial wave equation can be verified by simple substitution.

It is logical then to choose a transformation of variables similar to the argument found in the wave equation solution (I-1.63). Thus, we transform from the coordinates x_1 and t to the coordinates $\xi = x_1 - ct$ and $\eta = x_1 + ct$. The terms ξ and η are the natural coordinates (or characteristics) of the one-dimensional wave equation that form a grid in a new (transformed) coordinate system. This transformation in coordinates enables the wave equation to be written in the normal form

$$\frac{\partial^2 \psi}{\partial x_1^2} - \frac{1}{c^2}\frac{\partial^2 \psi}{\partial t^2} = 4\,\frac{\partial^2 \psi}{\partial \xi \partial \eta} = 0. \qquad\qquad (\text{I-1.64})$$

It is obvious that the differential equation in normal form [the second part of expression (I-1.64)] is a significant simplification with respect to solution than when in regular form [the first part of expression (I-1.64)]. The solution in normal form is readily seen to be Eq. (I-1.63).

Application of the method of characteristics to differential equations in which there are more than two independent variables is possible but involved, and does not, in general, yield any essential simplification of the equations. Nevertheless, even though its use thus appears limited for us. there accrues a decided increase in clarity of both the theory and of the physical ramification that is of value when we extend our treatment to three-dimensional considerations in Chapter I-2. The solutions there logically follow from the ideas developed here.

Another transformation technique associated with the method above and applicable to the wave equation can also be employed, although we do not require its inclusion in our work. The method is useful in problems in which c is not a constant but a slowly varying function of the space coordinates. In effect, the general solution to the wave equation can be modified with a transformation so as to separate the spatial dependency and the time dependency. For example, the first term in solution (I-1.63) can be assumed to be of the form

$$f = f[W(x_1) - c_0 t] \qquad\qquad (\text{I-1.65})$$

where c_0 is some reference constant, seen below to be interpreted as a reference velocity. Use of a solution of the type (I-1.65) enables one to transform the wave equation into a time-independent, first-order, partial differential conditional expression called the *eikonal equation*. Its solutions $W(x_1)$ (in three-dimensional space) represent surfaces along which, at a given instant of time, all points will be in phase. Such a surface is called a *wave surface* or *wave front*, the physical meaning of which is commented upon later in the section and in Chapter I-2. The normals to this surface are called *rays* and define the direction of propagation of the wave, specifically the direction of energy propagation.

In those situations where c is not a constant. the solution of the eikonal equation is not, in general, an exact solution of the wave equation. However in most practical mechanical wave propagation problems in which rapid changes in the physical properties do not occur, the eikonal equation is a good approximation. Refer to the text by Officer [16] for detailed discussion.

Although we do not do so, it is at this point that *Hugyens' principle* could be conveniently introduced and studied. In fact an alternate derivation of the eikonal equation can be obtained using physical arguments by invoking Hugyens' principle. This principle states that the position of the wave disturbance at a given time can be obtained by considering the effect by each point acting as a secondary source on the wave surface at a time an instant earlier. Hugyens' principle is useful in tracing the progress of elastic and nonelastic reflecting waves.

It is helpful to comment now on the physical meaning of a progressive wave from several points of view, first in association with characteristic theory second with emphasis upon a pictorial interpretation, and third with respect to its interrelation and connection with a standing wave.

As we have seen, the technique of transformation of coordinates has enabled us to introduce functions ξ and η (the characteristic curves) which are solutions of the unidimensional wave equation in the x_1, t plane. An equivalent statement is that, along the characteristic lines in the x_1, t plane, there are no changes in the quantities ξ and η. We may then write $x_1 - ct = $ constant and $x_1 + ct$ = constant, from which relations the expressions $dx_1/dt = \pm c$ follow. As noted in the Introduction, a wave may be described (ideally) as a configuration of the medium (such as the transverse shape of a wire or the distribution of density in a liquid) which moves through that medium with a finite velocity without giving the body as a whole any permanent displacement. Thus, these characteristic curves may be interpreted as the paths of disturbance waves that are moving at a velocity of $\pm c$ in a medium at rest. At each instant of time a wave front will correspond with a member of the characteristic family, and at a later time with some other member of the family. Hence, in a problem in which boundary conditions are imposed, a given boundary must cross both families of characteristics in order that f and g in solution (I-1.63), for example be specified for all values of ξ and η. When a boundary coincides with a characteristic, information is obtained only on that particular characteristic and none other in the same family.

The characteristic curves are quite naturally viewed as a net of interacting disturbances. The number of disturbances to be considered is arbitrary, depending only on the number of characteristics one desires to examine. Although in some instances an analytic solution can be obtained readily, *e.g.*, in a simple wave region, it becomes obvious that numerical treatments using computers are of great assistance in applying the theory of characteristics. For further discussion, the texts by Morse and Feshbach [17], Abbott [18], and Zel'dovich and Raizer [19] are of particular value; the reference by Courant and Friedricks [20] is also excellent, but greater emphasis is given there to the abstract theoretical aspects.

The general solution for a progressive wave in this case can be assigned a very simple interpretation that aids in its visualization. The interpretation is seen with particular ease in one space dimension. For illustration consider only the function f of Eq. (I-1.63). Then at any time t_1 ψ is a function of x_1 only, and can be represented by a certain curve such as shown in Fig. I.8; the specific

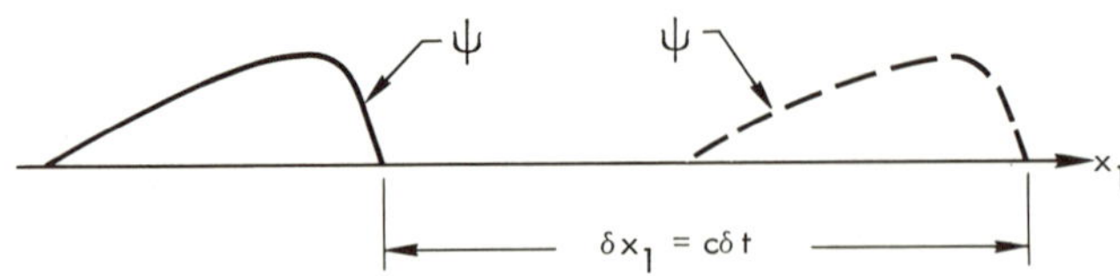

Fig. I.8. Progressive wave on one space dimension.

shape of the curve is given by f. After time is increased by an amount δt, the argument of the function f becomes $x_1 - c(t + \delta t)$. Since we are considering the unidimensional wave equation in which c is a constant and since we are dealing with propagation in an ideal elastic medium (obeying Hooke's law), the shape of the wave does not change during propagation.[10] The function f will remain unchanged, provided that simultaneously with the increase of t by δt the points along x_1 change by an amount $\delta x_1 = c\delta t$. This means that the curve ψ as shown for the time t in the figure can also be used for the time $t + \delta t$ if it is appropriately displaced in the x_1 direction. Hence, it is seen that f in the solution represents a wave moving in the direction of the positive x_1 axis with a constant velocity c. In the same manner, it can be shown that the function g represents a wave traveling in the direction of the negative x_1 axis. Thus, the solution (I-1.63) physically represents two waves progressing along the x_1 axis in opposite directions, each with constant velocity c.

The velocity of propagation of an elastic wave is not necessarily related to the material velocity; the latter, in Eq. (I-1.63) for example, is given by $\partial\psi/\partial t$. In fact, for these idealized linear waves, c is independent of the particle motion and remains constant irrespective of the shape and amplitude of the wave. It is noted that disturbances propagating with "large" amplitudes cannot necessarily be considered as moving with a constant velocity c and may be amplitude-dependent (see Chapter II-3).

It is possible to have elastic waves in three dimensions, other than plane, that maintain their shape but not their size. For if we consider a disturbance spreading from a point source, *i.e.*, a spherical wave, the deformation will depend on the value of the distance r from that point. Since $r^2 = x_1^2 + x_2^2 + x_3^2$, we have

$$\frac{\partial^2 \psi}{\partial x_1^2} = \frac{x_1^2}{r^2}\frac{\partial^2 \psi}{\partial r^2} + \frac{1}{r}\left(\frac{r^2 - x_1^2}{r^2}\right)\frac{\partial \psi}{\partial r}$$

with similar expressions for $\partial^2 \psi/\partial x_2^2$ and $\partial^2 \psi/\partial x_3^2$. Hence Eq. (I-1.59) becomes

$$\frac{\partial^2 \psi}{\partial t^2} = c^2\left(\frac{\partial^2 \psi}{\partial r^2} + \frac{2}{r}\frac{\partial \psi}{\partial r}\right)$$

or, since r is an independent variable,

$$\frac{\partial^2 (r\psi)}{\partial t^2} = c^2\frac{\partial^2 (r\psi)}{\partial r^2}. \qquad (I\text{-}1.66)$$

This is of the form of the classical wave equation and its general solution is

$$r\psi = F(r - ct) + G(r + ct) \qquad (I\text{-}1.67)$$

where F represents a spherical wave diverging from the origin of coordinates with the velocity c while G represents a converging spherical wave. The amplitude in both cases is inversely proportional to the distance r. Hence, although we consider a spherical wave in a perfectly elastic medium, the size, *i.e.*, the amplitude, varies as a function of position. Note that this simple analysis becomes invalid at the origin where a singularity exists.

The amplitude variation could also be predicted from physical reasoning. The area of the wave front of a diverging spherical wave increases as r^2. Thus the energy flow per unit area decreases as r^{-2}. As we will see later, energy flow is proportional to ψ^2. From solution (I-1.67) this flow would indeed be expected to decrease as r^{-2}, indicating that the form of the equation is correct.

Examination of a disturbance propagating with symmetry about an axis, *i.e.*, a cylindrical wave, leads to some interesting conclusions in comparison with those generated with regard to a spherical wave. Here we find that the amplitude decreases (approximately) as $r^{-1/2}$ rather than as r^{-1} for the spherical wave. Thus both the shape and size change during transmission of the cylindrical wave. Again, the amplitude factor for these axially symmetric waves could be developed from physical considerations using plausibility arguments.

It becomes reasonably clear now what is meant by a standing wave (generated by a separation of variables approach) and by a traveling wave (generated by a transformation of variables approach). (It is also obvious why we are interested in the latter.) In the comparison of the two wave patterns, a standing wave can be thought of as a superposition of two progressive waves moving in opposite directions that interfere constructively and destructively[11] to form the particular normal mode pattern. Thus, it is seen there exists a close correlation between problems in waves and problems in vibrations. Traditionally vibrations of systems or bodies (or, more specifically, consideration of normal modes of vibrations of finite systems) were used by the engineer, whereas the transmission of waves in bodies were of more interest to the mathematical physicist. This distinction in viewpoint no longer holds completely.

The use of a single simple harmonic progressive plane wave of infinite duration will be of considerable utility to us in the following chapters. An important type of periodic motion frequently occurring in nature can be closely approximated by such an idealized wave (or number of waves). A simple harmonic traveling wave is one in which the disturbance at any point in space varies sinusoidally with time and at any point in time varies sinusoidally in space. On the other hand, in a simple harmonic standing wave there are certain points (nodes) at which the distrubance vanishes for all time. The text by Braddick [21] presents these distinctions clearly.

It is necessary to be able to describe in some mathematical fashion this simple harmonic progressive plane wave train. A convenient and reasonable starting point from which to develop an amenable form is with the one-dimensional expressions (I-1.61) and (I-1.62). Since we are considering an idealized single harmonic wave of unique frequency, we can modify those expressions to read

$$\psi = A \, \sin\!\left[\frac{2\pi}{\Lambda}\,(x_1 - ct)\right] \qquad\qquad\qquad (\text{I-1.68})$$

where we have taken only the forward advancing wave. The quantity A is a constant, and the parameter Λ is the *wavelength*. The wavelength is the distance between two successively repeated points on the wave, *e.g.*, the crest or trough; equivalently, it is the distance a point on the wave moves for one cycle of the sine function. Thus in Eqs. (I-1.61) and (I-1.62), $\ell = \Lambda/2$. The term $2\pi/\Lambda$ is the number of waves per cycle and is called the *wave number*; it is usually given the symbol k [not to be confused with the coefficient of bulk modulus in Eq. (I-1.44)]. In an Eulerian frame of reference, the time to complete one full oscillation cycle is defined as the *period* τ; from Eq. (I-1.68) we write

$$kct = kc\tau = 2\pi$$

or

$$\tau = \frac{2\pi}{kc} = \frac{\Lambda}{c}. \tag{I-1.69}$$

Since the frequency ν is given as the number of cycles occurring per unit time, it follows that

$$\nu = \frac{1}{\tau} = \frac{c}{\Lambda}. \tag{I-1.70}$$

Hence, relationship (I-1.68) can be expressed as

$$\psi = A \sin \left[2\pi \frac{x_1}{\Lambda} - \nu t \right]$$

$$= A \sin (kx_1 - pt) \tag{I-1.71}$$

where $p = 2\pi\nu$ is the *circular frequency* and is defined as the number of radians per unit time. Since we are considering only one harmonic wave that is of infinite duration, we need not introduce a *phase shift* or *initial phase* here. The phase shift refers to where one begins "counting" a wave, necessary, for example, when treating interference or reflection phenomena of several waves together (see Chapter I-3).

For purposes of generality and mathematical convenience, Eq. (I-1.71) is often extended, first by using the superposition principle to include both sine and cosine terms, then by rewriting the trigonometric equation in terms of complex quantities, and finally by using Euler's formula[12] to express the result in exponential form (in Cartesian space). From these extensions, we have

$$\psi = U \exp [i (kx_i - pt)] \tag{I-1.72}$$

where U is a real coefficient related to the amplitude of the wave. Usually only the real part is required in the solution, but often an analysis can be made easier by using the complete exponential form, *e.g.*, in the determination of the phase relationship among the various mechanical and acoustic variables. We can further extend Eq. (I-1.72) to three dimensions as the need arises.

If the disturbance under analysis is not periodic or is of finite duration, then representation in the form (I-1.72) cannot strictly be given. However, an analysis based upon a combination of simple harmonic motion components each of different frequency is often possible. Discussion of such analyses is given in the following chapters, particularly in Chapter I-2, Section V.

It is obvious that understanding this idealized motion, such as specified by relationship (I-1.72), has both considerable basic and practical value.

V. Summary

The notion of stress can be introduced by considering the distribution of surface forces over a tetrahedron of infinitesimal volume imbedded in equilibrium in the interior of a continuous solid body. The specification of nine stress components S_{ij} are sufficient to characterize completely the state of stress at any point in the medium. The components S_{ii} (i not summed) are the normal stress components; the others are the shearing components. In general, these numbers are different at different points and constitute a stress field. The stress can be proved to be a second rank tensor. Moreover, it is symmetric so that among its components the relation $S_{ij} = S_{ji}$ holds, providing there exists no distribution of body torques within the body. This property of symmetry allows complete specification of stress at a point in a body with at most six independent numbers.

The notion of strain can be introduced by considering the relative displacement of two points in a deformable body. If δu_i is the relative displacement between any two points δx_j apart in a deformed body, and assuming the small strain approximation holds, we can write this relative displacement in rectilinear Cartesian coordinates from Taylor's series as

$$\delta u_i = \frac{\partial u_i}{\partial x_j}\, \delta x_j = \epsilon_{ij}\, \delta x_j + \omega_{ij}\, \delta x_j$$

where

$$\epsilon_{ij} = \frac{1}{2}\left(\frac{\partial u_i}{\partial x_j} + \frac{\partial u_j}{\partial x_i}\right)$$

and are defined as the components of strain; and

$$\omega_{ij} = \frac{1}{2}\left(\frac{\partial u_i}{\partial x_i} - \frac{\partial u_j}{\partial x_j}\right),$$

and are defined as the components of body rotation. The strain components ϵ_{ij} ($i = j$, i not summed) are tensions and/or compressions; the components ϵ_{ij} ($i \neq j$) are shearing strains. Strain has an arbitrary definition, but displacement is a consequence of the nature of the body and the boundary conditions imposed upon it and cannot be arbitrarily specified. Strain can be shown to be a symmetrical second rank tensor. The rotation tensor can also be shown to be of second rank, but it is found to be antisymmetrical, *i.e.*, $\omega_{ij} = -\omega_{ji}$.

The stress and strain fields are interrelated by means of some mathematical model. For small deformations, application of a generalization of Hooke's law of elasticity allows a connection between stress and strain that is satisfactory for the description of both the quasistatic and dynamic mechanical behavior of a great number of materials of physical interest, and yet is generally amenable to mathematical analysis. In three dimensions, this law can be written as

$$S_{ij} = C_{ijk\ell}\epsilon_{k\ell}, \tag{I-1.29}$$

where the numbers $C_{ijk\ell}$ form a fourth rank tensor and are called the elastic stiffness constants of the material. From symmetry and thermodynamical considerations, the number of stiffness constants is reduced to 21 in the most general anisotropic situation (triclinic).

For the most part, we consider bodies as isotropic (with properties identical in all directions) and homogeneous (with properties identical at all points). The condition of isotropy reduces the number of independent elastic constants to two, λ and μ, called Lamé's constants. Hooke's law then becomes

$$S_{ij} = \lambda\Delta\delta_{ij} + 2\mu\epsilon_{ij} \tag{I-1.32}$$

where δ_{ij} is the Kronecker delta, defined (in the Introduction) as

$$\delta_{ij} = \begin{cases} 0 \text{ if } i \neq j \\ 1 \text{ if } i = j \end{cases} \tag{5}$$

and where Δ is the dilatation defined,

$$\Delta = \epsilon_{ii}. \tag{I-1.33}$$

The equations of motion appropriate to continuous media in which all strains can be considered small are

$$\rho\frac{\partial^2 u_i}{\partial t^2} = \frac{\partial S_{ij}}{\partial x_j} \tag{I-1.49}$$

where ρ is the mass density, u_i are the displacements, and S_{ij} are the stress components. These relations hold irrespective of the stress-strain behavior of the medium, although the behavior must be consistent with certain physical and mathematical idealizations, mostly associated with the condition of small strain.

The field equations of motion in terms of the three displacement quantities u_i for an anisotropic, homogeneous, perfectly elastic solid under conditions of small displacements and negligible body forces are

$$\rho \frac{\partial^2 u_i}{\partial t^2} = \frac{1}{2} C_{ijk\ell} \left(\frac{\partial^2 u_k}{\partial x_\ell \partial x_j} + \frac{\partial^2 u_\ell}{\partial x_k \partial x_j} \right). \tag{I-1.52}$$

The field equations of motion in terms of the three displacement quantities for an isotropic, homogeneous, perfectly elastic solid under conditions of small displacement and negligible body forces are

$$\rho \frac{\partial^2 u_i}{\partial t^2} = (\lambda + \mu) \frac{\partial \Delta}{\partial x_i} + \mu \frac{\partial^2 u_i}{\partial x_j \partial x_j}. \tag{I-1.56}$$

Although relations (I-1.52) and (I-1.56) are couched in terms of a rectilinear Cartesian coordinate system, other systems can be used. The system selected depends primarily upon the governing physical geometry of a given problem.

In solutions of elastic stress wave problems, using Eqs. (I-1.52) or (I-1.56), we are led in the process to an expression in the form of the wave equation, a linear, homogeneous, second-order, partial differential relation with constant coefficients. The wave equation can be written

$$\frac{\partial^2 \psi}{\partial t^2} = c^2 \frac{\partial^2 \psi}{\partial x_i \partial x_i} \tag{I-1.59}$$

where $\psi = \psi(x_i, t)$ is the dependent variable and is a measure of some property of the disturbance such as displacement or velocity, and c is a physical constant which can be interpreted as the propagation velocity of the wave. Boundary conditions must be applied in addition to the differential equation in order to define properly and sufficiently a given problem.

There are two general alternate methods of solution to the wave equation boundary value problem: (1) the method of separation of variables in which the equation is separated into a set of ordinary differential equations, and (2) the method of transformation in which the equation is transformed to a new set of independent variables (natural coordinates) so as to simplify analysis. The method of separation lends itself to a standing wave or normal mode solution of the wave equation that is appropriate to problems in elastic vibration in systems of finite physical dimensions; the method of transformation (or method of characteristics) lends itself to a traveling wave or progressive wave solution of that equation which is appropriate to problems in elastic wave propagation. Both solutions are interrelated, although we are more concerned with the latter.

The general traveling wave solution of the wave equation in one space dimension is

$$\psi = f(x_1 - ct) + g(x_1 + ct) \qquad (\text{I-1.63})$$

where f and g are arbitrary functions of the arguments and are obtained from the boundary conditions of the problem. The solution represents two waves, one moving in the direction of the positive x_1 axis with a constant velocity c, and one traveling in the direction of the negative x_1 axis with the same velocity c.

The use of a single simple harmonic progressive wave train of infinite duration is of considerable mathematical utility. This periodic motion can be expressed as

$$\psi = U \exp [i(kx_1 - pt)] \qquad (\text{I-1.72})$$

where U is a real coefficient related to the amplitude of the wave, i is the imaginary unit, k is the wave number (defined as $2\pi/\Lambda$, in which Λ is the wavelength), and p is the circular frequency (defined as $2\pi\nu$, in which ν is the frequency).

Notes

[1] Although for some time agreement with experimental values using Newton's formula was known to be quite poor, it was not until 1816 that Laplace corrected the theoretical expression by considering the transmission of a wave in air as an adiabatic process.

[2] Some authors use the term stress vector to indicate traction. However, stress in general is not a vector, and to avoid any confusion we use the other terminology.

[3] The consideration of body motion (or dynamic behavior) at this point is in advance of our discussion. However, such consideration here is unavoidable if a general treatment is desired.

[4] When the strains are finite and if recourse is made to the definitions (I-1.18), it is no longer possible to give simple geometrical interpretations.

[5] Finite strain, as given by expressions (I-1.18), can also be shown to transform as a second rank tensor (see Wasley [12]).

[6] Using energy considerations (Wasley [12]), Poisson's ratio for a single-phase material can be shown to be bounded by the numbers 0 and $\frac{1}{2}$; i.e., $0 < \sigma < \frac{1}{2}$. For many structural materials the value of σ lies between 0.25 and 0.35; for a substance that is relatively fluid-like and incompressible (in an unconfined condition), σ approaches one-half.

[7] It is interesting to note the analogy here to the idealized concept of wave propagation in a rigid body commented upon in Section I in which the propagation velocity in the body is infinite.

[8] In Eq. (I-1.58), $\psi = \Delta$ and $c = [(\lambda + 2\mu)/\rho]^{1/2}$.

[9] This method is also useful when treating parabolic equations, e.g., the diffusion equation. However, in this situation there is only one real family of characteristics on which the solutions are constants, and this degeneracy causes the method to lose some of its value.

[10] Practically, the amplitude of the disturbance does decrease because of dissipation of energy.

[11]The term "interference" is somewhat misleading since the waves do not physically affect each other; the word, "interaction" is perhaps more appropriate. Interfere is a technical term used in physics which means the resultant effect produced by two or more waves (or influences) which arises as a consequence of linear response and the validity of the superposition principle, *cf.*, Newton's rings.

[12]$e^{i\theta} \equiv \exp(i\theta) = \cos\theta + i\sin\theta$.

References

[1] A. H. Cottrell, *The Mechanical Properties of Matter*, Wiley, New York, 1964.

[2] A. E. H. Love, *A Treatise on the Mathematical Theory of Elasticity*, 4th ed., Chap. III, pp. 1-3, Dover, New York, 1944.

[3] J. W. S. Rayleigh, *The Theory of Sound*, 2nd ed., Vol. I, pp. v-xxxii, Dover, New York, 1945.

[4] I. Todhunter and K. Pearson, *A History of the Theory of Elasticity*, Vol. II, Parts 1 and 2, Cambridge University Press, London, 1893.

[5] C. Truesdale, "Reactions of the History of Mechanics Upon Modern Research," *Proc. Fourth U.S. National Congress of Appl. Mech.*, p. 35, ASME, Berkeley, 1962.

[6] A. E. Green and W. Zerna, *Theoretical Elasticity*, Oxford University Press, London, 1954.

[7] I. I. Gol'denblat, *Some Problems of the Mechanics of Deformable Media*, P. Noordhoff, Groningen, The Netherlands, 1962.

[8] A. C. Eringen, *Mechanics of Continua*, Chaps. 1, 3, and 6, Wiley, New York, 1967.

[9] L. E. Malvern, *Introduction to the Mechanics of a Continuous Medium*, Chap. 3 and Sec. 5.3, Prentice-Hall, Englewood Cliffs, N. J., 1969.

[10] V. V. Novozhilov, *Foundations of the Nonlinear Theory of Elasticity*, pp. 1-16, Graylock Press, New York, 1953.

[11] I. S. Sokolnikoff, *Mathematical Theory of Elasticity*, 2nd ed., pp. 29-33, McGraw-Hill, New York, 1956.

[12] R. J. Wasley, *Propagation of Elastic Stress Disturbances in Deformable Solids*, Lawrence Livermore Laboratory, Report UCRL-14616, pp. 5-12, 147-158, 173-174, 1965.

[13] J. F. Nye, *Physical Properties of Crystals*, Chap. VIII, Oxford University Press, London, 1957.

[14] S. Timoshenko and J. N. Goodier, *Theory of Elasticity*, 2nd ed., pp. 6-10, McGraw-Hill, New York, 1951.

[15] R. V. Churchill, *Fourier Series and Boundary Value Problems*, McGraw-Hill, New York, 1941.

[16] C. B. Officer, *Introduction to the Theory of Sound Transmission*, Chap. 2, McGraw-Hill, New York, 1958.

[17] P. M. Morse and H. Feshbach, *Methods of Theoretical Physics*, Part I, pp. 676-692, McGraw-Hill, New York, 1953.

[18] M. B. Abbott, *An Introduction to the Method of Characteristics*, American Elsevier, New York, 1966.

[19] Ya. B. Zel'dovich and Yu. P. Razier, *Physics of Shock Waves and High-Temperature Hydrodynamic Phenomena*, Vol. I, pp. 15-44, Academic, New York, 1966.

[20] R. Courant and K. O. Friedricks, *Supersonic Flow and Shock Waves*, pp. 40-62, 75-78, Wiley-Interscience, New York, 1948.

[21] H. J. J. Braddick, *Vibrations, Waves, and Diffraction*, Chaps. 3 and 6, McGraw-Hill, New York, 1965.

WAVE PROPAGATION IN EXTENDED MEDIA

I. General

For solving the field equations of small motion for a homogeneous, perfectly elastic solid in which body forces are negligible, we require additional boundary conditions that relate to the specific geometry of the problem in question (Chapter I-1, Sections IV-B and IV-C). However, in the start of a motion resulting from a disturbance initiated at an internal point of the medium, the parts of the body near the boundary remain at rest, and the motion is the same whether the body is bounded or unbounded. Geometric complications such as wave reflections do not enter the unbounded problem. Accordingly, this simpler situation is treated first. These statements are not meant to imply that elastic (and nonelastic) waves in extended media are without practical importance. Their significance will become apparent in the following chapters, particularly in Chapters I-5 and II-3.

We reiterate that much of our elastic wave study is formulated upon the assumption that the motion is simple harmonic in nature and is thus describable by a harmonic progressive wave train of infinite duration (see Chapter I-1, Section IV-C).

II. Propagation in Anisotropic Media

We now discuss the propagation of small disturbances in anisotropic elastic solids. The topic is not difficult, and is of considerable importance in studies of crystalline materials. It will be seen, however, that any degree of anisotropy other than complete isotropy involves very cumbersome mathematical formulation and operation. For the reasons given in Chapter I-1, Section III-B, greater emphasis is placed upon transmission in isotropic media, and therefore, the following discussion is brief.

A straightforward formal procedure exists for examining the various kinds of waves that are possible in an extended medium (from the point of view of different discrete velocities). An expression for a simple harmonic wave in three dimensions is substituted directly into the expressions governing the propagation of elastic waves in anisotropic bodies. A determinantal relation is thereby obtained, enabling us to calculate the several propagation velocities. We employ essentially this technique in treating isotropic media in the next section.

However, such a general and direct approach gives greater detail, with accompanying unnecessary complexity, than is required at this time. Three simplifications are introduced that are helpful and illustrative both with respect to the actual mechanics and to the mathematics involved.

First, in the analyses we utilize, for the most part, plane waves [for example, a wave generalized to 3-space from Eq. (I-1.72)] propagating in a rectangular Cartesian coordinate system [Eq. (I-1.52)]. The solutions appropriate to such waves illustrate most of the important features of elastic wave transmission in solids.

Second, a rather specialized representation of a plane wave is employed instead of the complete harmonic description. The justification can be seen by recalling that, at this point, we are only concerned with the investigation of the various wave velocities possible and their associated directions, not with the wave shapes and their associated stresses and strains. Since any arbitrary plane disturbance (including a simple harmonic wave) is transmitted through an elastic medium without change in shape, *i.e.*, all parts of the given elastic plane disturbance are propagated at a constant velocity irrespective of its shape, the simplest description is advantageous.

Third, rather than using relations (I-1.52) directly in their final form, we obtain a more lucid presentation by retracing our steps somewhat and partially redeveloping (I-1.52) in terms of the plane wave description that we assume. Moreover, it is helpful to make the derivation in relation to, say, the x_1 component first for clarity and then to generalize using tensor notation.

Let us consider a plane wave advancing with a constant velocity c. A plane wave is one in which the displacement is constant over all points of a plane (termed a wave front in Chapter I-1, Section IV-C) perpendicular to the direction of propagation. We define the normal to the wave front by its direction cosines ℓ_i related, in respective order, to the coordinate axes x_i.[1] The wave front, or wave surface, is given by the equation $\ell_i x_i$ = constant.

Let **n** denote a unit vector along the normal to the advancing wave front and $\{\alpha\}$ the angles between **n** and the axes so that the relation $\ell_i = \cos \alpha_i$ exists. This expression is analogous to (3) in the Introduction in which $a_{ij} = \ell_i$. Double subscripts are not necessary here since one of the coordinate systems that we are "straddling" consists of only one axis, the normal. The condition

$$\sum_{n=1}^{3} \ell_i^2 = 1$$

follows directly from Eq. (4).

Now a unit distance, say s, along this normal has the value $\ell_i x_i$. Displacements along the normal are then functions of a single parameter,

$$\psi' = s - ct = \ell_i x_i - ct. \tag{I-2.1}$$

From Eq. (I-1.63) and the discussion in Section IV-C of the physical meaning of a progressive wave from the point of view of characteristic (transform) theory, it is clear that $\psi = f(\psi')$ and that now we are directing our attention only to the argument of the forward advancing wave along the direction of the vector $\mathbf{n}$. If we were to use the complete description of a simple harmonic plane wave, we would find at a given time the wave to be sinusoidal along the normal, and at a given position on the normal the disturbance to be uniformly periodic in time.

We start with our partial redevelopment of relations (I-1.52) using the definitions of small strain, Eqs. (I-1.17). It is interesting to find that, in this particular situation, the development is rendered somewhat less complicated with the use of the *engineer's* definition of strain,[2] that is, the second of Eqs. (I-1.17) without the factor 2. The choice is really immaterial as long as the reader is made cognizant of that choice, and we follow the lead of convention. Thus, we write

$$\epsilon_{11} = \frac{\partial u_1}{\partial x_1} = \ell_1 \frac{\partial u_1}{\partial \psi'},$$

and for the shearing strains $\hspace{6cm}$ (I-2.2)

$$\epsilon_{ij} = \ell_j \frac{\partial u_i}{\partial \psi'} + \ell_i \frac{\partial u_j}{\partial \psi'}.$$

With the substitution of expressions (I-2.2), the S_{11} component in the generalized Hooke's law (I-1.29) becomes, after rearranging and collecting terms,

$$S_{11} = (C_{1111}\ell_1 + C_{1112}\ell_2 + C_{1113}\ell_3) \frac{\partial u_1}{\partial \psi'}$$

$$+ (C_{1112}\ell_1 + C_{1122}\ell_2 + C_{1123}\ell_3) \frac{\partial u_2}{\partial \psi'}$$

$$+ (C_{1113}\ell_1 + C_{1123}\ell_2 + C_{1133}\ell_3) \frac{\partial u_3}{\partial \psi'}.$$

We can then calculate

$$\frac{\partial S_{11}}{\partial x_1} = \ell_1 \left[(C_{1111}\ell_1 + C_{1112}\ell_2 + C_{1113}\ell_3) \frac{\partial^2 u_1}{\partial \psi'^2} \right.$$

$$+ (C_{1112}\ell_1 + C_{1122}\ell_2 + C_{1123}\ell_3) \frac{\partial^2 u_2}{\partial \psi'^2}$$

$$\left. + (C_{1113}\ell_1 + C_{1123}\ell_2 + C_{1133}\ell_3) \frac{\partial^2 u_3}{\partial \psi'^2} \right].$$

Similar equations are obtained for $\partial S_{12}/\partial x_2$ and $\partial S_{13}/\partial x_3$. By differentiating displacement with respect to time and using relation (I-2.1), we get

$$\frac{\partial u_1}{\partial t} = - c \frac{\partial u_1}{\partial \psi'}$$

and

$$\frac{\partial^2 u_1}{\partial t^2} = c^2 \frac{\partial^2 u_1}{\partial \psi'^2}.$$

Thus the first expression of the equations of small motion (I-1.48) becomes

$$\lambda_{11} \frac{\partial^2 u_1}{\partial \psi'^2} + \lambda_{12} \frac{\partial^2 u_2}{\partial \psi'^2} + \lambda_{13} \frac{\partial^2 u_3}{\partial \psi'^2} = \rho c^2 \frac{\partial^2 u_1}{\partial \psi'^2},$$

together with two other similar equations. We can generalize this result and write in concise form

$$\lambda_{ik} \frac{\partial^2 u_k}{\partial \psi'^2} = \rho c^2 \frac{\partial^2 u_i}{\partial \psi'^2} \tag{I-2.3}$$

where

$$\lambda_{ik} = \frac{1}{2}(C_{ijmk}\ell_j\ell_m + C_{ijkn}\ell_j\ell_n)$$

and

$$\lambda_{ik} = \lambda_{ki}.$$

For illustration, we expand one of the λ_{ik} in terms of the elastic constants[3]

$$\lambda_{11} = C_{1111}\ell_1^2 + C_{1212}\ell_2^2 + C_{1313}\ell_3^2 + C_{1112}\ell_1\ell_2$$

$$+ C_{1213}\ell_2\ell_3 + C_{1113}\ell_1\ell_3.$$

The quantities λ_{ik} are not to be confused with the Lamé constant λ [see Eq. (I-1.32)].

We can now rewrite Eqs. (I-2.3) as a system of homogeneous linear simultaneous equations by simple transposition of the terms $\rho c^2(\partial^2 u_i/\partial \psi'^2)$. From algebra, such a system is consistent and can have nontrivial solutions only if the determinant of the coefficients vanishes. Hence, the condition is

$$\left| \lambda_{ik} - \delta_{ik}(\rho c^2) \right| = 0 \qquad (I\text{-}2.4)$$

where δ_{ik} is the unit tensor, Eq. (5). This determinant is called the *secular determinant* (or sometimes, the *characteristic equation of a square matrix*), and the third degree algebraic equation that it yields upon expansion is often called the *secular equation*.

Since the determinantal equation (I-2.4) is a cubic in c^2, it has three roots. Examination of this equation shows one that, if the λ_{ik} are real, the numbers $\{\rho c^2\}$ are all real. More specifically, it can be shown that these roots are positive and real because of the physical conditions imposed upon the elastic constants and because the wave velocities must be real. Thus, if the strains are small ("infinitesimal"), there are in general three plane waves, each with a distinct velocity, that can be transmitted in a completely anisotropic medium. The values of these velocities are functions of the elastic constants of the material, and the directions of propagation of the waves are defined by the direction cosines ℓ_i.

In the general situation for small strain, the directions of the particle motions accompanying the three plane waves are mutually perpendicular but are at some arbitrary directions relative to the propagation vector of the wave. Recent analytical work by Johnson [1] has examined this situation. If the direction of particle motion in a special instance is parallel to the direction of propagation, the wave is termed *longitudinal*; if normal, the wave is called *transverse*. Special directions exist in crystals for which the particle motions are either along or perpendicular to the direction of propagation; such directions are of practical importance in measuring the elastic constants of these crystals (see Chapter I-5, Section I-A). Usually, however, the wave train produces disturbances of the particles such that the resulting motion is some linear combination (because of the linearity of the governing equations) of longitudinal and transverse waves.

The problem of allowing finite strain from the propagation of waves in elastic media has also been studied (see Truesdell [2]). A general theory has been developed that shows that for anisotropic media there are nine different kinds of waves (insofar as velocities are concerned) that can travel along the principal axes of stress and strain; for isotropic media there are three different types (*i.e.*, velocities) possible. For infinitesimal strains, this general theory reduces to the above situation.

III. Propagation in Isotropic Media

The equations governing motion in isotropic media in terms of the three displacement quantities u_i are, in Cartesian tensor notation,

$$\rho \frac{\partial^2 u_i}{\partial t^2} = (\lambda + \mu) \frac{\partial \Delta}{\partial x_i} + \mu \frac{\partial^2 u_i}{\partial x_j \partial x_j}. \tag{I-1.56}$$

We treat the problem in two ways, the first being a formal extension of the technique advanced in the previous section, and the second being somewhat indirect but physically more illuminating. At present, we are still concerned with propagation of plane waves.

For isotropic materials, the determinantal equation (I-2.4) reduces directly to

$$\left| \ell_i \ell_k (\lambda + \mu) - \delta_{ik} (\mu - \rho c^2) \right| = 0. \tag{I-2.5}$$

For illustration, Eq. (I-2.5) may be partially expanded by writing

$$\left| \begin{matrix} \ell_1^2 (\lambda + \mu) + \mu - \rho c^2 & \ell_1 \ell_2 (\lambda + \mu) & \ell_1 \ell_3 (\lambda + \mu) \\ \cdots & \cdots & \cdots \end{matrix} \right| = 0.$$

Evaluation of the determinant and use of the relation for direction cosines

$$\sum_{n=1}^{3} \ell_i^2 = 1$$

leads to

$$(\lambda + 2\mu - \rho c^2)(\mu - \rho c^2)^2 = 0. \tag{I-2.6}$$

Thus for an extended isotropic elastic solid, in which small strains are allowed, there are two, and only two, distinct wave velocities possible (since two of the three roots are equal). We postpone further examination of the ramifications of this conclusion until discussion of the second technique of treatment has been given.

In Chapter I-1, Section IV-C, an example was presented to illustrate one method of generating the wave equation; specifically, we developed from the equations of motion (I-1.56) the wave equation in the form

$$\rho \frac{\partial^2 \Delta}{\partial t^2} = (\lambda + 2\mu) \frac{\partial^2 \Delta}{\partial x_j \partial x_j}. \tag{I-1.58}$$

Equations (I-1.58) show that the dilation Δ is propagated through the medium with velocity $c = [(\lambda + 2\mu)/\rho]^{1/2}$.

Now we can manipulate (I-1.56) in another fashion. We can differentiate the $i = 2$ component with respect to x_3 and the $i = 3$ component with respect to x_2. By subtracting the two, we have

$$\rho \frac{\partial^2}{\partial t^2}\left(\frac{\partial u_2}{\partial x_3} - \frac{\partial u_3}{\partial x_2}\right) = \mu \frac{\partial^2}{\partial x_i \partial x_i}\left(\frac{\partial u_2}{\partial x_3} - \frac{\partial u_3}{\partial x_2}\right)$$

or from definition (I-1.17),

$$\rho \frac{\partial^2 \omega_{23}}{\partial t^2} = \mu \frac{\partial^2 \omega_{23}}{\partial x_j \partial x_j}$$

where ω_{23} is the rotation about the x_1 axis. Similar expressions can be obtained for the other two rotations, and we write in general

$$\rho \frac{\partial^2 \omega_{ik}}{\partial t^2} = \mu \frac{\partial^2 \omega_{ik}}{\partial x_j \partial x_j}. \tag{I-2.7}$$

Hence the wave equations (I-2.7) show that the rotations are propagated through the medium with velocity $c = (\mu/\rho)^{1/2}$.

We can obtain yet additional information from expressions (I-1.56). First, if the dilatation is zero, the equations become

$$\rho \frac{\partial^2 u_i}{\partial t^2} = \mu \frac{\partial^2 u_i}{\partial x_j \partial x_j}. \tag{I-2.8}$$

Second, using potential theory, the theorem can be proved that if $\omega_{ij} = 0$, then the u_i are derivable from a scalar potential function ϕ and conversely. The hypothesis of the converse is: if $u_i = \partial \phi/\partial x_i$ the rotations vanish; such a conclusion is readily obtained from (I-1.17), with the usual assumption that the order of differentiation is immaterial. Thus from the definition $\Delta = \epsilon_{ii}$ [Eq. (I-1.33)] it follows that $\Delta = \partial^2 \phi/\partial x_j \partial x_j$ and $\partial \Delta/\partial x_i = \partial^2 u_i/\partial x_j \partial x_j$, and (I-1.56) becomes

$$\rho \frac{\partial^2 u_i}{\partial t^2} = (\lambda + 2\mu) \frac{\partial^2 u_i}{\partial x_j \partial x_j}. \tag{I-2.9}$$

Essentially, what we have done here is to split up our plane wave displacements into two parts: (1) a longitudinal part having zero rotation (or curl) which can be represented as the gradient of a scalar potential and (2) a transverse part having zero dilatation (or divergence) which can be represented as the curl of a vector potential.

It is noted that the use of potential functions is only a mathematical technique that is often helpful and convenient in certain problems to assist in obtaining solutions. The physical nature of the problem is not changed in any way – nor could we change it simply by mathematical manipulation. Upon obtaining the solution, we must generally determine whether it conforms satisfactorily with reality before we can make a judgment concerning the validity of the technique. We need not attach any physical meaning to these potential functions. More extensive use of potential functions is made in Chapter I-3, Section II.

Relations (I-2.6), (I-1.58), (I-2.7), (I-2.8), and (I-2.9) show that, in the interior of an elastic isotropic solid, a plane wave may be propagated with only two different velocities,[4] independent of the wave orientation. Waves involving no rotation travel with velocity

$$c_\ell = \left(\frac{\lambda + 2\mu}{\rho}\right)^{1/2},$$
(I-2.10)

and waves with no dilatation travel with velocity

$$c_s = \left(\frac{\mu}{\rho}\right)^{1/2}.$$
(I-2.11)

The first type of disturbance is transmitted with the higher velocity and is known as a *longitudinal* (hence the subscript ℓ) or *irrotational* wave; the second type is called a *shear* (hence the subscript s) or *equivoluminal* wave. Other names for the first type of wave are *dilatational*, *condensational*, or in seismology, *primary*, *push*, or *P*; other names for the second type are *transverse*, *distortional*, *solenoidal*, or in seismology, *secondary*, *shake*, or *SV* or *SH*, the latter depending on whether the displacements are in a vertical or horizontal plane, respectively. As previously indicated and detailed in Chapter I-2, Section IV, distinct directions of particle motions accompany these waves. Expression (I-2.10) will be found to be of significance in our consideration of nonelastic (shock) waves propagating in unbounded solids.

Equations (I-2.10) and (I-2.11) illustrate that, if the wave is either initially irrotational or initially equivoluminal, it will remain irrotational or equivoluminal, respectively, as long as the Lamé constants have the same values. If λ or μ have spatial variation, wave reflection will occur and a combination of longitudinal and transverse waves may result.

An important feature of these waves is that equivoluminal waves involve distortion without dilatation, but irrotational waves involve both phenomena. The velocity of distortional (equivoluminal) waves depends only upon the density and shear modulus of the medium, so it might appear at initial encounter that the velocity of dilatational (irrotational) waves should depend only upon the density and the bulk modulus, k. However, $k = \lambda + 2/3\mu$, [see (I-1.44)]. so that the velocity of dilatational waves is $[(k + 4/3\mu)/\rho]^{1/2}$, and the shear modulus as well as the bulk modulus is involved. A physical explanation for this situation is that the medium is not subjected to simple compression but rather to a combination of compression and shear. Imagine a small element of material, say a cube, selected from the unlimited medium through which a plane dilatational wave is progressing. Thus, a unit cross section taken normal to the direction of travel will not change during the passage of the wave. However, a dimension taken parallel to the direction of travel will be so altered. Consequently, the diagonal of the cube changes in length, and there is thus a change in *shape* of the small element as well as of its volume. Hence the resistance of the medium to shear as well as to its compressibility becomes involved. Note that if we consider an ideal isotropic fluid of unlimited extent the medium cannot support shear, and only one type of wave and, correspondingly, one distinct velocity of propagation are possible. The velocity in this case does indeed depend only upon the values of the bulk modulus and density, *i.e.*, $(k/\rho)^{1/2}$.

When treating the propagation of nonplanar waves in extended media, the same general procedures that we have discussed are relevant. Similar conclusions are reached with respect to possible velocities, that is, only two values of wave velocity can occur in unbounded, isotropic, elastic solids, and they are given by Eqs. (I-2.10) and (I-2.11). However, certain aspects of particle motion and deformation are different, and some comments are given in the next section.

IV. Direction of Particle Motion

Qualitative comments have been given in Chapter I-2, Section II regarding the orientations of the particle motions accompanying the three plane wave velocities possible in a general anisotropic medium. This section contains the necessary mathematics to determine quantitatively the directions of these particle motions (subject to the restriction of small strains), but it is applied only to isotropic media.

As before, let us take **n** as a unit vector along the normal to the given advancing plane wave front and $\{\alpha_k\}$ the angles between **n** and the coordinate axes so that $\ell_k = \cos \alpha_k$. Now let $\{\beta_k\}$ be the angles between a given vector displacement **U** of a particle on the wave front and the coordinate axes, so that $m_k = \cos \beta_k$. In Fig. I.9 one can recognize that the direction of the given displacement vector **U** $(= \vec{\text{U}})$ is the same for all particles on the front, but that this direction has no necessary connection with the direction of **n** $(= \vec{\text{n}})$. In other words, this wave is a combination of a longitudinal wave and a transverse wave.

The particle velocity of the point on the wave front can be written m_k $(\partial u_k / \partial t)$. It is proved below that the direction cosines m_k are related to the λ_{ik} constants given in expressions (I-2.3) by

$$\lambda_{ik} m_k = m_i \rho c^2. \tag{I-2.12}$$

In Eqs. (I-2.12), c^2 is the particular wave velocity under consideration (and in the case of completely anisotropic media, c^2 corresponds to three distinct

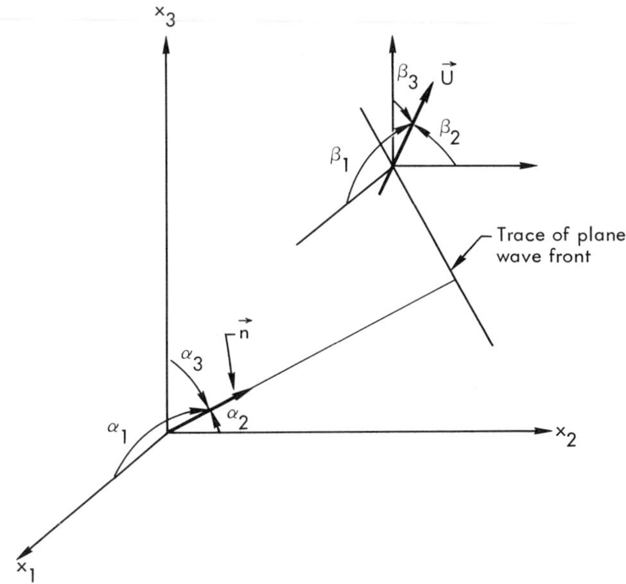

Fig. I.9. Direction of particle motion in a propagating plane wave.

values). These equations are particularly useful when considering methods for measuring elastic constants of nonisotropic crystals.

The starting point of the proof of the theorem contained in expressions (I-2.12) is the concise form of Eqs. (I-2.3)

$$\lambda_{ik} \frac{\partial^2 u_k}{\partial \psi'^2} = \rho c^2 \frac{\partial^2 u_i}{\partial \psi'^2},$$

$\qquad\qquad$ (I-2.3)

together with the general expression for an advancing plane wave solution taken in the direction[5] **n** [from Eqs. (I-1.68), (I-1.71), (I-1.72) and (I-2.1)],

$$\psi = u_k = U_k \exp\left[i(kx_n - pt)\right]$$
$$= U_k \exp\left[\frac{2\pi i}{\Lambda}(x_n - ct)\right]$$
$$= U_k \exp\left(\frac{2\pi i}{\Lambda}\psi'\right),$$

$\qquad\qquad$ (I-2.13)

in which the U_k are components of the displacement amplitude with respect to the axes. The U_k may not be functions of the position coordinates if consistency with the requirement of a plane wave solution is to be achieved. Equation (I-2.13) inserted into (I-2.3) yields

$$\lambda_{ik} U_k = \rho c^2 U_i, \tag{I-2.14}$$

with suitable change of subscripts. A division of this result by the magnitude of **U**

$$|\mathbf{U}| = (U_k U_k)^{1/2}$$

gives

$$\lambda_{ik} m_k = m_i \rho c^2, \tag{I-2.12}$$

with the use of the definition $U_k = |\mathbf{U}| \cos \beta_k = |\mathbf{U}| m_k$. Since the particle velocity has components $\partial u_k / \partial t$ this definition is consistent with the definition of the m_k given above, and Eq. (I-2.12) is just the required result.

Substituting the values of λ_{ik} from (I-2.5) and the values of c from (I-2.10) and (I-2.11) into (I-2.12), we obtain for extended isotropic media that the relations $m_1 = \ell_1$, $m_2 = \ell_2$, $m_3 = \ell_3$ hold for c_ϱ and $m_k \ell_k = 0$ for c_s. Thus c_ϱ corresponds to a wave in which the particle motion is in the direction of propagation (hence the term "longitudinal wave"), and c_s corresponds to a wave for which the direction of particle motion is perpendicular to the direction of propagation (hence the term "shear wave"). It is also apparent that these conditions are valid irrespective of the orientations of the waves, *i.e.*, only two directions of particle motion are possible in isotropic bodies.

When investigating nonplanar waves, the geometry introduces somewhat increased complexity into the analysis. For example, consider a spherical elastic wave propagating in an extended isotropic solid. It is found (see Lindsay [3]) that the displacement direction at a point on an irrotational spherical wave corresponds to the direction of wave propagation at that point. However, the particle motion associated with a solenoidal spherical wave is not necessarily transverse as it is in the equivalent plane wave.

V. Pulse Propagation

Under some conditions of propagating mechanical elastic disturbances in solids, adequate analysis is not possible using the concept of a simple harmonic progressive wave of infinite extent. In treating aperiodic transient disturbances or disturbances of finite duration, new techniques are required. The understanding of pulse propagation is especially significant in certain experimental work (such as ultrasonic pulse transmission for nondestructive testing discussed in Chapter I-5 and as the study of shock waves covered in Chapter II-3) inasmuch as harmonic wave trains of very long ("infinite") duration cannot be physically produced. Thus, it is desirable to have some acquaintance, even though brief, with mathematical approaches useful in pulse description and analysis.

The subject is easiest to introduce when the medium that is transmitting the pulse is elastic, unbounded, and allows propagation of waves of various frequencies at the same velocity. Moreover, to this point we have restricted the investigation to propagation velocities and particle motion directions [refer to Eqs. (I-2.4), (I-2.10), (I-2.11), and (I-2.12)]. We have not yet concerned ourselves with the other physical ramifications of the disturbance, such as the amplitude variation of stress and displacement. The effects of extending these conditions are investigated in subsequent chapters.

A standard method of solution is analysis by Fourier series, according to which any traveling wave[6] that is reasonably well-behaved, periodic, and of extended duration propagating in an elastic (unbounded) medium can be resolved into an equivalent set of discrete simple harmonic motion components with a suitable distribution of frequencies. By expressing the wave within an exact mathematical context, we are able to generate a completely analytical solution. Such an equivalent representation is a key operation in many theoretical disciplines in physics and engineering.

In most practical situations, a few terms in the Fourier series suffice for adequate description of the wave. We may concern ourselves with periodic functions of either time or (multidimensional) space, *i.e.*, the same formalism is appropriate in both cases. Sinusoidal terms (or related exponentials with imaginary arguments) are usual wave functions used for such representation, but are not the only ones possible. For example, Bessel functions are used in the description of waves on a circular membrane. Indeed, according to generalized Fourier methods, any complete set of functions may be used. The functions selected depend upon the geometry of the solid in which the wave is being propagated and upon the geometry of the wave itself, *i.e.*, planar or nonplanar.

We do not raise detailed questions here concerning the conditions placed upon the function to be represented. It is sufficient to state that convergency conditions are satisfied if the function and its first derivative (either space or time depending upon the function) are continuous except for a finite number of discontinuities within each period. These conditions are not stringent requirements, and they are satisfied for essentially all functions useful in mechanics. Details of these standard techniques of Fourier synthesis can be found in many books, *e.g.*, the text by Churchill [4] .

If we have a quantity which is aperiodic (isolated) but we are only concerned with its behavior over a limited range of some variable, we may still utilize Fourier series analysis and relate the quantity to a set of harmonic components, with this limited range taken as the longest period in the series. Of course the set of components combine to form a function that repeats periodically for all time, but this is of no practical consequence.

Extending this approach a bit, in many practical situations we can adequately reproduce aperiodic pulses using a series analysis by taking longer and longer ranges of representation. Exact mathematical correspondence with the physical shape of the pulse is generally impossible with this method, even with many terms in the series, and a decision must be made by the analyst whether the resolution achieved is acceptable for the problem under consideration.

An exact mathematical description of a completely aperiodic pulse can be developed by yet further extension of the principles of Fourier series. Essentially, the range of representation is increased indefinitely by spacing the Fourier terms so close in frequency space that a continuous frequency spectrum results. The problem is then reduced to the determination of the Fourier transform of the initial conditions (of the wave) and the evaluation of the resulting integral equation. Analytical evaluation of such equations can be achieved by several techniques, all of which are rather sophisticated, *e.g.*, contour integration in the complex plane (and thus the use of complex frequencies), convolution operations, the approximate methods of stationary phase and "saddle point" integration, and use of the Dirac δ-function. The literature on Fourier transform theory and the Fourier integral theorem is extensive, and two good references are those by Morse [5] and by Stratton [6] .

It is a consequence of Fourier series analysis that the individual harmonic waves comprising the disturbance are each of different frequencies. Under certain conditions, these waves can separate as they move along, the separation depending upon the characteristics of the individual component waves. This discussion leads us to the separate phenomena of *phase*, *group*, and *signal velocities*, discussed in Chapter I-4.

VI. Summary

The propagation of small disturbances in anisotropic media can be examined by direct substitution of an expression representing a simple harmonic wave into the field equations of motion. Although this is essentially the technique employed in treating wave propagation in isotropic media, two simplifications are introduced in order to emphasize the mechanics involved: A specialized representation of a plane wave front is used in conjunction with a step by step development of the governing equations. This method provides the equations of small motion

$$\lambda_{ik} \frac{\partial^2 u_k}{\partial \psi'^2} = \rho c^2 \frac{\partial^2 u_i}{\partial \psi'^2} \qquad\qquad (I\text{-}2.3)$$

where the independent variable ψ' is defined as $(\ell_i x_i - ct)$ (ℓ_i being the direction cosines of the wave front normal), and the coefficients are given by

$$\lambda_{ik} = \frac{1}{2} \left(C_{ijmk}\ell_j\ell_m + C_{ijkn}\ell_j\ell_n \right).$$

By rewriting expressions (I-2.3), a system of linear homogeneous simultaneous equations can be formed. Such a system can be solved nontrivially only if the determinant of the coefficients vanishes. Hence, the condition on (I-2.3) is

$$\left| \lambda_{ik} - \delta_{ik}\,(\rho c^2) \right| = 0. \qquad\qquad (I\text{-}2.4)$$

Condition (I-2.4) is called the secular determinant. The three roots of (I-2.4) are positive and real and represent the three wave velocities that can be possibly transmitted in a completely anisotropic medium under the restriction of infinitesimal strains.

When treating isotropic elastic materials, two methods are appropriate to the investigation of the propagation of plane waves. First, the determinantal equation (I-2.4) directly reduces to

$$\left| \ell_i\ell_k\,(\lambda + \mu) - \delta_{ik}\,(\mu - \rho c^2) \right| = 0. \qquad\qquad (I\text{-}2.5)$$

Evaluation of the determinant leads to the solutions

$$c_\ell = \left(\frac{\lambda + 2\mu}{\rho} \right)^{1/2} \qquad\qquad (I\text{-}2.10)$$

and

$$c_s = \left(\frac{\mu}{\rho}\right)^{1/2}, \tag{I-2.11}$$

showing that, in the interior of an elastic isotropic solid, a planar wave (or nonplanar wave for that matter) may be transmitted with only two different velocities (assuming that the limitation of infinitesimal strains is imposed).

The second method essentially performs various manipulations with the equations of motion (I-1.56) and examines the consequences of allowing first the dilatation and then the rotation to become zero. This latter approach enables conclusions to be obtained concerning the physical nature of (I-2.10) and (I-2.11). Thus, solution (I-2.10) represents a disturbance traveling with no rotation. It is known as a longitudinal wave (hence the subscript ℓ); other names are irrotational, dilatational, and P waves. Equation (I-2.11) represents a wave propagating with no dilatation. It is known as a shear wave (hence the subscript s); other names are equivoluminal, transverse, distortional, and SV or SH waves (the latter depending upon whether the displacements are in a vertical or horizontal plane, respectively). An important feature of these disturbances is that plane shear waves involve distortion without dilatation, but that plane longitudinal waves involve both phenomena.

It is proved that the direction cosines m_k for the particle motions are related to the λ_{ik} constants in expressions (I-2.3) by

$$\lambda_{ik} m_k = m_i \rho c^2. \tag{I-2.12}$$

By applying Eqs. (I-2.12) to isotropic bodies, we find that c_ℓ corresponds to a wave in which the particle motion is in the direction of propagation (hence the term "longitudinal wave"), and c_s corresponds to a wave for which the direction of particle motion is perpendicular to the direction of propagation (hence the term "shear wave"). Moreover, these conditions are the only ones possible in the plane wave situation.

Under some conditions of elastic wave propagation in solids, *e.g.* pulse transmission, adequate analysis is not possible using the concept of a simple harmonic progressive wave of infinite extent. Then a standard method of solution applies the principles of Fourier series analysis over a limited range of some space or time variable that is representative of the wave. Although this approach is only approximate, often in many practical situations one can adequately reproduce isolated pulses using a series analysis by taking longer and longer ranges of representation. Exact mathematical description of a completely aperiodic pulse can be developed by extending the principles of Fourier series analysis and using Fourier transform theory and integral theorem solutions.

Notes

[1] Direction cosines have been described in Section IV in the Introduction in regard to the transformation of coordinate systems and the mathematical definition of tensor quantities.

[2] The engineer's definition of strain also means that the reference length is taken as the *original* length rather than the *deformed* length [see Eq. (I-1.19)]. Such distinction is of no significance in small strain. Another definition suitable for treatment of large strains is provided by Eq. (II-3.1).

[3] With the definition of strain as given by Eqs. (I-1.17), we find that the coefficients λ_{ik} cannot be formed as conveniently as above. For example, $\lambda_{ik} \neq \lambda_{ki}$ and

$$\lambda_{11} = C_{1111}\ell_1^2 + \frac{1}{2}C_{1212}\ell_2^2 + \frac{1}{2}C_{1313}\ell_3^2 + \frac{3}{2}C_{1112}\ell_1\ell_2$$
$$+ C_{1213}\ell_2\ell_3 + \frac{3}{2}C_{1113}\ell_1\ell_3.$$

[4] We have previously mentioned that, for isotropic media in which we allow finite strain, three different velocities are possible.

[5] We are not allowed to give here as simplified a representation of a plane wave as we did in Chapter I-2, Section II.

[6] Standing waves are also amenable to Fourier series analysis.

References

[1] J. N. Johnson, *J. Appl. Phys*, **42**, 5522 (1971).

[2] C. Truesdell, *Arch. Rat. Mech. Anal.*, **8**, 263 (1961).

[3] R. B. Lindsay, *Mechanical Radiation*, pp. 152-153, McGraw-Hill, New York, 1960.

[4] R. V. Churchill, *Fourier Series and Boundary Value Problems*, McGraw-Hill, New York, 1965.

[5] P. M. Morse, *Vibration and Sound*, pp. 16-17, 42-52, McGraw-Hill New York, 1948.

[6] J. A. Stratton, *Electromagnetic Theory*, pp. 284-297, McGraw-Hill, New York, 1941.

WAVE PROPAGATION IN REFLECTION AND REFRACTION SEMI-EXTENDED MEDIA:

I. General

We now examine the transmission of elastic disturbances in the simplest bounded medium we can envision: a semi-infinite solid with a plane boundary. Such a body is called a "semi-extended" medium. It is one step removed from the idealized unbounded configuration that we treated in Chapter I-2. We are approaching a more "practical" environment, one that has geometrical limitations, and consequently one that engenders greater mathematical difficulty.

Waves can propagate either near[1] and parallel to this plane-bounding surface or at some arbitrary angle to it. Under the former condition, the disturbances are called *surface waves*; under the latter, we are led to the phenomenon of *wave reflection*. Both situations are examined.

Most of the restrictions introduced previously are maintained in this chapter: The medium is homogeneous and isotropic, the strains are small, and all response is elastic.

II. Surface Waves

There are various types of elastic waves that can propagate parallel to the bounding surface in semi-extended, isotropic,[2] homogeneous solids. Although we will mention several kinds, we consider primarily the transmission of plane

Rayleigh waves. The importance of these disturbances is discussed in Chapter I-5.

Rayleigh waves are similar to shallow depth gravity surface waves in liquids in that they propagate near the surface, their amplitudes attenuate exponentially with distance from this surface, and their particle paths are ellipses.

Two specific requirements are imposed when the propagation of classical Rayleigh surface waves is discussed. As indicated, the boundary must be plane so that the problems of reflections or scattering do not arise. The boundary must also be free of stress, *i.e.*, a "free surface." These conditions allow a relatively simple and straightforward, yet useful, analytical solution. In theory, a free surface is taken to be a solid-vacuum interface, but in practice a solid-gas interface is permissible if the pressure of the gas is restricted to a reasonable value, say less than 10 atmospheres.

It is instructive to develop the mathematics necessary for an understanding of Rayleigh waves. Much of this mathematics is also relevant to the treatment of the other kinds of surface waves we mention in this section. The general procedure is to determine the solution of the equations of small motion (I-1.56) using the expression for a progressive continuous simple harmonic plane wave and the boundary condition of a stress-free interface.

For ease of representation and mathematical treatment, the boundary is taken to be the $x_1 x_2$ plane with the x_3 axis directed positive toward the interior of the medium. We take the plane wave to be traveling in the positive x_1 direction. Since we are then considering a two-dimensional problem (in the $x_1 x_3$ plane) the displacements will be independent of x_2, that is, at a given point (x_1, x_3) the displacements will be identical regardless of the x_2 coordinate selected.

The mathematical procedures are simplified by the separation of the effects of dilatation and rotation produced by the disturbance in the medium. To accomplish this separation, we enlarge upon the ideas advanced in Section III of Chapter I-2. Specifically, two new variables are introduced via potential theory. We define two displacement potential functions ϕ and χ, such that

$$u_1 = \frac{\partial \phi}{\partial x_1} + \frac{\partial \chi}{\partial x_3}$$

and   (I-3.1)

$$u_3 = \frac{\partial \phi}{\partial x_3} - \frac{\partial \chi}{\partial x_1}.$$

The dilatation can then be written as

$$\Delta = \epsilon_{jj} = \frac{\partial u_1}{\partial x_1} + \frac{\partial u_3}{\partial x_3} = \frac{\partial^2 \phi}{\partial x_j \partial x_j}$$

which is in terms of ϕ only, and the rotation in the $x_1 x_3$ plane as

$$2\omega_{13} = \frac{\partial u_1}{\partial x_3} - \frac{\partial u_3}{\partial x_1} = \frac{\partial^2 \chi}{\partial x_j \partial x_j}$$

which is in terms of χ only. Upon substitution of these potential functions (I-3.1) into the $i = 1$ component of the equations of small motion in an isotropic medium (I-1.56), and with appropriate shifts in the order of differentiation, we obtain

$$\rho \frac{\partial}{\partial x_1} \left(\frac{\partial^2 \phi}{\partial t^2} \right) + \rho \frac{\partial}{\partial x_3} \left(\frac{\partial^2 \chi}{\partial t^2} \right) = (\lambda + \mu) \frac{\partial}{\partial x_1} \left(\frac{\partial^2 \phi}{\partial x_j \partial x_j} \right)$$
$$+ \mu \frac{\partial^2 u_1}{\partial x_j \partial x_j}.$$

By adding and subtracting to the above expression the term

$$\mu \Delta = \mu \left(\frac{\partial u_1}{\partial x_1} + \frac{\partial u_3}{\partial x_3} \right),$$

it follows that

$$\rho \frac{\partial}{\partial x_1} \left(\frac{\partial^2 \phi}{\partial t^2} \right) + \rho \frac{\partial}{\partial x_3} \left(\frac{\partial^2 \chi}{\partial t^2} \right) = (\lambda + 2\mu) \frac{\partial}{\partial x_1} \left(\frac{\partial^2 \phi}{\partial x_j \partial x_j} \right)$$
$$+ \mu \frac{\partial}{\partial x_3} \left(\frac{\partial^2 \chi}{\partial x_j \partial x_j} \right).$$

In a similar fashion for the $i = 3$ component, we get

$$\rho \frac{\partial}{\partial x_3}\left(\frac{\partial^2 \phi}{\partial t^2}\right) - \rho \frac{\partial}{\partial x_1}\left(\frac{\partial^2 \chi}{\partial t^2}\right) = (\lambda + 2\mu)\frac{\partial}{\partial x_3}\left(\frac{\partial^2 \phi}{\partial x_j \partial x_j}\right)$$
$$- \mu \frac{\partial}{\partial x_1}\left(\frac{\partial^2 \chi}{\partial x_j \partial x_j}\right).$$

Since the Lamé constants and the density are not functions of position (as yet), the two equations will be satisfied if

$$\frac{\partial^2 \phi}{\partial t^2} = \left(\frac{\lambda + 2\mu}{\rho}\right)\frac{\partial^2 \phi}{\partial x_j \partial x_j} = c_\varrho^2 \frac{\partial^2 \phi}{\partial x_j \partial x_j}$$

and (I-3.2)

$$\frac{\partial^2 \chi}{\partial t^2} = \left(\frac{\mu}{\rho}\right)\frac{\partial^2 \chi}{\partial x_j \partial x_j} = c_s^2 \frac{\partial^2 \chi}{\partial x_j \partial x_j}$$

with the use of expressions (I-2.10) and (I-2.11).

Let us consider as a trial solution of wave equations (I-3.2) a sinusoidal wave of infinite duration propagating in the x_1 direction (thus disallowing the possibility of the occurrence of reflections). From Eq. (I-1.72) we can write

$$\phi = M(x_3) \exp[i(kx_1 - pt)]$$

and (I-3.3)

$$\chi = N(x_3) \exp[i(kx_1 - pt)]$$

where M and N are functions that determine the change of amplitude of the waves with depth. Note that in the sense that the amplitude (and hence displacement) is allowed to vary as a function of depth, the disturbance is not necessarily plane. By using Eq. (I-1.70), the velocity is represented as $c_r = \Lambda \nu = p/k$, in which all parameters here refer to and are characteristic of the Rayleigh wave. We take (or rather, assert) the velocity to be constant as before, but different from either the longitudinal (c_ϱ) or the shear (c_s) velocities.

Upon substitution of the expression for ϕ from (I-3.3) into the appropriate relation (I-3.2), we obtain a second-order, ordinary, linear differential equation with constant coefficients,

$$\frac{d^2M}{dx_3^2} - \left(k^2 - \frac{p^2}{c_\varrho^2}\right) M = 0. \tag{I-3.4}$$

There are two linearly independent solutions to Eq. (I-3.4). One of them corresponds to a disturbance that increases with increasing x_3, which means the amplitude becomes divergently large at $x_3 = \infty$ with the attendant difficulty that the amount of energy to produce such a disturbance also becomes infinite. This solution does not correspond to physical reality, and can be discarded. The solution of the differential equation (I-3.4) in which we are interested is

$$M = a \, \exp\left[-\left(k^2 - \frac{p^2}{c_\varrho^2}\right)^{1/2} x_3\right] = a \, \exp\left(-\xi_\varrho x_3\right)$$

where a is a constant relating to a reference amplitude of the disturbance and depending upon initial conditions, and where the introduction of the auxiliary symbol ξ_ϱ is done strictly for purposes of simplification in writing. Similarly, after substitution of the expression for χ from (I-3.3) into the second of (I-3.2), we obtain the relevant solution

$$N = b \, \exp\left[-\left(k^2 - \frac{p^2}{c_s^2}\right)^{1/2} x_3\right] = b \, \exp\left(-\xi_s x_3\right)$$

where b is a constant, again depending upon initial conditions, and the auxiliary symbol ξ_s is introduced for reasons similar to those for ξ_ϱ above.

The assumed solutions (I-3.3) become

$$\phi = a \, \exp\left[-\xi_\varrho x_3 + i\left(kx_1 - pt\right)\right]$$

and $\tag{I-3.5}$

$$\chi = b \, \exp\left[-\xi_s x_3 + i\left(kx_1 - pt\right)\right].$$

Note that the coefficients of x_3 in Eq. (I-3.5) do indeed serve as modification factors of the wave amplitude with respect to depth [see relations (I-3.3)].

The boundary conditions are now established at the interface so as to satisfy the requirement that it be stress-free. Hence we have $S_{3i} = 0$ at the surface, $x_3 = 0$. First, let us write from Eqs. (I-1.32) and (I-1.17),

$$S_{33} = \lambda\Delta + 2\mu\,\frac{\partial u_3}{\partial x_3}.$$

With the help of (I-3.1), the above expression becomes

$$S_{33} = \lambda\,\frac{\partial^2\phi}{\partial x_j\partial x_j} + 2\mu\left(\frac{\partial^2\phi}{\partial x_3^2} - \frac{\partial^2\chi}{\partial x_1\partial x_3}\right) = (\lambda + 2\mu)\,\frac{\partial^2\phi}{\partial x_3^2}$$

$$+ \lambda\,\frac{\partial^2\phi}{\partial x_1^2} - 2\mu\,\frac{\partial^2\chi}{\partial x_1\partial x_3}.$$

Now substituting from the assumed solutions (I-3.5), we obtain at $x_3 = 0$

$$a\left[(\lambda + 2\mu)\xi_\varrho^2 - \lambda k^2\right] + i2b\mu k\xi_s = 0. \tag{I-3.6}$$

Second, we have

$$S_{31} = \mu\left(\frac{\partial u_1}{\partial x_3} + \frac{\partial u_3}{\partial x_1}\right),$$

and from (I-3.1) we obtain

$$S_{31} = \mu\left(2\frac{\partial^2\phi}{\partial x_1\partial x_3} + \frac{\partial^2\chi}{\partial x_3^2} - \frac{\partial^2\chi}{\partial x_1^2}\right).$$

Again from (I-3.5), we have at $x_3 = 0$,

$$-i2ak\xi_\varrho + b\left(2k^2 - p^2/c_s^2\right) = 0. \tag{I-3.7}$$

Note that we need not consider S_{32}. Since there is no dependence on displacement in the x_2 direction, S_{32} vanishes trivially.

We can eliminate the constants a and b from the simultaneous equations (I-3.6) and (I-3.7) and get

$$4\mu k^2 \xi_\varrho \xi_s = \left[(\lambda + 2\mu)\, \xi_\varrho^2 - \lambda k^2 \right] \left(2k^2 - p^2/c_s^2 \right).$$

By squaring both sides of the preceding expression, dividing the result by $\mu^2 k^8$, and noting that

$$\frac{c_s^2}{c_\varrho^2} = \frac{\mu}{\lambda + 2\mu},$$

we obtain the result

$$\left(p^2/c_s^2 k^2 \right)^3 - 8 \left(p^2/c_s^2 k^2 \right)^2 + \left(24 - 16 c_s^2/c_\varrho^2 \right) \left(p^2/c_s^2 k^2 \right)$$

$$+ 16 \left(c_s^2/c_\varrho^2 - 1 \right) = 0. \tag{I-3.8}$$

This expression is termed the *propagation condition*. The term *frequency equation* is also sometimes employed in this connection. However, such nomenclature may cause some confusion to the reader because of the term's use in association with the solution of vibration problems using the separation of variables technique.

Equation (I-3.8) is cubic in $p^2/c_s^2 k^2$. Its solution has three roots, only one of which corresponds to a wave in the real plane and which has physical significance to us at this time.[3] Note that if the value of Poisson's ratio σ is known, the equation may be solved numerically through the use of the expression

$$\frac{c_s^2}{c_\varrho^2} = \frac{1 - 2\sigma}{2(1 - \sigma)}. \tag{I-3.9}$$

Note also that the determination of the solution enables us to find specifically the velocity of the Rayleigh surface waves by the relationship

$$\frac{p^2}{c_s^2 k^2} = \frac{c_r^2}{c_s^2} \tag{I-3.10}$$

providing the velocity of the distortion wave in an extended medium (with the same Poisson's ratio) is known. Of particular importance is the fact that c_r is independent of the frequency, $\nu = p/2\pi$, and depends only upon the elastic constants of the material. This independence shows that providing λ and μ remain constants, *i.e.*, Hooke's law applies and no reflections occur, a plane Rayleigh wave will travel in the x_1 direction without change in velocity or in form.

We do not limit our attention to velocity calculations as we did in Chapter I-2. We also examine the displacements and stresses that result from the passing of the Rayleigh wave. Thus from Eqs. (I-3.1) and (I-3.5), it follows that

$$u_1 = \frac{\partial \phi}{\partial x_1} + \frac{\partial \chi}{\partial x_3} = [iak \exp(-\xi_\varrho x_3) - b\xi_s \exp(-\xi_s x_3)] \exp i(kx_1 - pt). \qquad \text{(I-3.11)}$$

If we use (I-3.7), we expand the result by Euler's formula [see Eq. (I-1.72)], and use only the real part of the resulting solution, the above expression reduces to

$$u_1 = -ak \left[\exp(-\xi_\varrho x_3) - 2\xi_\varrho \xi_s \left(2k^2 - \frac{p^2}{c_s^2} \right)^{-1} \exp(-\xi_s x_3) \right] \sin(kx_1 - pt). \qquad \text{(I-3.12)}$$

As in Chapter I-1, we introduced the trial solutions as complex functions for convenience. We are not reducing the utility of our solution (I-3.11) here by our simplification to solution (I-3.12). However, to avoid confusion, the reader should exercise care in reading the two equations. The displacements u_1 in (I-3.11) is complex whereas that in (I-3.12) is real, and strictly the two should be distinguished from each other, perhaps by a superscript. In most texts, such notational clarification is not made, and the reader is expected to understand what is intended.

In a similar fashion, we obtain

$$u_3 = -a\xi_\varrho \left[\exp(-\xi_\varrho x_3) - 2k^2 \left(2k^2 - \frac{p^2}{c_s^2} \right)^{-1} \exp(-\xi_s x_3) \right] \cos(kx_1 - pt). \qquad \text{(I-3.13)}$$

Equations (I-3.12) and (I-3.13) define completely the solution because of the manner in which we have selected the coordinate system for this two-dimensional problem, *i.e.*, $u_2 = 0$ everywhere.

Parametric equations for an ellipse are

$$y_1 = C_1 \sin \theta$$

and   (I-3.14)

$$y_3 = C_3 \cos \theta$$

where y_1 and y_3 are the coordinates of a point on the ellipse consistent with the parametric angle θ, and C_1 and C_3 are the lengths of the semiaxes of the ellipse. Comparing equations (I-3.14) with solutions (I-3.12) and (I-3.13), θ represents the arguments $(kx_1 - pt)$, y_1 and y_3 are identified with the displacements u_1 and u_3, respectively, and the C's correspond to the coefficients of the trigonometric functions. Hence the expressions for u_1 and u_3 show that the path of any particle in the medium is an ellipse. For any physical problem it can be shown that $C_1 < C_3$; it follows then that the major axis is perpendicular to the bounding surface.[4] As indicated, the rate at which the amplitude of the wave decays with depth depends upon the values of the attenuation factors in the solutions, $i.e.$, the coefficients of x_3.

Let us now consider some numerical examples. The method of approach is as follows: We assume a value for Poisson's ratio that is representative of the material we wish to study; the propagation condition relationship (I-3.8) is then solved with the use of expression (I-3.9), and the ratio of the velocities of the (real) Rayleigh wave c_r to the shear wave c_s for the same material is obtained by Eq. (I-3.10); the variation of the amplitude of the displacements and stresses with depth and wavelength (both more conveniently obtained in nondimensional form in terms of reference values) are calculated, respectively, from (I-3.12) and (I-3.13) and from a combination of (I-1.32) and (I-1.17).

Figure I.10 shows results for the amplitude ratios of the displacements for three values of Poisson's ratio calculated in accordance with the above method. The value $\sigma = 0$ represents one possible material extreme, a laterally inflexible solid (that is isotropic and single phase); $\sigma = 0.5$ represents the other possible extreme, a perfect (incompressible) fluid. The value $\sigma = 0.25$ (which results when $\lambda = \mu$) is often used for glass. The curves are given in nondimensional form in which $u_{1/0}$ and $u_{3/0}$ refer to the displacements in the x_1 and x_3 directions at $x_3 = 0$, respectively. Also listed on the figure for the three values of σ are the ratio between the maximum displacements at the surface (i.e., between the minor and the major axes of the ellipse), the ratio between c_r and c_s and the ratio between c_s and c_ϱ.

The variation of the amplitudes of the stresses with depth and wavelength can be calculated in the fashion indicated above. The application of the pertinent

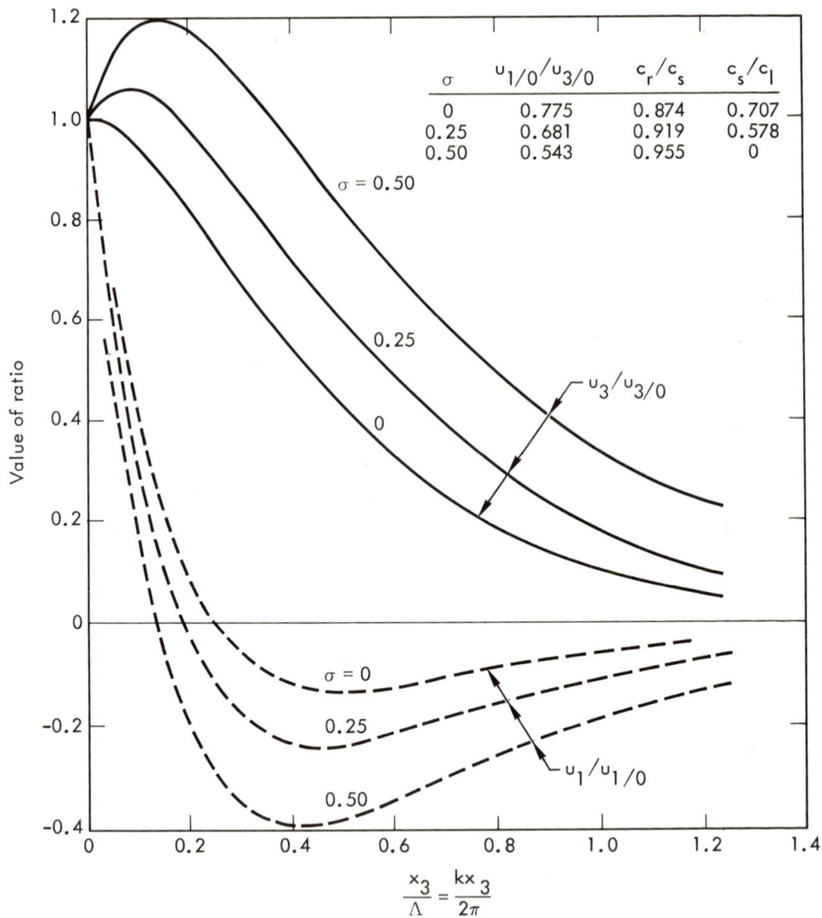

σ	$u_{1/0}/u_{3/0}$	c_r/c_s	c_s/c_l
0	0.775	0.874	0.707
0.25	0.681	0.919	0.578
0.50	0.543	0.955	0

Fig. I.10. Displacement ratios associated with Rayleigh waves.

equations is straightforward. The texts by Kolsky [2] and Volterra and Zachmanoglou [3] contain some results.

Several observations are possible from the examination of Fig. I.10 and the expressions associated with its construction. We find the velocity of propagation of Rayleigh waves is smaller than that of either of the unbounded (body) elastic waves; more specifically, $c_r < c_s < c_\varrho$.[5]

The amplitude of the displacement u_1 along the direction of propagation x_1 is attenuated rapidly with increasing values of kx_3, becoming zero for the depth approximately equal to 0.2 wavelength (the exact value depending upon the value of σ). At this approximate depth, u_1 is zero for all values of x_1 and t, and one concludes that there is a plane below (and parallel with) the surface in which there is no motion parallel to the surface. For greater depths the amplitude once again assumes finite values; the vibrations take place in the opposite phase because of the reversal of sign, but again approach zero as depth increases. This result is related to the splitting of the vector displacement field **u** so that its dilatation (or divergence) is related only to one of the potential functions ϕ, and its rotation (or curl) is related only to the other, χ.

The rate at which the amplitude of the motion in the x_3 direction u_3 is attenuated with depth is shown in the figure. As the depth increases, the amplitude of the vibration first increases (except in the limiting case where $\sigma = 0$), reaching a maximum at about 0.15 wavelength (the exact value again depending upon the value of σ), and then decreases continuously. Where the depth equals the wavelength, this amplitude has decreased to about a quarter of its value at the surface. Note that there is no finite depth at which the motion in a direction normal to the surface vanishes.

Since the wave number k is directly proportional to the frequency of the vibrations of surface waves [see Eqs. (I-1.69) and (I-1.70)], it is apparent from Fig. I.10 that Rayleigh waves of high frequency will be attenuated with depth more rapidly than those of low frequency. Upon further examination, it also follows that the greater the frequency, the less the depth at which u_1 vanishes; in this regard, the behavior is analogous to that of the *skin effect* of high-frequency alternating currents moving in solid conductors.

III. Reflection and Refraction of Elastic Waves

We now investigate the other orientation condition of elastic wave travel in semi-extended isotropic media, namely, the propagation of a plane wave moving at some arbitrary angle to the plane bounding surface. Under these circumstances, we encounter the phenomenon of wave reflection.

Elastic waves are reflected at the boundaries of solids, but not usually in the simple way that they are reflected at the boundaries of containers of fluids. Generally, upon reflection of a pure dilatational or distortional wave, a disturbance is generated which is a mixture of both types of wave, the amplitude of each type depending upon the angle of incidence and upon the elastic constants.

The discussion of wave reflection and refraction, particularly the situation of normal incidence on a plane surface or interface, is of central importance in the study of shock waves and dynamic fracture — even though the treatment is restricted in this chapter to elastic phenomena. There are also certain properties of reflected and refracted waves of considerable useful application in dynamic elastic study (see Chapter I-5, Section II).

A. Reflection at a Free Boundary – Normal Incidence

An elastic wave will be reflected when it reaches a free surface of the material in which it is traveling. To eliminate the possibility of refracted waves, a free boundary is taken here to be a stress-free surface in a vacuum. In other words, we do not allow any disturbance to be excited on the nonsolid side of the boundary by the impinging wave; thus a solid-gas interface is not possible as before.

The simplest reflection occurs when the wave strikes the surface normally. If the incident wave is longitudinal with a given sense (*i.e.*, sign), then the reflected wave must also be longitudinal but with its sense opposite to the incident wave. The reason for the change of sense is that the stress normal to the surface must be zero. This boundary condition is analogous to having a vacuum on one side of the bounding surface.

As an illustrative example, consider a section of a "wave" of extremely long period in which the stress rises instantaneously from zero to its full value, *i.e.*, the stress is a step function. Figure I.11 depicts such a wave in longitudinal compression. This compression wave will be reflected in tension with an equal but opposite stress, thereby canceling the incident compressional stress so that the material traversed by the reflected wave is stress-free. To visualize reflections of this sort, it is helpful to introduce the concept of having outside the material a "virtual" disturbance of the same shape as the incident wave, but producing a stress of opposite sense. The position of the virtual disturbance moving toward the free surface must be such that both the incident wave and the virtual wave strike the surface at the same time. The incident wave passes out of the material while the other enters into it. This crossing of the two waves occurs without any distortion. The linearity of the wave equation permits this visualization, since linearity means the equation contains no terms relating to interaction of waves.

For one cycle of a saw-toothed compression wave, Fig. I.12 shows that the stresses no longer exactly cancel, and that there is a net tension which is a function of the distance traversed by the reflected wave. Such residual transient tension stresses can cause fracture or *spall* of the material if the tension exceeds the dynamic tensile strength of the material. One takes advantage of this condition in the Hopkinson pressure bar technique (Chapter II-1, Section I-A).

A transverse disturbance that impinges normally upon a free surface can be examined in a similar fashion. It is found (next section) that the resulting reflection will also be a transverse wave normal to the surface. The extension to the problem of reflection and transmission of a plane wave normally incident on a plane interface between two solid media is discussed in Section III-C in this chapter.

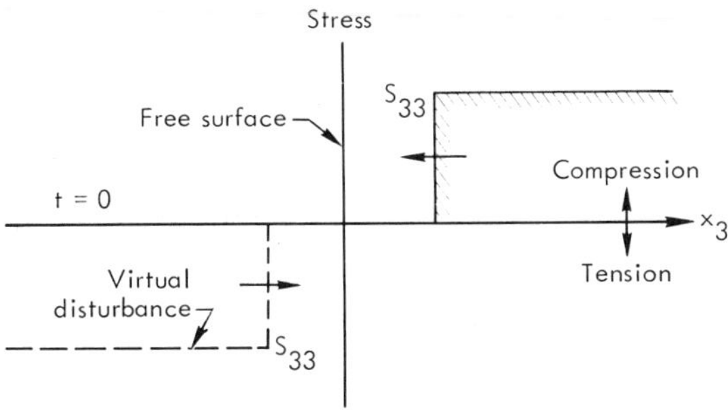

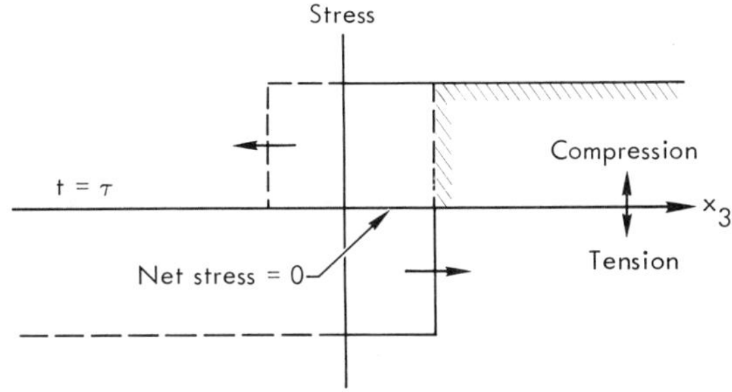

Fig. I.11. Reflection of a step function stress disturbance normally incident upon a plane free surface.

An excellent introduction to this topic is the work by Rinehart [4].

B. *Reflection at a Free Boundary – Oblique Incidence*

When an elastic wave strikes a plane free surface obliquely, reflection is more complicated because, except for certain exceptional circumstances to be treated later, the energy of the incident wave is partitioned in general into two reflected waves instead of only one.

Let us consider first a longitudinal plane wave reflecting at a plane free surface. We want to determine the kind, amplitude, and direction of the reflected waves. There are obviously several possible combinations of these reflected disturbances that can be investigated. One can laboriously examine each of the possibilities chosen, using random selection, but with some logic (and prior knowledge) we "assume" the combination to consist of a longitudinal wave and a shear wave. It will be seen that these two reflected disturbances do indeed satisfy all the boundary conditions.

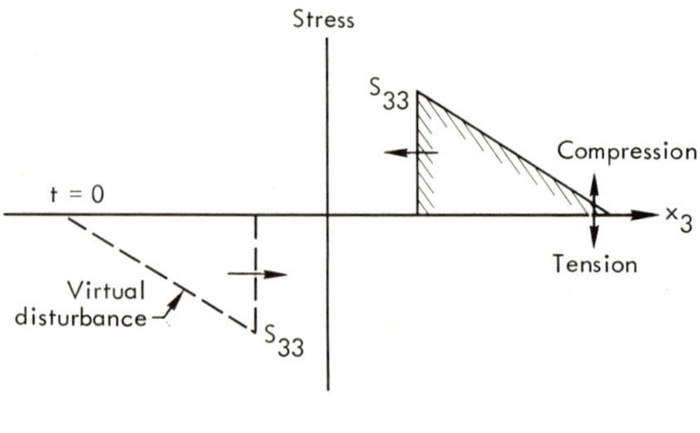

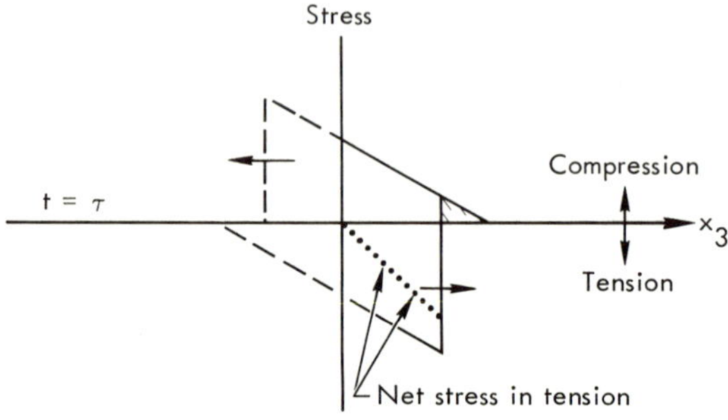

Fig. I.12. Reflection of a saw-toothed stress disturbance normally incident upon a plane free surface.

To maintain consistency with the previous configuration considerations, we take the incident longitudinal wave to lie in the $x_1 x_3$ plane (*i.e.*, it is two-dimensional), its maximum amplitude to be A_{LI}, and its direction of propagation to be at an angle α_{LI} with the x_3 axis so that its travel is a function of increasing x_1 and decreasing x_3. The subscripts *I, R, L* and *S* will be used for incident, reflected, longitudinal, and shear, respectively. Figure I.13 illustrates the geometry; note that the boundary is the $x_1 x_2$ plane and the x_3 axis is its normal. The reflected longitudinal wave is labeled with its maximum amplitude A_{LR} and the propagation direction is shown to be at an angle of α_{LR} with the x_3 axis; the corresponding values for the reflected shear wave are A_{SR} and α_{SR}, respectively. The angle α_{SR} will be shown to be less than the angle α_{LR}.

As illustrated in Fig. I.13, an incident simple harmonic wave in which the displacement is normal to the advancing wave front can be written from Eqs. (I-1.72) and (I-2.13) as

$$u_{LI} = A_{LI} \exp \left[i(k_{LI} x_1 \sin \alpha_{LI} - k_{LI} x_3 \cos \alpha_{LI} - pt) \right] \qquad (\text{I-3.15})$$

where u_{LI} is the normal displacement, $k_{LI} \sin \alpha_{LI}$ and $k_{LI} \cos \alpha_{LI}$ are, respectively, the x_1 and x_3 components of the wave number of the incident disturbance. Since we are confining our comments at present to propagation of a wave in the real domain, we need not continue the use of complex quantities. Hence, we use only the real part of Eq. (I-3.15) and write

$$u_{LI} = A_{LI} \cos (k_{LI}x_1 \sin \alpha_{LI} - k_{LI}x_3 \cos \alpha_{LI} - pt). \qquad \text{(I-3.16)}$$

Relationship (I-3.16) permits some simplification in the following mathematical treatment.

The reflected longitudinal wave can be expressed as

$$u_{LR} = A_{LR} \exp [i(k_{LR}x_1 \sin \alpha_{LR} + k_{LR}x_3 \cos \alpha_{LR} - pt + \phi_1))] \qquad \text{(I-3.17)}$$

in which u_{LR} is the displacement perpendicular to the wave front and ϕ_1 is the phase shift, a constant which allows for any change in phase of the wave upon reflection. Since $k_{LR} = k_{LI} = p/c_\varrho$, we can state for waves in real space

$$u_{LR} = A_{LR} \cos (k_{LI}x_1 \sin \alpha_{LR} + k_{LI}x_3 \cos \alpha_{LR} - pt + \phi_1). \qquad \text{(I-3.18)}$$

Similarly, the reflected shear wave is given in the real domain as

$$u_{SR} = A_{SR} \cos (k_{SR}x_1 \sin \alpha_{SR} + k_{SR}x_3 \cos \alpha_{SR} - pt + \phi_2) \qquad \text{(I-3.19)}$$

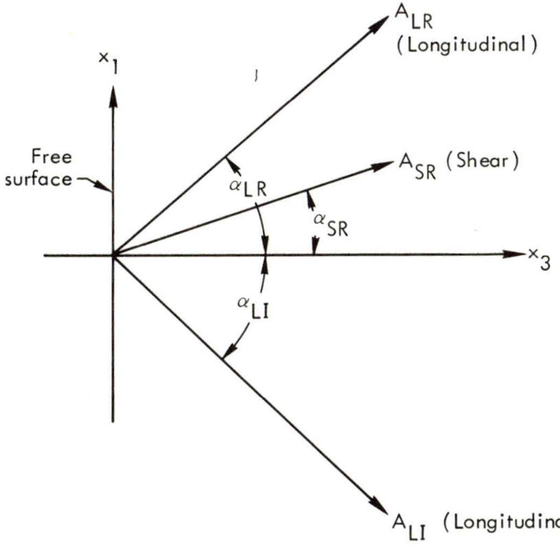

Fig. I.13. Reflection of a longitudinal wave obliquely incident upon a plane free surface.

where u_{SR} is the displacement in the $x_1 x_3$ plane parallel to the wave front, $k_{SR} = p/c_s$, and ϕ_2 is the phase shift constant.

Together with the theory of linear dependence of functions, two of the six stress components are enough to determine four unknowns which appear in the problem. These four unknowns are the amplitudes of the two reflected waves and the angles relative to the boundary normal at which they are reflected. Thus, at the free boundary, $x_3 = 0$, both the stress normal to the surface S_{33} and the shearing stress, $S_{31} = S_{13}$, on the surface must be zero for all values of x_1 and t. These boundary conditions enable the four unknowns to be determined. Using Eqs. (I-1.32) and (I-1.17), we have at $x_3 = 0$,

$$S_{33} = \lambda \Delta + 2 \mu \frac{\partial u_3}{\partial x_3} = 0$$

$$(\text{I-3.20})$$

$$S_{31} = \mu \left(\frac{\partial u_3}{\partial x_1} + \frac{\partial u_1}{\partial x_3} \right) = 0$$

in which u_3 and u_1 are the *net* displacements in the x_3 and x_1 directions produced by the incident and reflected waves, and where $\Delta = \partial u_3/\partial x_3 + \partial u_1/\partial x_1$. Each net displacement u_i is the sum of contributions made severally by the three waves and, because of superposition, can be expressed in terms of a linear combination of the displacements of the individual waves. Thus

$$u_3 = -u_{LI} \cos \alpha_{LI} + u_{LR} \cos \alpha_{LI} - u_{SR} \sin \alpha_{SR}$$

and

$$(\text{I-3.21})$$

$$u_1 = u_{LI} \sin \alpha_{LI} + u_{LR} \sin \alpha_{LI} + u_{SR} \cos \alpha_{SR}$$

where the sign convention for the waves is taken in an arbitrary, but consistent, fashion.

The first conditions (I-3.20) (after differentiation and rearrangement of terms) yields

$$-A_{LI} k_{LI} (\lambda + 2\mu \cos^2 \alpha_{LI}) \sin(k_{LI} x_1 \sin \alpha_{LI} - pt)$$

$$-A_{LR} k_{LI} [\lambda \cos(\alpha_{LI} - \alpha_{LR}) + 2\mu(\cos \alpha_{LI} \cos \alpha_{LR})]$$

$$\sin(k_{LI} x_1 \sin \alpha_{LR} - pt + \phi_1) + A_{SR} k_{SR} (\mu \sin 2\alpha_{SR})$$

$$\times \sin(k_{SR} x_1 \sin \alpha_{SR} - pt + \phi_2) = 0. \qquad (\text{I-1.22})$$

The second of (I-3.20) gives

$$A_{LI}k_{LI}(\sin 2\alpha_{LI}) \sin(k_{LI}x_1 \sin \alpha_{LI} - pt)$$

$$-A_{LR}k_{LI} \sin(\alpha_{LI} + \alpha_{LR}) \sin(k_{LI}x_1 \sin \alpha_{LR} - pt + \phi_1)$$

$$-A_{SR}k_{SR}(\cos 2\alpha_{SR}) \sin(k_{SR}x_1 \sin \alpha_{SR} - pt + \phi_2) = 0. \qquad \text{(I-3.23)}$$

Equations (I-3.22) and (I-3.23) can only be satisfied for all x_1 and t if

$$k_{LI} \sin \alpha_{LI} = k_{LI} \sin \alpha_{LR} = k_{SR} \sin \alpha_{SR} \qquad \text{(I-3.24)}$$

and if the phase constants, ϕ_1 and ϕ_2, are taken from some combination of the numbers 0 and π. The four possible combinations of these numbers will give equivalent results since the selection of a value of π for a phase constant of a wave will simply change the sign of the amplitude of that wave. Hence, a complete specification of the problem is given by selecting $\phi_1 = \phi_2 = 0$.

From Eqs. (I-3.24) and (I-3.9), it follows that

$$\alpha_{LI} = \alpha_{LR} \qquad \text{(I-3.25)}$$

and

$$\frac{\sin \alpha_{LI}}{\sin \alpha_{SR}} = \frac{c_\varrho}{c_s} = \left(\frac{\lambda + 2\mu}{\mu}\right)^{1/2} = \left[\frac{2(1 - \sigma)}{1 - 2\sigma}\right]^{1/2}. \qquad \text{(I-3.26)}$$

The similarity between elastic mechanical waves and optical waves can be seen by observing that expressions (I-3.25) and (I-3.26) have, respectively, the same forms as the law of reflection and the law of refraction (Snell's law).[6] A dilatation wave is reflected at the angle of incidence, and a distortion wave is reflected at an angle similar to that for the refraction of light. One can read in relationship (I-3.26) that the "refractive ratio" is $[(\lambda + 2\mu)/\mu]^{1/2}$. Note also that $\alpha_{SR} < \alpha_{LR}$.

The relations among the amplitudes of the waves become, from (I-3.22) and (I-3.25),

$$(A_{LI} + A_{LR}) (\lambda + 2\mu \cos^2\alpha_{LI}) - A_{SR} \frac{c_\varrho}{c_s} \mu \sin 2\alpha_{SR} = 0;$$

and from (I-3.26),

$$(A_{LI} + A_{LR}) \sin \alpha_{LI} \cos 2\alpha_{SR} - A_{SR} \sin \alpha_{SR} \sin 2\alpha_{SR} = 0. \qquad \text{(I-3.27)}$$

From Eqs. (I-3.23) and (I-3.25), we get

$$(A_{LI} - A_{LR}) \sin 2\alpha_{LI} - A_{SR} \frac{c_\varrho}{c_s} \cos 2\alpha_{SR} = 0$$

and using (I-3.26) with an appropriate trigonometric identity, if follows that

$$2(A_{LI} - A_{LR}) \cos \alpha_{LI} \sin \alpha_{SR} - A_{SR} \cos 2\alpha_{SR} = 0. \tag{I-3.28}$$

We may obtain the amplitudes of the two reflected waves in terms of the amplitude of the incident wave from expressions (I-3.27) and (I-3.28). Of particular importance is the fact that these two equations are applicable to harmonic waves of any frequency, and consequently, because of the principle of superposition, they will also be true for elastic waves of any shape.

The dependence of the angle of reflection of the shear wave α_{SR} upon the angle of the incident dilatation wave α_{LI} is shown in Fig. I.14. Expression (I-3.26), in which Poisson's ratio is a parameter, is used to determine these values. We use the same values for σ as used in Fig. I.10. As expected, in a perfect (incompressible) fluid there is no reflected shear wave for any angle of incidence of the longitudinal wave.

Figure I.15, with the same parametric values of Poisson's ratio, shows the ratios of the reflected amplitudes, A_{LR} and A_{SR}, to the incident amplitude A_{LI} as functions of the angle of incidence α_{LI}. Expressions (I-3.26), (I-3.27), and (I-3.28) were used to determine values for this figure.

It is appropriate to summarize the several important conclusions which can be

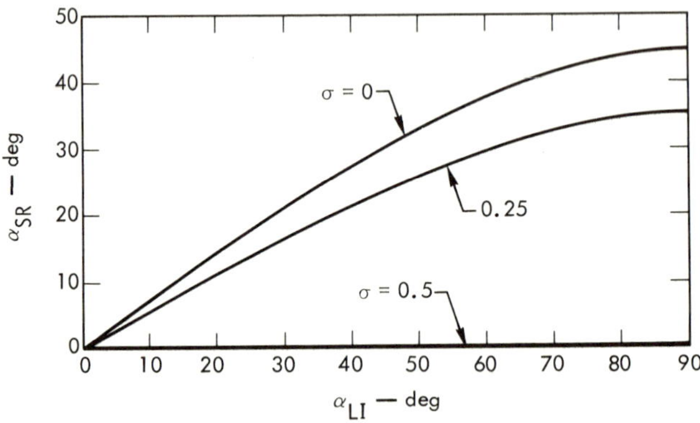

Fig. I.14. Angle of reflection of shear wave α_{SR} as a function of angle of incidence of a longitudinal wave α_{LI} and of Poisson's ratio σ.

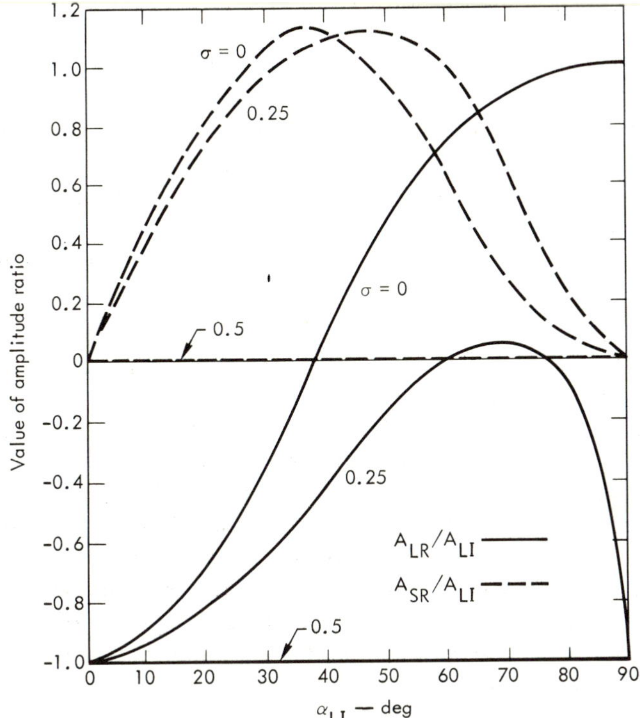

Fig. I.15. Ratios of reflected displacement amplitudes to incident displacement amplitude as functions of angle of incidence of a longitudinal wave α_{LI} and of Poisson's ratio σ.

drawn by examining expressions (I-3.26) - (I-3.28) and Figs. I.14 and I.15:

1. For all values of Poisson's ratio σ, $(A_{LR}/A_{LI}) = -1$ and $(A_{SR}/A_{LI}) = 0$ at normal incidence ($\alpha_{LI} = \alpha_{SR} = 0$). Thus, there is no reflected distortion wave, and the amplitude of the reflected dilatation wave is equal to that of the incident wave apart from a phase change of π upon reflection. The same result was obtained in Section III-4 in this chapter.

2. For the value $\sigma = 0.5$, a medium behaves as a perfect isotropic, incompressible fluid, and there is no reflected distortion wave. The energy is transferred to a reflected dilatation wave of equal but opposite amplitude for all angles of incidence.

3. A phase change of π occurs in the reflected longitudinal wave for all angles of incidence only when $\sigma < 0.263$. Expression (I-3.26) enables this condition on Poisson's ratio to be translated into the requirement that $(\lambda/\mu) > 1.110$.

4. The reflected transverse wave is always in phase with the incident longitudinal wave.

5. For $\sigma < 0.263$, there are two angles of incidence at which the reflected longitudinal wave has an amplitude of zero. At these angles all the energy goes into shear waves. Since σ is a property of materials, it follows that the vanishing of A_{LR} cannot occur for all media.[7]

6. At grazing incidence ($\alpha_{LI} = \pi/2$) no shear wave is reflected, and $(A_{LR}/A_{LI}) = -1$ for all $\sigma > 0$.

7. The amplitude ratio A_{SR}/A_{LI} can exceed unity for certain values of σ and α_{LI}.

To this point, we have made only general statements regarding the planes of vibration of the several waves. We have relied upon our intuition and assumed that the various disturbances have movement only in the $x_1 x_3$ plane [see the comments associated with Eqs. (I-3.20)]. However, although true, such a condition is not obvious, especially with regard to the reflected transverse wave and the polarization direction of its vibrations. The condition can be proved in a straightforward, although somewhat tedious, manner. The proof is obtained by first assuming the propagation directions of the two reflected waves to be quite general, assigning two angles each for their specification, and then imposing the three boundary conditions at $x_3 = 0$, $S_{i2} = 0$. Development of the complete proof provides the reader with an excellent opportunity to improve his ability to visualize and describe mathematically problems in three dimensions. For details, see Wasley [5].

Associated with the propagation of waves in elastic solid media is material motion and velocity. For example, when a longitudinal wave such as we have been investigating strikes a free surface, normally or obliquely, the surface will be displaced. The direction of this displacement relative to the normal to the boundary is termed the *angle of emergence* α_e and is shown in Fig. I.16. The particle velocity is represented by v, Σv_1 being the sum of the particle velocity components of the three waves parallel to the surface and Σv_3 being the sum of the particle velocity components perpendicular to the surface. Note that in the figure we are considering a phase change of π in the reflected longitudinal wave with respect to the incident longitudinal wave. Hence, Fig. I.16 is strictly applicable to materials in which $\sigma > 0.263$ for certain values of α_{LI} (see Fig. I.15). However, the essential ideas remain the same for all values of Poisson's ratio.

Evidently,

$$\alpha_e = \tan^{-1} \frac{\Sigma v_1}{\Sigma v_3}$$

and   (I-3.29)

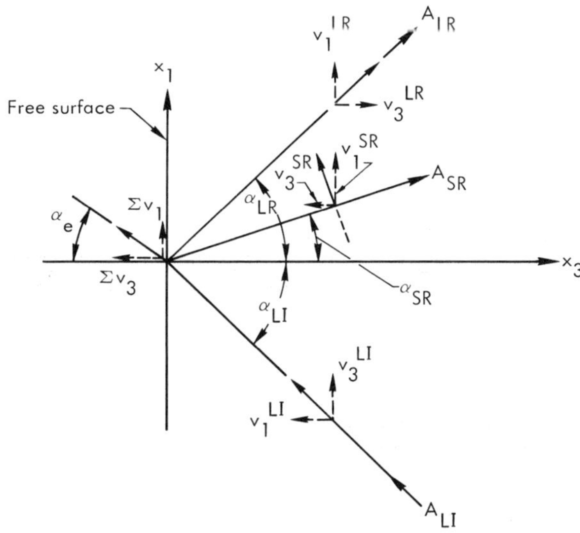

Fig. I.16. Motion of a plane free surface caused by an obliquely incident longitudinal wave.

$$v_e = \left[(\Sigma v_1)^2 + (\Sigma v_3)^2\right]^{1/2}$$

where v_e is the resultant particle velocity at the free surface. Recall we have asserted that the vibrations of the reflected shear wave generated by the oblique reflection of the longitudinal wave are polarized in the plane of incidence, and hence the particle velocity associated with this shear wave is normal to its direction of propagation and is also in the $x_1 x_3$ plane. If the incident longitudinal wave falls normally on the surface ($\alpha_e = 0$), the direction of surface motion is parallel to the x_3 axis.

It is instructive to examine the angle of emergence with respect to its relationship with the incident wave and to the parameters of the elastic medium. We have from Eqs. (I-3.16), (I-3.18), and (I-3.19)

$$v_j^{(i)} = \sum_i v_j^{(i)} = \sum_i \frac{\partial u_j^{(i)}}{\partial t} = \sum_i T_j^{(i)} \frac{\partial u^{(i)}}{\partial t} = \sum_i T_j^{(i)} A^{(i)} \frac{\partial}{\partial t} (\cos \zeta^{(i)})$$

where T, A and ζ are, respectively, the appropriate trigonometric function, amplitude and argument of the ith wave ($i = LI, LR$ and SR) in the jth direction ($j = 1$ representing the x_1 direction and $j = 3$ the x_3 direction). From relationships (I-3.29) and (I-3.24) we can write

$$\tan \alpha_e = \frac{\displaystyle\sum_i T_1^{(i)} A^{(i)} \sin \zeta^{(i)}}{\displaystyle\sum_i T_3^{(i)} A^{(i)} \sin \zeta^{(i)}}$$

$$= \frac{A_{LI} \sin \alpha_{LI} \sin \zeta_{LI} + A_{LR} \sin \alpha_{LI} \sin \zeta_{LR} + A_{SR} \cos \alpha_{SR} \sin \zeta_{SR}}{-A_{LI} \cos \alpha_{LI} \sin \zeta_{LI} + A_{LR} \cos \alpha_{LI} \sin \zeta_{LR} - A_{SR} \sin \alpha_{SR} \sin \zeta_{SR}}.$$

At $x_3 = 0$,

$$\tan \alpha_e = \frac{A_{LI} \sin \alpha_{LI} + A_{LR} \sin \alpha_{LI} + A_{SR} \cos \alpha_{SR}}{-A_{LI} \cos \alpha_{LI} + A_{LR} \cos \alpha_{LI} - A_{SR} \sin \alpha_{SR}}$$

$$= \frac{\sin \alpha_{LI}(1 + A_{LR}/A_{LI}) + (A_{SR}/A_{LI}) \cos \alpha_{SR}}{\cos \alpha_{LI}(-1 + A_{LR}/A_{LI}) - (A_{SR}/A_{LI}) \sin \alpha_{SR}}.$$

Now using expressions (I-3.27) and (I-3.28), we have

$$2(1 - A_{LR}/A_{LI}) \cos \alpha_{LI} \sin \alpha_{SR} = (A_{SR}/A_{LI}) \cos 2\alpha_{SR}$$

and

$$(1 + A_{LR}/A_{LI}) \sin \alpha_{LI} \cos 2\alpha_{SR} = (A_{SR}/A_{LI}) \sin \alpha_{SR} \sin 2\alpha_{SR}$$

and it follows that

$$\tan \alpha_e = \frac{(A_{SR}/A_{LI}) \sin \alpha_{SR} \tan 2\alpha_{SR} + (A_{SR}/A_{LI}) \cos \alpha_{SR}}{-\dfrac{(A_{SR}/A_{LI}) \cos 2\alpha_{SR}}{2 \sin \alpha_{SR}} - (A_{SR}/A_{LI}) \sin \alpha_{SR}}$$

$$= -\frac{2 \sin^2\alpha_{SR} \tan 2\alpha_{SR} + 2 \sin \alpha_{SR} \cos \alpha_{SR}}{\cos 2\alpha_{SR} + 2 \sin^2 \alpha_{SR}} = -\tan 2\alpha_{SR}.$$

Thus, we see that the simple expression

$$\alpha_e = -2\alpha_{SR} \tag{I-3.30}$$

exists, and from Eq. (I-3.26), the angle of emergence is a function of both the angle of incidence and Poisson's ratio. The negative sign in expression (I-3.30) arises because of the wave directions and the sign conventions chosen.

It is interesting now to consider the division of energy of the incident wave between the two reflected waves. Using Fig. I.17, let us write the energy $E^{(i)}$ of the wave filling a cylinder as

$$E^{(i)} = \int_0^t \rho\, e^{(i)}\, c^{(i)}\, S^{(i)}\, dt \tag{I-3.31}$$

where, as before, $i = LI$, LR and SR; $e^{(i)}$ is the energy density of the ith wave; $c^{(i)}$ is the propagation velocity of the ith wave (either c_ϱ or c_s); and $S^{(i)}$ is the cross-sectional area perpendicular to the propagation direction. The last quantity can be represented by

$$S^{(i)} = S \cos \alpha^{(i)}$$

where S is the area of the wave disturbance on the free surface ($x_3 = 0$). Eviden-

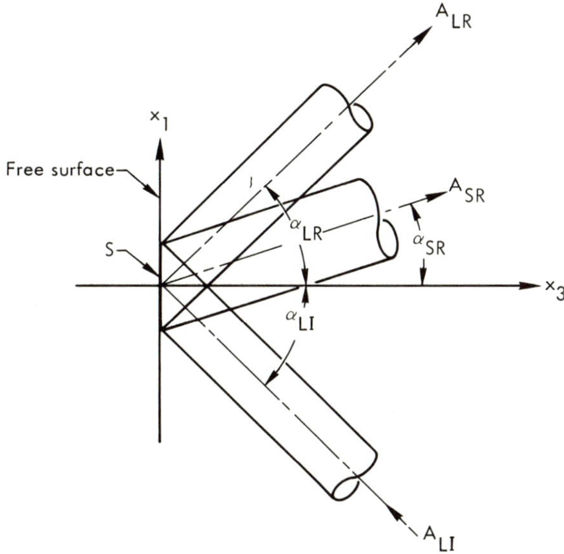

Fig. I.17. Division of energy of a reflecting longitudinal wave obliquely incident upon a plane free surface.

tly from Fig. I.17, all three waves have identical areas S. From expression (I-3.31) we have over the time interval δt

$$E^{(i)} = \rho e^{(i)} c^{(i)} S \cos \alpha^{(i)} \delta t \qquad\qquad (I\text{-}3.32)$$

where we assume none of the quantities vary with time; expressions for the energies of each of the three waves follow directly.

It is first necessary to illustrate that the energy density of a wave is proportional to the square of the amplitude of the wave. A simple plausibility argument can be given. We represent a wave in the real plane moving in the x_3 direction as

$$u_3 = A \sin pt$$

where A is the maximum amplitude. The kinetic energy per unit mass T can be seen to be

$$T = \frac{1}{2}\rho \dot{u}_3^2 = \frac{1}{2}\rho A^2 p^2 \cos^2 pt.$$

The potential energy per unit mass V can be seen to be

$$V = \frac{1}{2}ku_3^2 = \frac{1}{2} kA^2 \sin^2 pt$$

where k is a constant representing the (linear) resistance to movement, *i.e.*, the force equals ku_3. Note that both T and V are proportional to A^2; and in fact, for simple harmonic motion, it can be shown that the average kinetic energy per unit mass (per cycle) is equal to the average potential energy per unit mass (per cycle).

From the above, we may write

$$E^{(i)} \propto (A^{(i)})^2 c^{(i)} S \cos \alpha^{(i)} \delta t. \qquad\qquad (I\text{-}3.33)$$

Thus, the energy flux, *i.e.*, the energy per unit area per unit time, is

$$\frac{E^{(i)}}{S \cos \alpha^{(i)} \delta t} \propto (A^{(i)})^2 c^{(i)}. \qquad\qquad (I\text{-}3.34)$$

It can be seen that the energy flux of a distortion wave is less than that of a dilatation wave of the same amplitude, the ratio being c_s/c_ϱ. Note also that since the distortion wave is reflected at a smaller angle than the angle of incidence, the cross section of a reflected beam of distortional waves will be greater than the cross section of the incident beam, and the energy density will be lower (providing again the two waves have the same displacement amplitudes).

For any given incident dilatation wave, the sum of the energies of the two reflected waves is equal to the energy of the incident wave. We may write

$$E_{LI} = E_{LR} + E_{SR}$$

and from expressions (I-3.32), (I-3.25), and (I-3.26), we can readily show

$$1 = (A_{LR}/A_{LI})^2 + (A_{SR}/A_{LI})^2 \frac{\sin 2\alpha_{SR}}{\sin 2\alpha_{LI}} \tag{I-3.35}$$

or in the symmetrical form,

$$A_{LI}^2 \sin 2\alpha_{LI} = A_{LR}^2 \sin 2\alpha_{LR} + A_{SR}^2 \sin 2\alpha_{SR}.$$

It is appropriate to check whether the ratios A_{LR}/A_{LI} and A_{SR}/A_{LI} obtained from (I-3.27) and (I-3.28) satisfy the conservation of energy expressed by (I-3.35). We have arrived at Eq. (I-3.35) by a line of argument which is quite independent of the particular values of the ratios, and it is just this independence of the two arguments which makes the check a sensible thing to do.

Solving Eqs. (I-3.27) and (I-3.28), we have[8]

$$\frac{A_{LR}}{A_{LI}} = \frac{2 \cos \alpha_{LI} \sin^2\alpha_{SR} \sin 2\alpha_{SR} - \cos^2 2\alpha_{SR} \sin \alpha_{LI}}{2 \cos \alpha_{LI} \sin^2\alpha_{SR} \sin 2\alpha_{SR} + \cos^2 2\alpha_{SR} \sin \alpha_{LI}}$$

and   (I-3.36)

$$\frac{A_{SR}}{A_{LI}} = \frac{2 \sin 2\alpha_{LI} \cos 2\alpha_{SR} \sin \alpha_{SR}}{2 \cos \alpha_{LI} \sin^2\alpha_{SR} \sin 2\alpha_{SR} + \cos^2 2\alpha_{SR} \sin \alpha_{LI}}.$$

Let us introduce the following abbreviations:

$$a \equiv 2 \cos \alpha_{LI} \sin^2\alpha_{SR} \sin 2\alpha_{SR},$$

$$b \equiv \cos^2 2\alpha_{SR} \sin \alpha_{LI},$$

and

$$c \equiv \sin 2\alpha_{LI} \cos 2\alpha_{SR} \sin \alpha_{SR},$$

so that relationships (I-3.36) can be written in the simple manner

$$\frac{A_{LR}}{A_{LI}} = \frac{a - b}{a + b}$$

and   (I-3.37)

$$\frac{A_{SR}}{A_{LI}} = \frac{2c}{a + b} \, .$$

Now we compute the last term of Eq. (I-3.35) and find

$$\left(\frac{A_{SR}}{A_{LI}}\right)^2 \frac{\sin 2\alpha_{SR}}{\sin 2\alpha_{LI}} = \frac{4c^2 \sin 2\alpha_{SR}}{(a + b)^2 \sin 2\alpha_{LI}}$$

$$= \frac{4 \sin 2\alpha_{LI} \cos^2 2\alpha_{SR} \sin^2\alpha_{SR} \sin 2\alpha_{SR}}{(a + b)^2} \, .$$ (I-3.38)

Upon inserting the first of Eqs. (I-3.37) and Eqs. (I-3.38) into the energy relation (I-3.35), we obtain

$$1 = \frac{a^2 - 2ab + b^2}{(a + b)^2} + \frac{4ab}{(a + b)^2} = \frac{(a + b)^2}{(a + b)^2} = 1$$

and the check is complete.

Next, let us consider a shear wave incident obliquely on a plane free surface. We ask the same questions posed earlier in our treatment of an incident longitudinal wave, namely, the kind, amplitude, and direction of the reflected waves, the planes of vibration of the various waves, the direction of movement of the free boundary, and the division of energy among the several disturbances. Since similar techniques used with the previous problem are also appropriate for investigation of the phenomena associated with an incident wave of distortion, we omit many of the details and proceed in essentially outline form to the results. (See Wasley [5] for additional discussion.)

We examine a plane shear wave incident upon a free surface at angle β_{SI} with the normal to that surface. As before, we take the x_3 axis along the normal and let the wave propagate in the $x_1 x_3$ plane. Figure I.18 shows the geometry; note the similarity to Fig. I.13.

Although the polarization direction of the vibrations of the incident wave is known to be in a plane normal to the direction of propagation, it is necessary to make a further statement regarding this polarization direction. That is, the displacements of the impinging shear wave can have their directions, in general, at

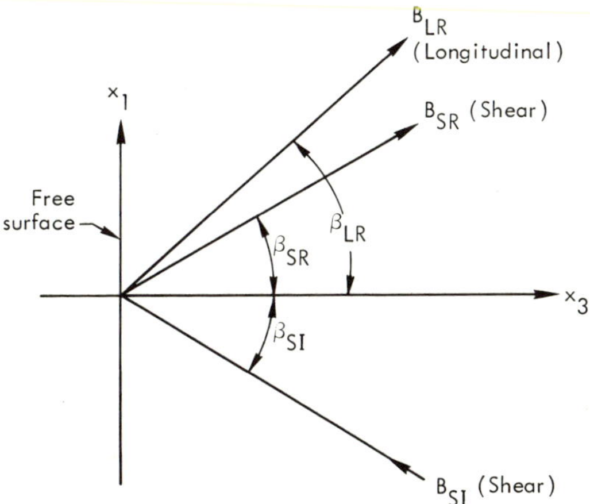

Fig. I.18. Reflection of a shear wave obliquely incident upon a plane free surface.

some arbitrary angle with the x_2 axis ($\neq \pi/2$). However, these displacements may be resolved into two components, those displacements parallel to the x_2 axis and those perpendicular to that axis. It is sufficient to determine the conditions for the reflection of a shear wave for these two special vibration directions, since by superposition the reflection condition for an arbitrary direction of vibration may be synthesized from a combination of these results.

Let us first take the vibration direction of the incident shear wave to be parallel to the x_2 axis. We "assume" that only a distortion wave is reflected here, write the appropriate expressions for representation of the two waves [see Eqs. (I-3.15) - (I-3.19)], form the net displacements produced by these waves [see Eqs. (I-3.21)], and determine whether the relevant boundary conditions at $x_3 = 0$

$$S_{3i} = 0 \tag{I-3.39}$$

are satisfied. Since there is no motion of the incident wave in either the x_1 or x_3 direction, we may reason on the basis of our previous efforts and the boundary conditions $S_{31} = 0$ and $S_{33} = 0$ that the reflected wave will have no vibrations parallel to the $x_1 x_3$ plane and that the problem is two-dimensional. Thus the net displacements can be expressed in a particularly simple fashion in this situation.

The above procedure results in an equation which can be satisfied for all values of x_3 and t if, indeed, only a reflected shear wave is generated and is transmitted with a change in phase at an angle equal to the angle of incidence (*i.e.*, $\beta_{SI} = \beta_{SR}$).

For the second situation, let us take the vibration direction of the incident shear wave to be in the $x_1 x_3$ plane. The treatment is analogous to that previously described. The boundary conditions of which we make use are $S_{33} = 0$ and $S_{31} = 0$ at $x_3 = 0$. It is now found that these conditions can only be satisfied by assuming that a longitudinal wave as well as a shear wave is reflected. The wave of distortion is reflected at an angle equal to the angle of incidence, as in the situation above, and the longitudinal disturbance is reflected at an angle β_{LR} (shown in Fig. I.18) where

$$\frac{\sin \beta_{LR}}{\sin \beta_{SI}} = \frac{c_\ell}{c_s}. \tag{I-3.40}$$

It follows that $\beta_{SI} = \beta_{SR} < \beta_{LR}$. Note the similarity between Eqs. (I-3.40) and (I-3.26). In fact, Fig. I.14 can be used to illustrate the dependence of the angle of reflection of the dilatation wave β_{LR} upon the angle of the incident shear wave β_{SI}; for this purpose we substitute β_{LR} for α_{LI} and β_{SI} for α_{SR}.

From these conclusions, we may readily determine the connection among the various amplitudes, B_{SI}, B_{SR}, and B_{LR}, for any angle of incidence.[9] Thus

$$(B_{SI} - B_{SR}) \sin 2\beta_{SI} \sin \beta_{SI} + B_{LR} \sin \beta_{LR} \cos 2\beta_{SI} = 0 \tag{I-3.41}$$

and

$$(B_{SI} + B_{SR}) \cos 2\beta_{SI} + 2B_{LR} \sin \beta_{SI} \cos \beta_{LR} = 0 \tag{I-3.42}$$

or if written in terms of the amplitude ratios,

$$\frac{B_{SR}}{B_{SI}} = -\frac{\cos^2 2\beta_{SI} \sin \beta_{LR} - 2 \sin^2 \beta_{SI} \sin 2\beta_{SI} \cos \beta_{LR}}{\cos^2 2\beta_{SI} \sin \beta_{LR} + 2 \sin^2 \beta_{SI} \sin 2\beta_{SI} \cos \beta_{LR}}$$

and

$$\tag{I-3.43}$$

$$\frac{B_{LR}}{B_{SI}} = -\frac{2 \sin \beta_{SI} \sin 2\beta_{SI} \cos 2\beta_{SI}}{\cos^2 2\beta_{SI} \sin \beta_{LR} + 2 \sin^2 \beta_{SI} \sin 2\beta_{SI} \cos \beta_{LR}}.$$

Figure I.19, with Poisson's ratio as a parameter, shows the ratios of the reflected stress amplitudes, B_{SR} and B_{LR}, to the incident stress amplitude as functions of the angle of incidence, β_{SI}. Expressions (I-3.40) and (I-3.43) were used to determine values for this figure.

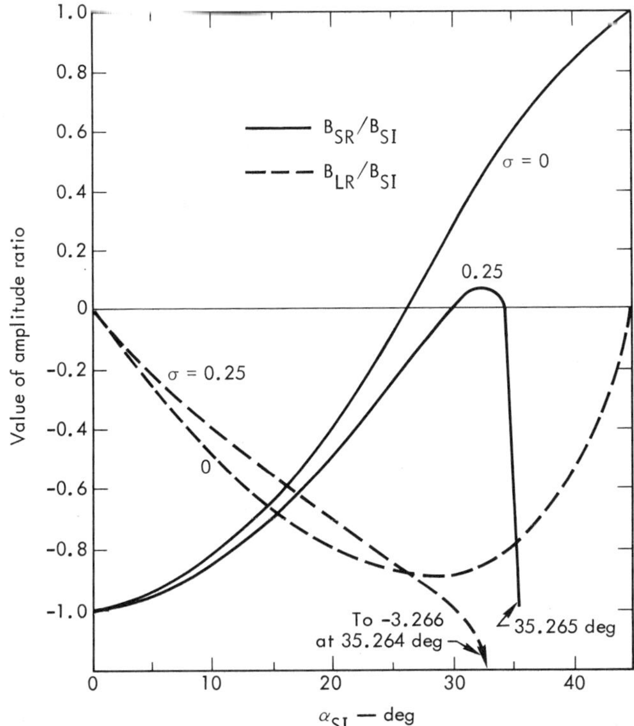

Fig. I.19. Ratios of reflected displacement amplitudes to incident displacement amplitude as functions of angle of incidence of a shear wave α_{SI} and of Poisson's ratio σ.

We summarize several important conclusions which can be formed by examining the two cases considered above:

1. In general, if the direction of vibration of the incident shear wave is parallel to the x_2 axis, only a shear wave is reflected. If the direction is perpendicular to the x_2 axis, both a shear and longitudinal wave are reflected. In both situations, $\beta_{SI} = \beta_{SR}$.

2. At normal incidence for $\sigma < 0.5$, Fig. I.19 illustrates that no dilatation wave is reflected, and the reflected distortion wave is π out of phase with the incident wave. This lack of phase coincidence is similar to the comparable situation of an incident longitudinal disturbance (see Fig. I.15).

3. In Fig. I.19, both amplitude ratios collapse to a point for $\sigma = 0.5$ at $\beta_{SI} = 0$. The graph of B_{SR}/B_{SI} consists of a point at -1, and the graph of B_{LR}/B_{SI} consists of a point at 0. However, this is of only academic interest, since

for an isotropic incompressible fluid where $\sigma = 1/2$ there is no shear wave possible.

4. As in the instance of an incident longitudinal wave, for $\sigma < 0.263$ there are two angles of incidence for which the amplitude of the reflected shear wave is zero and all the energy is converted into a longitudinal wave. For $\sigma < 0.263$, $(B_{SR}/B_{SI}) < 0$ and there is a change in phase.

5. Since $\sin \beta_{LR}$ cannot be larger than unity, we note from expression (I-3.40) and from Fig. I.19 that there is an angle beyond which a longitudinal wave cannot be reflected. This critical angle is called the angle of *total reflection* and is discussed at the end of Section III-C. For this critical angle, the shear wave is reflected at the angle of incidence and is equal in amplitude to the incident distortion wave.

6. The amplitude ratio B_{LR}/B_{SI} can exceed unity for certain values of σ and β_{SI}.

We shall not pursue discussions of the division of energy and of the direction of movement of the free boundary. The methods of approach are similar to those given earlier for an incident longitudinal wave.

C. Reflection and Refraction at a Plane Interface

In this section we are concerned with phenomena at a plane interface between two elastic isotropic media which have abruptly different physical properties. The treatment of the interface problem is similar to that previously described for reflection at a free boundary. It is enough to present only the general procedures involved and to give results of use to us.

We require that the two media do not move relative to each other at the interface. Under these conditions, a disturbance incident obliquely upon this interface will generate, in general, four waves; two of these waves (one longitudinal and one shear) are refracted into the second medium, and two (again one longitudinal and one shear) are reflected from the boundary. This condition of continuous slip-free contact can be expressed in terms of the normal and tangential displacements and stresses across the boundary. Thus, at the interface there are now four separate boundary conditions.[10] Let us consider first a longitudinal simple harmonic wave propagating in the $x_1 x_3$ plane and incident on the boundary at an angle α_{LI} with the normal. As before, we have a two-dimensional problem. Although the geometry will follow the lead established in Section III-B and illustrated in Fig. I.13, it must be extended to include specification of the refracted waves, which in turn requires notational distinction between the two media. Hence, the medium in which the incident disturbance is traveling will be denoted by the subscript 1 and the other medium by the subscript 2. The amplitude of the refracted longitudinal wave is labeled A_{LRf} and its propagation direction (with respect to the normal to the interface) α_{LRf}.

Similarly, the refracted shear wave is given by the quantities A_{SRf} and α_{SRf}.

Using the above boundary conditions and the theory of linear dependence of functions, it can be shown that the five waves satisfy the angular conditions

$$\frac{\sin \alpha_{LI}}{c_{\varrho 1}} = \frac{\sin \alpha_{LR}}{c_{\varrho 1}} = \frac{\sin \alpha_{SR}}{c_{s1}} = \frac{\sin \alpha_{LRf}}{c_{\varrho 2}} = \frac{\sin \alpha_{SRf}}{c_{s2}}. \tag{I-3.44}$$

It is again instructive to point out the similarity in form between the expressions for mechanical wave reflection and refraction [Eqs. (I-3.44)] and the equivalent expressions relating to optical reflection and refraction. Many of the principles in the two disciplines are the same.

If we now substitute suitable expressions for the waves into the boundary conditions, we obtain four relations among the amplitudes. In summary, we have

$$(A_{LI} + A_{LR}) \sin \alpha_{LI} + A_{SR} \cos \alpha_{SR} - A_{LRf} \sin \alpha_{LRf} + A_{SRf} \cos \alpha_{SRf} = 0$$

$$(A_{LI} - A_{LR}) \cos \alpha_{LI} + A_{SR} \sin \alpha_{SR} - A_{LRf} \cos \alpha_{LRf} - A_{SRf} \sin \alpha_{SRf} = 0$$

$$(A_{LI} + A_{LR}) c_{\varrho 1} \cos 2\alpha_{SR} - A_{SR} c_{s1} \sin 2\alpha_{SR}$$

$$- A_{LRf} c_{\varrho 2}(\rho_2/\rho_1) \cos 2\alpha_{SRf} - A_{SRf} c_{s2}(\rho_2/\rho_1) \sin 2\alpha_{SRf} = 0$$

and $\tag{I-3.45}$

$$[(A_{LI} - A_{LR}) \sin 2\alpha_{LI} - A_{SR}(c_{\varrho 1}/c_{s1}) \cos 2\alpha_{SR}] \rho_1 c_{s1}^2$$

$$- [A_{LRf}(c_{\varrho 1}/c_{\varrho 2}) \sin 2\alpha_{LRf} - A_{SRf}(c_{\varrho 1}/c_{s2}) \cos 2\alpha_{SRf}] \rho_2 c_{s2}^2 = 0$$

where the densities ρ_1 and ρ_2 can be expressed if desired in terms of $\lambda_1, \lambda_2, \mu_1, \mu_2, c_{\varrho 1}, c_{\varrho 2}, c_{s1}$, and c_{s2} through relations (I-2.10) and (I-2.11). As before, the above equations enable the amplitudes of the reflected and refracted waves to be given in terms of the amplitude of the incident longitudinal wave as a function of angle of incidence and Poisson's ratio. A graph similar to Fig. I.15 can be constructed.

A special case of significant practical interest, in both the elastic and non-elastic cases, is the situation in which the initial longitudinal wave is incident normally upon the boundary. We find for this condition that only normal longitudinal waves result, and that

$$\frac{A_{LR}}{A_{LI}} = \frac{\rho_2 c_{\varrho 2} - \rho_1 c_{\varrho 1}}{\rho_2 c_{\varrho 2} + \rho_1 c_{\varrho 1}}$$

and   (I-3.46)

$$\frac{A_{LRf}}{A_{LI}} = \frac{2\rho_2 c_{\varrho 2}}{\rho_2 c_{\varrho 2} + \rho_1 c_{\varrho 1}}.$$

The amplitude ratio in the first equation of (I-3.46) is called the *reflection factor* while that ratio in the second of (I-3.46) is named the *transmission factor*. Note that the amplitude of the reflected stress wave depends upon the quantity $(\rho_2 c_{\varrho 2} - \rho_1 c_{\varrho 1})$ and that *no* wave will be reflected at normal incidence when the product of the density and longitudinal sound velocity is the same for the two media. This product ρc is known as the *characteristic acoustic impedance* of the medium. It is a general characteristic of wave motion that reflection (either normal or oblique) takes place if, and only if, the propagating waves encounter a change of impedance. When the acoustic impedance of the second medium is greater than that of the first, the amplitude of the displacement upon reflection is of the same sense as that of the incident wave. When the impedance of the second medium is less than that of the first, $(A_{LR}/A_{LI}) < 0$, as expected from earlier considerations.

Let us now examine the characteristics of an elastic shear wave, simple harmonic in nature, reflecting and refracting at a plane interface. We again take the incident wave to be propagating in the $x_1 x_3$ plane. The geometry is similar to that illustrated in Fig. I.18 except that extension is made, as above, to the specification of the refracted waves. As in the situation of a shear wave impinging upon a free surface, two cases must be distinguished regarding the polarization direction of vibration.

In the situation where the vibration direction of the incident wave is parallel to the x_2 axis, it is found that motion of all the waves that are generated also occurs in a direction parallel to that axis and that no longitudinal waves can develop. By procedures previously described, it follows that the reflected shear wave will be propagated at an angle equal to the angle of incidence and that the refracted shear wave will be propagated at an angle β_{SRf} with the interface normal, where

$$\frac{\sin \beta_{SRf}}{\sin \beta_{SI}} = \frac{c_{s2}}{c_{s1}}.$$ (I-3.47)

Upon substitution of suitable expressions for the waves into the appropriate boundary conditions, we obtain the relations

$$B_{SI} + B_{SR} - B_{SRf} = 0$$

and   (I-3.48)

$$(B_{SI} - B_{SR}) \rho_1 \sin 2\beta_{SI} - B_{SRf}\rho_2 \sin 2\beta_{SRf} = 0.$$

When the vibrations of the incident wave are perpendicular to the x_2 axis, four additional waves result at the interface. The boundary conditions and the theory of linear dependence of functions require the relationships among the respective propagation direction angles to conform to a sine law similar to (I-3.44), namely,

$$\frac{\sin \beta_{SI}}{c_{s1}} = \frac{\sin \beta_{SR}}{c_{s1}} = \frac{\sin \beta_{LR}}{c_{\ell 1}} = \frac{\sin \beta_{LRf}}{c_{\ell 2}} = \frac{\sin \beta_{SRf}}{c_{s2}}.$$ (I-3.49)

The four boundary conditions in this instance lead to the relations

$$(B_{SI} + B_{SR}) \cos \beta_{SI} + B_{LR} \sin \beta_{LR} - B_{LRf} \sin \beta_{LRf} - B_{SRf} \cos \beta_{SRf} = 0$$

$$(B_{SI} - B_{SR}) \sin \beta_{SI} + B_{LR} \cos \beta_{LR} + B_{LRf} \cos \beta_{LRf} - B_{SRf} \sin \beta_{SRf} = 0$$

$$(B_{SI} + B_{SR})c_{s1} \sin 2\beta_{SI} - B_{LR}c_{\ell 1} \cos 2\beta_{SI}$$

$$+ B_{LRf}c_{\ell 2}(\rho_2/\rho_1) \cos 2\beta_{SRf} - B_{SRf}c_{s2}(\rho_2/\rho_1) \sin 2\beta_{SRf} = 0$$

and   (I-3.50)

$$[(B_{SI} - B_{SR}) \cos 2\beta_{SI} - B_{LR}(c_{s1}/c_{\ell 1}) \sin 2\beta_{LR}] \, \rho_1 c_{s1}$$

$$- [B_{LRf}(c_{s2}/c_{\ell 2}) \sin 2\beta_{LRf} + B_{SRf} \cos 2\beta_{SRf}] \, \rho_2 c_{s2} = 0$$

where the densities ρ_1 and ρ_2 can again be expressed in terms of the Lamé constants and the acoustic velocities using (I-2.10) and (I-2.11).

We will not consider specifically the division of energy among the waves and movement of the interface. Similar methods, previously explained, may be used to treat these subjects.

There are two points that should be mentioned before leaving the topic of elastic wave reflection and refraction. First, it is emphasized again that because it is not necessary to specify the frequency of the harmonic waves and since nowhere is it stated that the propagation velocity is frequency-dependent, the equations are applicable for consideration of elastic waves of reasonably arbitrary shape by use of Fourier analysis and the principle of superposition.

Second, the condition arises that there is a limiting angle of incidence for the impinging wave which makes the angle of the reflected or refracted wave equal to $\pi/2$, *i.e.*, the wave propagates along the boundary. For angles of incidence greater than this critical angle, the simple relationships given previously no longer apply, and a phenomenon results which is similar to that called *total reflection* in optics. There are several cases that need to be examined. When a shear wave is incident upon a free boundary at an angle whose sine is greater than $c_s/c_\varrho = [\mu/(\lambda + 2\mu)]^{1/2}$, a longitudinal wave cannot be reflected. Note that an incident longitudinal disturbance on a free surface will not raise problems associated with total reflection since c_s is always less than c_ϱ. When extending the examination to a longitudinal wave incident upon an interface between two media, we find from (I-3.44) that total reflection[11] of a refracted longitudinal wave can occur if $\alpha_{LI} > \sin^{-1}(c_{\varrho 1}/c_{\varrho 2})$. Similar difficulties arise when a shear wave is incident upon an interface between two media. In the case where the vibration direction of the incident wave is parallel to the x_2 axis (using the same geometry as before), total reflection of the refracted shear wave can occur if $\beta_{SI} > \sin^{-1}(c_{s1}/c_{s2})$. Where this incident shear wave vibration direction is normal to the x_2 axis, the condition of total reflection can occur for the refracted longitudinal wave if $\beta_{SI} > \sin^{-1}(c_{s1}/c_{\varrho 2})$ and again for the refracted shear wave if $\beta_{SI} > \sin^{-1}(c_{s1}/c_{s2})$.

When the angle of incidence does exceed the critical angle, the treatment becomes relatively sophisticated. Complex quantities must be introduced to account for sine and cosine values greater than unity. It is found that instead of the generation of the usual reflected or refracted plane wave, a disturbance results which possesses distinctly different properties. For example, the amplitude of the disturbance decreases exponentially with distance from the interface and generally causes complicated phase changes in the other waves generated. Moreover, when total reflection occurs, the results are found to include terms that cause dispersion, and the analysis of waves of arbitrary shape becomes a more delicate matter.

D. *Reflection and Refraction at Curved Surfaces: Scattering*

A more general condition of reflection and refraction of elastic waves occurs when restrictions of the planarity of the wave and/or of the interfaces are removed. When mechanical radiation impinges upon an uneven surface, it is said to *scatter*. More specifically, a scattered wave is defined as the difference between the actual wave and the undisturbed wave which would exist if (usually finite) perturbations were not present.

The general method of approach to solve problems in elastic scattering involves the superposition of potential fields to account separately for the incident wave field and the scattered wave field. For the elastic wave-scattering situation, there is the additional complication that at least two potential functions are necessary to describe the incident wave field, whereas the bulk acoustic problem requires only one. The potential functions are required to satisfy the wave equation (or the corresponding Helmholtz equation). Thus, in the case of a bulk acoustic wave scattered by some obstacle, for example, the boundary value problem is to find a solution of the wave (or Helmholtz) equation consistent with the scattering potential (or a prescribed normal derivative) on the boundary of the obstacle.

The details of solution can become fairly involved and sophisticated, requiring analytical techniques such as integral equation representation and Green's function theory to assist in the formation for the scattering potentials in terms of the incident potentials. Further discussion here becomes too specialized. The reference by Banaugh [6], in particular, gives a good presentation of the various methods of solution, including the use of digital computers. Comments on application are given in Chapter I-5.

IV. Summary

To examine the propagation of elastic disturbances in a semi-infinite homogeneous isotropic solid with a plane boundary (a semi-extended medium), it is necessary to stipulate the direction of propagation of the disturbances. Waves can either propagate parallel and near to the boundary or at some arbitrary angle to it. Under the first condition, the disturbances are called surface waves. Under the second, the phenomenon of wave reflection occurs.

There are several kinds of elastic surface waves that can be transmitted in semi-extended solids. One of the most important types is plane Rayleigh waves.

The general mathematical procedure for studying Rayleigh waves is to determine the solution of the equations of small motion (I-1.56) using the expression for a progressive continuous simple harmonic plane wave and the boundary condition of a stress-free interface. A two-dimensional coordinate system is used in formulating the problem. The technique allows us to obtain the propagation condition, which, in turn, enables us to find the velocity of the Rayleigh surface waves by the relationship

$$\frac{p^2}{c_s^2 k^2} = \frac{c_r^2}{c_s^2} \tag{I-3.10}$$

providing the shear wave velocity c_s and Poisson's ratio σ are known.

The displacements, u_1 and u_3, follow from the above and are written

$$u_1 = -ak\left[\exp(-\xi_\varrho x_3)-2\xi_\varrho\xi_s\left(2k^2-\frac{p^2}{c_s^2}\right)^{-1}\exp(-\xi_s x_3)\right]\sin(kx_1-pt) \qquad (\text{I-3.12})$$

and

$$u_3 = -a\xi_\varrho\left[\exp(-\xi_\varrho x_3)-2k^2\left(2k^2-\frac{p^2}{c_s^2}\right)^{-1}\exp(-\xi_s x_3)\right]\cos(kx_1-pt) \qquad (\text{I-3.13})$$

where $\xi_\varrho = \left(k^2-p^2/c_\varrho^2\right)^{1/2}$ and $\xi_s = \left(k^2-p^2/c_s^2\right)^{1/2}$. Equations (I-3.12) and (I-3.13) define completely the solution because of the two dimensional formulation, i.e., $u_2 = 0$ everywhere.

When a plane wave propagates at some arbitrary angle to the bounding surface, wave reflection occurs. The simplest reflection occurs when the wave strikes the free surface normally. If the incident wave is longitudinal with a given sense, then the reflected wave is normal and must also be longitudinal but with its sense and direction opposite to the incident wave. A transverse disturbance that impinges normally upon a free surface is found to produce a reflection that is also a transverse wave normal to the surface.

When an elastic wave strikes a plane free surface obliquely reflection is more complicated because, in general, the energy of the incident wave is partitioned into two reflected waves instead of only one.

A longitudinal plane wave that is obliquely incident upon a plane free surface results in a reflected longitudinal and a reflected shear wave. The conditions of reflection direction are

$$\alpha_{LI} = \alpha_{LR} \qquad (\text{I-3.25})$$

$$\frac{\sin\alpha_{LI}}{\sin\alpha_{SR}} = \frac{c_\varrho}{c_s} = \left(\frac{\lambda+2\mu}{\mu}\right)^{1/2} \qquad (\text{I-3.26})$$

where α is the propagation direction angle of the particular wave in question, and the subscripts LI, LR and SR refer to the incident longitudinal, reflected longitudinal and reflected shear waves, respectively. The relations among the amplitudes of the waves are

$$\frac{A_{LR}}{A_{LI}} = \frac{2 \cos \alpha_{LI} \sin^2\alpha_{SR} \sin 2\alpha_{SR} - \cos^2 2\alpha_{SR} \sin \alpha_{LI}}{2 \cos \alpha_{LI} \sin^2\alpha_{SR} \sin 2\alpha_{SR} + \cos^2 2\alpha_{SR} \sin \alpha_{LI}}$$

and $(I\text{-}3.36)$

$$\frac{A_{SR}}{A_{LI}} = \frac{2 \sin 2\alpha_{LI} \cos 2\alpha_{SR} \sin \alpha_{SR}}{2 \cos \alpha_{LI} \sin^2\alpha_{SR} \sin 2\alpha_{SR} + \cos^2 2\alpha_{SR} \sin \alpha_{LI}}$$

With regard to the planes of vibration of the various waves, the following theorem can be proved: If a longitudinal wave that is incident on a free surface produces no displacement parallel to the coordinate axis x_2 (see Fig. I.13), then the reflected waves, one longitudinal and one transverse, also do not produce displacement parallel to x_2 and the problem is two-dimensional. The direction of displacement of the free surface is found to be in a very simple relation with the propagation direction of the reflected shear wave, namely

$$\alpha_e = -2\alpha_{SR} \qquad (I\text{-}3.30)$$

when α_e is the angle of emergence and is taken with respect to the normal as before.

A shear wave that is obliquely incident upon a plane free surface can be described mathematically in a similar fashion. For the more important situation in which the vibration direction of the incident shear wave is in the x_1x_3 plane, it is found that

$$\beta_{SI} = \beta_{SR}$$

$$\frac{\sin \beta_{LR}}{\sin \beta_{SI}} = \frac{c_\varrho}{c_s} \qquad (I\text{-}3.40)$$

$$\frac{B_{SR}}{B_{SI}} = - \frac{\cos^2 2\beta_{SI} \sin \beta_{LR} - 2 \sin^2\beta_{SI} \sin 2\beta_{SI} \cos \beta_{LR}}{\cos^2 2\beta_{SI} \sin \beta_{LR} + 2 \sin^2\beta_{SI} \sin 2\beta_{SI} \cos \beta_{LR}}$$

and $(I\text{-}3.43)$

$$\frac{B_{LR}}{B_{SI}} = - \frac{2 \sin \beta_{SI} \sin 2\beta_{SI} \cos 2\beta_{SI}}{\cos^2 2\beta_{SI} \sin \beta_{LR} + 2 \sin^2\beta_{SI} \sin 2\beta_{SI} \cos \beta_{LR}} .$$

The notation is comparable to that given earlier.

Using similar techinques, extension can be made of the above concepts and results to reflection and refraction at a plane slip-free interface between two different elastic isotropic media. Under these conditions, a disturbance incident obliquely upon this interface will generate, in general, four waves; two of these waves (one longitudinal and one shear) are refracted into the second medium, and two (again one longitudinal and one shear) are reflected from the boundary.

A special case of significant practical interest is the situation in which an initial longitudinal wave is incident normally upon the boundary. We find for this condition that only normal longitudinal waves result and that

$$\frac{A_{LR}}{A_{LI}} = \frac{\rho_2 c_{\varrho 2} - \rho_1 c_{\varrho 1}}{\rho_2 c_{\varrho 2} + \rho_1 c_{\varrho 1}}$$

and   (I-3.46)

$$\frac{A_{LRf}}{A_{LI}} = \frac{2\rho_2 c_{\varrho 2}}{\rho_2 c_{\varrho 2} + \rho_1 c_{\varrho 1}}$$

where A_{LRf} refers to the amplitude of the refracted wave, subscript 1 refers to the medium in which the incident disturbance propagates, and subscript 2 refers to the other medium; ρc is known as the acoustic impedance.

When treating interfaces that are not plane, the phenomenon of scattering results and rather involved mathematical techniques are required for analysis.

Notes

[1] What we mean by "near" will soon become clear; otherwise the disturbances are body waves (Chapter I-2).

[2] The propagation of surface waves in anisotropic media is a subject that is highly specialized, although not difficult. This subject is not presently of enough concern to warrant its inclusion here (see Musgrave [1]).

[3] The other two roots are of concern when studying certain reflection phenomena.

[4] The relative positions of the major and minor axes are reversed between Rayleigh waves and shallow depth gravity surface waves in liquids.

[5] We remind the reader the three velocities are found to be constants for any one homogeneous, elastic material.

[6] Expression (I-3.26) is used in the design of explosive plane wave generating systems which, in turn, are used in shock wave experiments (see Chapters II-1 and II-3).

[7] As stated in Section II, a common material for which $\sigma < 0.263$ is glass with $\sigma = 0.25$.

[8] Equations (I-3.36) are forms convenient to use in obtaining values for plotting Fig. I.15.

[9] Again, when treating shear waves, the sign convention for the various disturbances must be taken in a careful and consistent manner.

[10] Note that four waves are generated with four boundary conditions, whereas before (Section III-B), two waves are generated with two boundary conditions; such a one-to-one correspondence is in general always true.

[11] The apparent inconsistency with respect to the application of the word "reflection" to a refracted wave arises from the origin of the term "total reflection" in optics. There it is found generally for light (electromagnetic) waves, that penetration into the second medium (medium 2 in our geometry) is only a few wavelengths. The title of total reflection is thus justified in optics, but the carryover of the expression to elastic mechanical waves is not entirely appropriate.

References

[1] M. J. P. Musgrave, "Elastic Waves in Anisotropic Media," in *Progress in Solid Mechanics*, Vol. II (I. N. Sneddon and R. Hill, eds.), North-Holland, Amsterdam, 1961.

[2] H. Kolsky, *Stress Waves in Solids*, Oxford University Press, London, 1953; pp. 16-23, Dover, New York, 1963.

[3] E. Volterra and E. C. Zachmanoglou, *Dynamics of Vibrations*, pp. 539-545, Charles E. Merrill Books, Columbus, Ohio, 1965.

[4] J. S. Rinehart, "On Fractures Caused by Explosions and Impacts," in *Quarterly of the Colorado School of Mines*, Golden, Colorado, 1960.

[5] R. J. Wasley, *Propagation of Elastic Stress Disturbances in Deformable Solids*, University of California Lawrence Livermore Laboratory, Report UCRL-14616, pp. 40-56, 175-184, 1965.

[6] R. P. Banaugh, *Scattering of Acoustic and Elastic Waves by Surfaces of Arbitrary Shape*, University of California Lawrence Livermore Laboratory, Report UCRL-6779, pp. 1-9, 99-107, 1962.

WAVE PROPAGATION IN CIRCULAR CYLINDRICAL RODS

I. General

Bounded media, such as cylinders and plates, will also allow the transmission of stress disturbances. The only difference between the propagation of elastic stress disturbances in unbounded and bounded media is geometrical. In theory, the transmission of such disturbances can be treated by solving the equations of small motion with the appropriate boundary conditions. It is apparent, however, that unless this bounded geometry is simple enough, a precise analysis of the problem rapidly becomes computationally laborious. Thus it is not surprising that very few "exact" analyses exist.

A very important analytical exact solution does exist for trains of progressive simple harmonic waves of infinite duration propagating in uniform, isotropic, solid circular cylinders of infinite length. This solution is called the *Pochhammer-Chree* analysis. Exact treatments are also available for other bounded geometries, *e.g.*, the infinite plate. We will examine the Pochhammer-Chree solution for elastic wave travel in cylindrical bars in some detail, and will mention some of the main features of exact analysis of wave and pulse propagation in other geometries. Many parts of the various treatments are common to each other. Because of the complexity of the exact analyses, we will also examine selected elementary and approximate methods of solution.

We are primarily interested in the propagation of elastic waves in cylinders because of its relevance and the understanding it contributes to the subject treated in Chapter II-2. Other practical applications of the above analyses are mentioned briefly in Chapter I-5.

II. Propagation in Circular Cylinders of Infinite Length

A. Exact Solution; Pochhammer-Chree Analysis

1. Governing Equations

We begin with the equations of small motion previously derived for a Hookean elastic isotropic medium in a rectilinear Cartesian coordinate system. It is necessary to transform these equations to the proper system so as to enable convenient treatment of the propagation of disturbances in a circular cylinder, namely, transformation to a system of cylindrical coordinates. We outline two (although not completely independent) methods of transformation. Inserting the details of the transformations and manipulations is a very useful exercise. The reader can refer to Section IV in the Introduction and to the appropriate references.

The equations of motion are written in two forms. First, in terms of Cartesian tensor (or indicial) notation, the expressions are

$$\rho \frac{\partial^2 u_i}{\partial t^2} = (\lambda + \mu) \frac{\partial \Delta}{\partial x_i} + \mu \frac{\partial^2 u_i}{\partial x_j \partial x_j}$$

$$= (\lambda + \mu) \frac{\partial}{\partial x_i} \left(\frac{\partial u_j}{\partial x_j} \right) + \mu \frac{\partial^2 u_i}{\partial x_j \partial x_j}. \tag{I-1.56}$$

Symbolic (in this case, vector) notation is used for the second representation. The reason here is to show the power of symbolic methods (although not necessarily their simplicity) in certain situations. There is also an element of tradition in the use of vector notation for such purposes, even though there is some limitation associated with its use with second and higher rank tensors. Thus, we write

$$\rho \frac{\partial^2 \mathbf{u}}{\partial t^2} = (\lambda + \mu) \, \nabla (\nabla \cdot \mathbf{u}) + \mu \nabla^2 \mathbf{u} \tag{I-4.1}$$

where $\mathbf{u}$ is the displacement vector, and the meaning of the symbolic operators can be obtained directly from the corresponding relations (I-1.56).

The two methods of transformation considered are: (1) rewriting expressions (I-1.56) in cylindrical coordinates using the appropriate transformation laws of tensor analysis and the rules governing chain differentiation; and (2) transformation of expression (I-4.1) into a cylindrical coordinate system using well known definitions of the vector and scalar operators. The methods are interrelated — the differences being in definition and manipulation — and the results obtained are the same.

Since we are introducing a curvilinear coordinate system it is, *in general*, necessary to distinguish between two types of tensor transformation behavior, namely, *covariant* and *contravariant* behavior.[1] The two types of transformation are defined by certain rules governing the transformation of the components from one coordinate system to another. Velocity is an example of a contravariant tensor of first rank. The stress tensor can be written in either contravariant or covariant second-rank form, depending upon the nature of the initial reference condition selected.

For complex situations such as the use of nonorthogonal curvilinear coordinate systems, the application of the theory of relativity, or where reliance must be placed entirely upon abstract mathematics, the more intricate contravariant and covariant notation is usually necessary. Fortunately, however, there is a significant simplification that one can often introduce when working in many areas of mechanics, electric field theory, and similar disciplines that obviates the requirement for the more complicated tensor formalism. It can be shown that the Cartesian notation can be applied to any mutually orthogonal curvilinear coordinate system, provided the base vectors are taken as unit length vectors lying along the directions of the curvilinear axes. A base vector system is taken here to mean any system of n noncoplanar vectors passing through a common origin upon which an n-dimensional coordinate system is referenced. Even if no restrictions were placed upon the coordinate system, either with respect to orthogonality of base vectors or coincidence of the two, one can still retain Cartesian notation as long as a specific problem is treated and each calculation is checked for correspondence with physical reality. One must be extremely careful in any transformations made in such a situation, however, and each case must be examined separately.

The first of the two transformation methods makes use of, specifically: (1) the simple Cartesian notation via the above base vector restriction, (2) the tensor (vector) transformation law,

$$u_i = \frac{\partial x_i}{\partial x_j'} u_j',$$

(I-4.2)

and (3) the rules of chain differentiation.

Although there is no objection to transforming the equations of motion written in the form of (I-1.56), an equivalent form can be developed which is somewhat more convenient (and conventional) for further analysis. It involves simply a regrouping of the displacement terms in a fashion so that they form certain components of the rotation tensor $\omega_{k\ell}$.

To make this regrouping, we need to define another tensor that is similar in some respects to the unit tensor. The third-rank tensor ϵ_{ijk} is called the *alternating tensor* and is defined as

$$\epsilon_{ijk} = \begin{cases} 0 \text{ if any two of the subscripts are equal,} \\ +1 \text{ if the subscripts are in cyclic order,} \\ -1 \text{ if the subscripts are in noncyclic order.} \end{cases} \qquad \text{(I-4.3)}$$

It can be shown (most easily by expansion; see also Chou and Pagano [1]) that the relationship

$$\frac{\partial^2 u_i}{\partial x_j \partial x_j} = \frac{\partial}{\partial x_i}\left(\frac{\partial u_\ell}{\partial x_\ell}\right) - 2\partial\epsilon_{ijk}\frac{\partial\omega_j}{\partial x_k} \qquad \text{(I-4.4)}$$

holds, where the notation used for the rotation quantities has been modified to indicate the specific components remaining. The nonvanishing components are defined as

$$\omega_1 \equiv \omega_{23} = -\omega_{32},$$

$$\omega_2 \equiv \omega_{31} = -\omega_{13}, \qquad \text{(I-4.5)}$$

and

$$\omega_3 \equiv \omega_{12} = -\omega_{21}.$$

It is helpful when considering Eqs. (I-4.4) and (I-4.5) to recall that associated with the second rank antisymmetric rotation tensor in three dimensions, there is a tensor of first rank (a vector) related to the displacement function u_k by the expression $\epsilon_{ijk}(\partial u_k / \partial x_j)$.

Using relations (I-4.3) - (I-4.5), we can rewrite Eqs. (I-1.56) as

$$\rho\frac{\partial^2 u_i}{\partial t^2} = (\lambda + 2\mu)\frac{\partial}{\partial x_i}\left(\frac{\partial u_\ell}{\partial x_\ell}\right) - 2\mu\epsilon_{ijk}\frac{\partial\omega_j}{\partial x_k}. \qquad \text{(I-4.6)}$$

That Eqs. (I-4.6) are equivalent to (I-1.56) can be verified by the expansion and recollection of terms.

To proceed with the transformation, it is most convenient to find first the functional relationship between the particular curvilinear coordinates of interest and the Cartesian coordinates x_i, and then make the transformations accordingly. In circular cylindrical coordinates the variables are the radius r, the angle θ, and the dimension along the centerline (axis) of the cylinder z. The coordinate r is the distance from the z axis. In terms of the curvilinear coordinates, we have the relations

$$x_1 = r \cos \theta,$$

$$x_2 = r \sin \theta, \qquad (I\text{-}4.7)$$

and

$$x_3 = z.$$

It is recognized that there is some sacrifice of consistency (and generality) here. It is not necessary to particularize the notational description of the problem as yet. Indeed, although we could take $r = x_1'$, $\theta = x_2'$ and $z = x_3'$ and continue the treatment on that basis, we adopt Eqs. (I-4.7) for our development of the Pochhammer-Chree and related analyses.

The displacements in (r, θ, z) can now be denoted by u_r, u_θ, and u_z. Making use of rule (I-4.2), we can write

$$u_1 = \frac{\partial x_1}{\partial r} u_r + \frac{\partial x_1}{r\partial \theta} u_\theta + \frac{\partial x_1}{\partial z} u_z,$$

$$u_2 = \frac{\partial x_2}{\partial r} u_r + \frac{\partial x_2}{r\partial \theta} u_\theta + \frac{\partial x_2}{\partial z} u_z, \qquad (I\text{-}4.8)$$

and

$$u_3 = \frac{\partial x_3}{\partial r} u_r + \frac{\partial x_3}{r\partial \theta} u_\theta + \frac{\partial x_3}{\partial z} u_z.$$

Note that in Eqs. (I-4.8) we use the differential quantity $r\,\partial\theta$ instead of $\partial\theta$ alone. This procedure enables both proper dimensions to be maintained and a direction to be readily associated with u_θ. Upon substitution into expressions (I-4.8) the appropriate derivatives of relations (I-4.7) for the direction cosines, we obtain

$$u_1 = u_r \cos \theta - u_\theta \sin \theta,$$

$$u_2 = u_r \sin \theta + u_\theta \cos \theta, \qquad (I\text{-}4.9)$$

and

$$u_3 = u_z.$$

One observes that Eqs. (I-4.9) can also be formed by simply taking the components of u_r, u_θ, and u_z in the x_1, x_2, and x_3 directions, respectively, and summing suitably.

Chain differentiation allows us to write

$$\frac{\partial}{\partial x_1} = \frac{\partial r}{\partial x_1}\frac{\partial}{\partial r} + r\frac{\partial \theta}{\partial x_1}\frac{1}{r}\frac{\partial}{\partial \theta} = \cos\theta\,\frac{\partial}{\partial r} - \frac{\sin\theta}{r}\frac{\partial}{\partial \theta},$$

$$\frac{\partial}{\partial x_2} = \frac{\partial r}{\partial x_2}\frac{\partial}{\partial r} + r\frac{\partial \theta}{\partial x_2}\frac{1}{r}\frac{\partial}{\partial \theta} = \sin\theta\,\frac{\partial}{\partial r} + \frac{\cos\theta}{r}\frac{\partial}{\partial \theta}, \qquad (I\text{-}4.10)$$

and

$$\frac{\partial}{\partial x_3} = \frac{\partial}{\partial z},$$

where we have used relations (I-4.7) in the form

$$\theta = \tan^{-1}\frac{x_2}{x_1},$$

and   (I-4.11)

$$r^2 = x_1^2 + x_2^2.$$

We will show only the transformation of the dilatation, $\Delta = \partial u_j/\partial x_j$, in expressions (I-4.6) to indicate the method used since the actual mechanics of transformation after the procedure has been set up is a relatively straightforward exercise in algebra. Using Eqs. (I-4.9) and (I-4.10), the dilatation becomes, in cylindrical coordinates,

$$\Delta = \frac{\partial u_1}{\partial x_1} + \frac{\partial u_2}{\partial x_2} + \frac{\partial u_3}{\partial x_3}$$

$$= \cos\theta\,\frac{\partial}{\partial r}(u_r\cos\theta - u_\theta\sin\theta) - \frac{\sin\theta}{r}\frac{\partial}{\partial \theta}(u_r\cos\theta - u_\theta\sin\theta)$$

$$+ \sin\theta\,\frac{\partial}{\partial r}(u_r\sin\theta + u_\theta\cos\theta) + \frac{\cos\theta}{r}\frac{\partial}{\partial \theta}(u_r\sin\theta + u_\theta\cos\theta) + \frac{\partial u_z}{\partial z}$$

$$= \frac{1}{r} \frac{\partial(ru_r)}{\partial r} + \frac{1}{r} \frac{\partial u_\theta}{\partial \theta} + \frac{\partial u_z}{\partial z}. \tag{I-4.12}$$

Several steps have been omitted in the calculus and algebra.

The other terms in Eqs. (I-4.6) [or (I-1.56)] transform in a similar manner, and we find the equations of small motion are expressed in cylindrical coordinates as

$$\rho \frac{\partial^2 u_r}{\partial t^2} = (\lambda + 2\mu) \frac{\partial \Delta}{\partial r} - \frac{2\mu}{r} \frac{\partial \omega_z}{\partial \theta} + 2\mu \frac{\partial \omega_\theta}{\partial z},$$

$$\rho \frac{\partial^2 u_\theta}{\partial t^2} = \frac{(\lambda + 2\mu)}{r} \frac{\partial \Delta}{\partial \theta} - 2\mu \frac{\partial \omega_r}{\partial z} + 2\mu \frac{\partial \omega_z}{\partial r}, \tag{I-4.13}$$

and

$$\rho \frac{\partial^2 u_z}{\partial t^2} = (\lambda + 2\mu) \frac{\partial \Delta}{\partial z} - \frac{2\mu}{r} \frac{\partial(r\omega_\theta)}{\partial r} + \frac{2\mu}{r} \frac{\partial \omega_r}{\partial \theta},$$

where Δ is given by Eqs. (I-4.12), and

$$\omega_r = \frac{1}{2} \left(\frac{1}{r} \frac{\partial u_z}{\partial \theta} - \frac{\partial u_\theta}{\partial z} \right),$$

$$\omega_\theta = \frac{1}{2} \left(\frac{\partial u_r}{\partial z} - \frac{\partial u_z}{\partial z} \right), \tag{I-4.14}$$

and

$$\omega_z = \frac{1}{2r} \left(\frac{\partial(ru_\theta)}{\partial r} - \frac{\partial u_r}{\partial \theta} \right).$$

By analogy with expressions (I-4.5), ω_r is taken equivalent to ω_1, ω_θ is equivalent to ω_2, and ω_z is equivalent to ω_3.

We now transform the vector form of the equations of motion (I-4.1) into a cylindrical coordinate representation. As before, it is helpful to transform (I-4.11) first into a more convenient form for later work. By noting the following vector relations (see Hildebrand [2]),

$$\nabla^2 \mathbf{u} = \nabla(\nabla \cdot \mathbf{u}) - \nabla \times \nabla \times \mathbf{u} \tag{I-4.15}$$

and

$$\omega = \frac{1}{2} \nabla \times \mathbf{u},$$

we can write

$$\rho \frac{\partial^2 \mathbf{u}}{\partial t^2} = (\lambda + 2\mu) \nabla (\nabla \cdot \mathbf{u}) - \mu \nabla \times \nabla \times \mathbf{u}$$

$$= (\lambda + 2\mu) \nabla (\nabla \cdot \mathbf{u}) - 2\mu \nabla \times \omega. \tag{I-4.16}$$

Equation (I-4.16) is entirely equivalent to (I-4.6). Thus expressions (I-1.56), (I-4.1), (I-4.6) and (I-4.16) are all statements of the equations of small motion.

As an example, similar to above, we may transform the dilatation, $\Delta = \nabla \cdot \mathbf{u}$, using relations and definitions developed for vector representation. Equation (I-4.12) is again the result.

In the solution of Eqs. (I-4.13), certain of the Hooke's law relations are used as boundary conditions. Specifically, in (I-1.32) we need S_{11}, S_{12}, and S_{13} written as appropriate functions of strain and transformed to cylindrical coordinates. A formal method of approach is to transform separately the stress components and the strain components from a Cartesian to a cylindrical system using the rules on tensor operation outlined above and in the Introduction. The results are inserted into Eqs. (I-1.32) and recombined (by transposition and substitution) into the desired forms. The transformations are not difficult, but involve tedious trigonometric manipulation. Several examples are presented in detail by Mason [3].

We may write by analogy with the symbols of stress used in the Cartesian notation that

$$S_{11} \overset{\triangle}{=} S_{rr}$$

$$S_{12} = S_{21} \overset{\triangle}{=} S_{r\theta} = S_{\theta r}, \tag{I-4.17}$$

and

$$S_{13} = S_{31} \overset{\triangle}{=} S_{rz} = S_{zr}.$$

We use the same analogy to express ϵ_{rr}, $\epsilon_{r\theta}$, and ϵ_{rz}. The relations between the stress and strain components in isotropic media are to be found using the procedure outlined above

$$S_{rr} = \lambda\Delta + 2\mu\epsilon_{rr} - \lambda\Delta + 2\mu\frac{\partial u_r}{\partial r},$$

$$S_{r\theta} = 2\mu\epsilon_{r\theta} = \mu\left[\frac{1}{r}\frac{\partial u_r}{\partial\theta} + r\frac{\partial}{\partial r}\left(\frac{u_\theta}{r}\right)\right], \qquad (\text{I-4.18})$$

and

$$S_{rz} = 2\mu\epsilon_{rz} = \mu\left(\frac{\partial u_r}{\partial z} + \frac{\partial u_z}{\partial r}\right).$$

Solutions of the equations of motion (I-4.13) are investigated using expressions for sinusoidal waves of infinite·duration propagating along an isotropic solid cylinder (*i.e.*, in the z direction) of infinite length. The displacement of each point is assumed to be a simple harmonic function of z and t. From Eq. (I-1.72), we can write

$$u_r = U_r(r,\theta)\exp\left[i(kz - pt)\right],$$

$$u_\theta = U_\theta(r,\theta)\exp\left[i(kz - pt)\right], \qquad (\text{I-4.19})$$

and

$$u_z = U_z(r,\theta)\exp\left[i(kz - pt)\right],$$

where U_r, U_θ, and U_z are real factors related to the amplitude of the waves. Note that the amplitude factors are functions of both r and θ. They differ in this regard from the expression for a plane wave (see (I-2.13)) with a constant amplitude factor, and from the expression for a Rayleigh wave (see (I-3.3)), the amplitude of which varies with depth only. We note, as we have in the situation for a Rayleigh wave, that in the sense the amplitudes (and hence displacements) are allowed to vary as functions of r and θ, the disturbances are not necessarily plane. No account is taken for a phase constant ϕ in Eqs. (I-4.19); we assume our waves to begin at a time where $\phi = 0$.

When Eqs. (I-4.13) and (I-4.19) are combined and the boundary conditions of vanishing of the three stress components (I-4.18) at the surface of the circular cylinder are imposed, one obtains expressions involving the frequency and wave length of the vibrations, the radius of the cylinder, the elastic constants of the material, and its density. These expressions are termed the *dispersion equations* (or *propagation conditions* or *frequency equations*), and form the basis of the Pochhammer-Chree treatment.

2. Particular Solutions

We now consider three particular types, *i.e.*, *modes*, of wave propagation in infinitely long circular cylindrical bars, namely, *longitudinal*, *torsional*, and *transverse* (or *flexural*). Each of these modes is identified with a specific dispersion equation, each equation having an infinite number of roots. Each root corresponds to a particular deformation pattern that exists independently of any other. The first root is termed the *fundamental*, the second root the *first harmonic*, and so on.

This rather specific approach was the one taken originally by Pochhammer and Chree. However, the problem can be considered more generally, and the results of such an investigation are presented after the section on transverse vibrations. Although the Pochhammer-Chree analysis has associated with it an infinite set of dispersion equations, *i.e.*, a frequency spectrum, the three modes indicated above correspond to the best known and the most important types of vibration, and in which the bulk of the activity is concentrated. More complex modes and higher harmonics are interesting, but are presently primarily of academic concern in that they correspond to no physically significant motion, *i.e.*, such vibrations are apparently rarely excited.

a. Longitudinal Mode. We examine first the propagation of longitudinal waves. Longitudinal motion of a cylinder is characterized by the periodic extension and contraction of its elements without lateral displacement of its axis. We assume that each particle vibrates in its rz plane so that the displacement u_θ vanishes. The condition is also imposed that the motion be symmetrical about the axis of the cylinder so that u_r and u_z are independent of θ. It then follows that $\partial u_z/\partial\theta = \partial u_\theta/\partial z = \partial(ru_\theta)/\partial r = \partial u_r/\partial\theta = 0$, and thus from (I-4.14) that $\omega_r = \omega_z = 0$. Using the assumed solutions (I-4.19), the first and third of the equations of motion (I-4.13) becomes[2]

$$\rho\, p^2 u_r = -(\lambda + 2\mu)\frac{\partial\Delta}{\partial r} - i\,2\mu k\omega_\theta$$

and
$$\tag{I-4.20}$$
$$\rho\, p^2 u_z = -i(\lambda + 2\mu)\,k\Delta + \frac{2\mu}{r}\frac{\partial(r\omega_\theta)}{\partial r}.$$

The technique of solution now becomes strictly mathematical. We obtain two differential equations, one in which the dependent variable is the dilatation Δ and one in which the dependent variable is the component of rotation ω_θ. We do this by eliminating first Δ and then ω_θ from (I-4.20) by using definitions (I-4.12) and (I-4.14). After some manipulation, we get

$$\frac{\partial^2 \Delta}{\partial r^2} + \frac{1}{r}\frac{\partial \Delta}{\partial r} + \left[\frac{\rho\, p^2}{(\lambda + 2\mu)} - k^2\right]\Delta = 0 \tag{I-4.21}$$

and

$$\frac{\partial^2 \omega_\theta}{\partial r^2} + \frac{1}{r}\frac{\partial \omega_\theta}{\partial r} - \frac{\omega_\theta}{r^2} + \left[\frac{\rho p^2}{\mu} - k^2\right]\omega_\theta = 0. \tag{I-4.22}$$

(It is a good exercise for the reader to check that the above two equations are dimensionally correct.)

By rewriting Eq. (I-4.21), we may obtain Bessel's equation of zero order, the solutions of which are well known. Let $g = [\rho p^2/(\lambda + 2\mu) - k^2]^{1/2}$, and we can then write

$$\frac{\partial^2 \Delta}{\partial (gr)^2} + \frac{1}{(gr)}\frac{\partial \Delta}{\partial (gr)} + \Delta = 0 \tag{I-4.23}$$

from which the solution that is finite at the axis, $r = 0$, is

$$\Delta = G\, J_0(gr) \tag{I-4.24}$$

where J_0 represents the Bessel function of the first kind, of order zero, and G is a function of z and t but independent of r.

In a similar manner, expression (I-4.22) can be transformed into Bessel's equation of order one. Let $h = [\rho p^2/\mu - k^2]^{1/2}$, and we can obtain

$$\frac{\partial^2 \omega_\theta}{\partial (hr)^2} + \frac{1}{(hr)}\frac{\partial \omega_\theta}{\partial (hr)} + \left(1 - \frac{1}{(hr)^2}\right)\omega_\theta = 0. \tag{I-4.25}$$

Again, the solution which is finite at the axis, $r = 0$, is

$$\omega_\theta = H\, J_1(hr) \tag{I-4.26}$$

where J_1 represents the Bessel function of the first kind, of order one, and H is a function of z and t but independent of r.

Now we can combine the assumed solutions (I-4.19) with expressions (I-4.12) and (I-4.14) and write

$$\Delta = \frac{\partial u_r}{\partial r} + \frac{u_r}{r} + \frac{\partial u_z}{\partial z}$$

$$= \left[\frac{\partial U_r(r)}{\partial r} + \frac{U_r(r)}{r} + ikU_z(r) \right] \exp\left[i(kz - pt) \right] \qquad (\text{I-4.27})$$

and

$$\omega_\theta = \frac{1}{2} \left(\frac{\partial u_r}{\partial z} - \frac{\partial u_z}{\partial r} \right)$$

$$= \frac{1}{2} \left[ikU_r(r) - \frac{U_z(r)}{r} \right] \exp\left[i(kz - pt) \right]. \qquad (\text{I-4.28})$$

Thus we have two expressions for Δ [(I-4.24) and (I-4.27)] and two for ω_θ [(I-4.26) and (I-4.28)]. To satisfy these equations, $U_r(r)$ and $U_z(r)$ must be in the form[3]

$$U_r(r) = A \frac{\partial}{\partial r} J_0(gr) + BkJ_1(hr) \qquad (\text{I-4.29})$$

and

$$U_z(r) = iAkJ_0(gr) + i\frac{B}{r} \frac{\partial}{\partial r} \left[rJ_1(hr) \right] \qquad (\text{I-4.30})$$

where A and B are constants.

The next step toward developing the dispersion equation relevant to the propagation of longitudinal waves is to incorporate solutions (I-4.19) and conditions (I-4.29) and (I-4.30) into the appropriate boundary equations. We require the bounding surface of our cylinder at $r = R$ to be free from tractions. Then from (I-4.18) one can read

$$S_{rr}\Big|_{r=R} = 0 = \lambda \left(\frac{\partial u_r}{\partial r} + \frac{u_r}{r} + \frac{\partial u_z}{\partial z} \right)_{r=R} + 2\mu \left(\frac{\partial u_r}{\partial r} \right)_{r=R}$$

$$= A \left[2\mu \frac{\partial^2 J_0(gR)}{\partial R^2} - \frac{\lambda \rho p^2 J_0(gR)}{\lambda + 2\mu} \right] + 2B\mu k \frac{\partial J_1(hR)}{\partial R} \qquad (\text{I-4.31})$$

and

$$S_{rz}\bigg|_{r=R} = 0 = \mu\left(\frac{\partial u_r}{\partial z} + \frac{\partial u_z}{\partial r}\right)_{r=R}$$

$$= 2Ak\frac{\partial J_0(gR)}{\partial R} + B\left(2k^2 - \frac{\rho p^2}{\mu}\right)J_1(hR) \qquad (\text{I-4.32})$$

where $\partial/\partial r\big|_{r=R}$ has been simplified to $\partial/\partial R$. Use is made here of Bessel's equations of zero and first order in their respective forms,

$$\frac{\partial^2 J_0(gr)}{\partial r^2} + \frac{1}{r}\frac{\partial J_0(gr)}{\partial r} + g^2 J_0(gr) = 0 \qquad (\text{I-4.33})$$

and

$$\frac{\partial^2 J_1(hr)}{\partial r^2} + \frac{1}{r}\frac{\partial J_1(hr)}{\partial r} + \left(h^2 - \frac{1}{r^2}\right)J_1(hr) = 0. \qquad (\text{I-4.34})$$

The dispersion or frequency equation is obtained by eliminating the constants A and B from Eqs. (I-4.31) and (I-4.32),

$$\left(\frac{\rho p^2}{\mu} - 2k^2\right)J_1(hR)\left[2\mu\frac{\partial^2 J_0(gR)}{\partial R^2} - \frac{\lambda\rho p^2 J_0(gR)}{\lambda + 2\mu}\right]$$

$$+ 4k^2\mu\frac{\partial J_0(gR)}{\partial R}\frac{\partial J_1(hR)}{\partial R} = 0. \qquad (\text{I-4.35})$$

This relationship is written in terms of six parameters, namely, λ, μ, ρ, R, p (or $\nu = p/2\pi$), and k (or $\Lambda = 2\pi/k$). However, using a nondimensional form, the number of variables may be effectively reduced to three, c_p/c_0, R/Λ, and Poisson's ratio σ, where c_p is the phase velocity (we are more specific below with what is meant by the adjective, phase) and $c_0 = (E/\rho)^{1/2}$, the velocity of waves traveling in a bar with infinitely long wavelength (to be considered in detail when discussing elementary methods of solution). Thus, for a given value of σ, an equation involving only c_p/c_0 and R/Λ is obtained, and a numerical solution is straightforward, although laborious. As stated, the equation has an infinite number of roots, each root corresponding to a different harmonic excited in a cylinder.

The dispersion, specifically called the *acoustic* or *mechanical* dispersion, arises physically in this situation from interaction of the wave with the boundary of the system, causing the continual partial conversion of shear waves to longitudinal waves and vice versa. The analytical prediction of the dispersion by the Pochhammer-Chree frequency equation for longitudinal vibration (I-4.35), in the form of the variation of wave velocity with wavelength (or wave frequency), has only relatively recently been investigated in any detail. To illustrate the dispersion for such waves, Fig. I.20 plots the fundamental and the first two harmonics over the most interesting range of wavelengths for Poisson's ratio equal to 0.25. The result of the variation of σ upon the dispersion of the fundamental is also shown in the figure. It is interesting to note that in the limit of a laterally inflexible medium, solution of the frequency equation in the fundamental mode predicts a discontinuity in slope at $R/\Lambda = 0.293$.

The values of c_ϱ/c_0, c_s/c_0, and c_r/c_0 for $\sigma = 0.25$ are also indicated in Fig. I.20. These terms are obtained with the following relations (using results

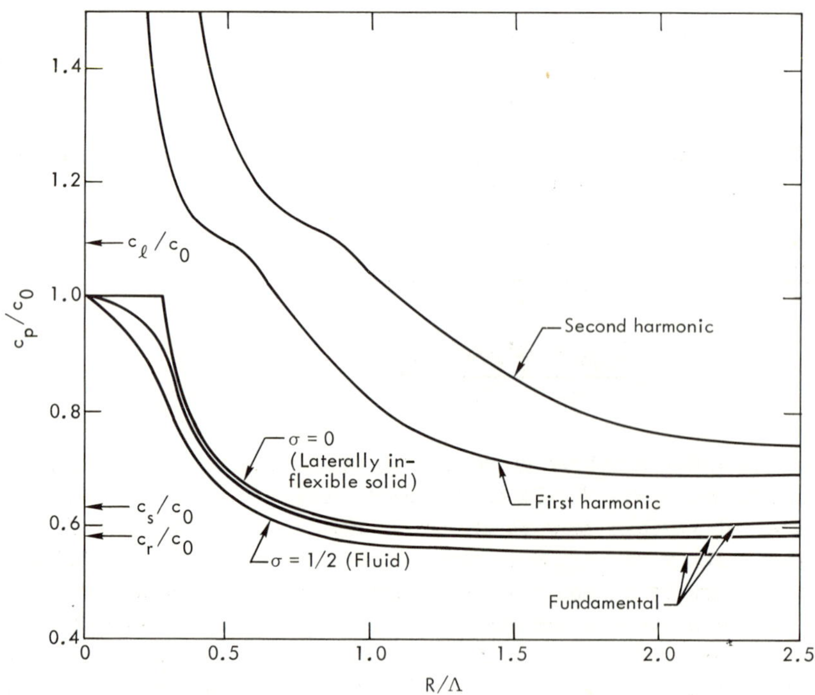

Fig. I.20. Phase velocity of longitudinal waves in a solid circular cylinder of radius R and Poisson's ratio 0.25 (except where noted).

from Chapter I-1, Section III-B, Chapter I-2, Section III and Chapter I-3, Section II):

$$\frac{c_\varrho}{c_0} = \left(\frac{\lambda + 2\mu}{E}\right)^{1/2} = \left[\frac{1-\sigma}{(1+\sigma)(1-2\sigma)}\right]^{1/2} = \underset{\sigma = 0.25}{} 1.095$$

$$\frac{c_s}{c_0} = \left(\frac{\mu}{E}\right)^{1/2} = \left[\frac{1}{2(1+\sigma)}\right]^{1/2} = \underset{\sigma = 0.25}{} 0.633 \qquad (I\text{-}4.36)$$

and

$$\frac{c_r}{c_0} = \left(\frac{c_r}{c_s}\right)\left(\frac{c_s}{c_0}\right) = \underset{\sigma = 0.25}{} (0.919)(0.633) = 0.582.$$

It may be seen from the lower curve in the figure that in the range $0 < R/\Lambda < 0.1$, the phase velocity does not decrease significantly from c_0 and can be taken to be that value in many situations. At the other extreme, the phase velocity ratio is predicted to approach c_r/c_0 when R/Λ tends to infinity.

It has been found experimentally (see Mason [4]) that some of the energy of an extensional pulse in a cylinder is propagated with velocity "near" to the dilatational velocity of the medium c_ϱ. However, it is also noted that the phase velocity may vary over a wide range, both below and well above c_ϱ (if one considers the higher harmonics). There are several questions that can be raised on the basis of these two observations. For example, what is the meaning of phase velocity? How is mechanical energy, *i.e.*, a signal, transmitted in cylinders? Is this transmission related to the phase velocity? Can the Pochhammer-Chree analysis predict the velocities of energy propagation? These questions suggest a brief but important digression.

Until now, there has been no requirement to raise these questions since in our previous investigations the velocity was a constant (or constants) independent of frequency, and all disturbances were transmitted without change in shape. We spoke indifferently of *wave* velocity and *phase* velocity. Such cannot now be the situation when we consider dispersive media, and we examine more explicitly the meaning of *phase* velocity and the associated concepts of *group* and *signal* velocities.

The most common concept of the velocity of propagation of waves is that of phase velocity. By definition, this velocity refers to the rate at which a specific element on a periodic simple harmonic wave of infinite duration, *e.g.*, a crest, moves in Eulerian space (see Fig. I.8). More simply, we stand at a certain position in a laboratory frame of reference and count the number of (simple harmonic) waves that pass us in a given amount of time. If displacement of the medium is represented by the function $u_j = U_j \exp\left[i(kz - pt)\right]$, then the surfaces of constant state or phase are defined by

$$kz - pt = \text{constant} \tag{I-4.37}$$

and these surfaces are propagated with the constant (phase) velocity

$$c_p = \frac{dz}{dt} = \frac{p}{k}. \tag{I-4.38}$$

The existence of such large phase velocities as can be noted in the harmonics in Fig. I.13 has been interpreted by some writers (see, for example, Green [5]) as being due to the wave front intersecting the axis of the cylinder obliquely. The wave front thus appears to travel a greater distance down the cylinder in a given time than the actual distance moved by the front perpendicular to itself.

The fact that we consider a single simple harmonic wave of infinite duration implies that this wave travels at a constant velocity and that it has associated with it a constant frequency. However, if this disturbance is not periodic or is of finite duration, representation cannot be specified in the form above. Then, if the disturbance is examined by means of a Fourier analysis, it is found that the individual harmonic waves constituting the disturbance are each of different frequencies. If the medium is dispersive, these waves travel at different velocities and separate as they move along, the degree of separation depending upon the characteristics of the individual component waves. This discussion leads us to the concept of *group* velocity.

To introduce the properties of such a disturbance in a dispersive medium, one method is to superpose two harmonic waves propagating in the same direction, but which differ very slightly in frequency and wave number (a complete and very lucid, discussion is given by Brillouin [6]). For simplicity, we take

$$u_1 = U \cos (kz - pt)$$

and   (I-4.39)

$$u_2 = U \cos \left[(k + \delta k)z - (p + \delta p)t\right].$$

If we add and rewrite the result by means of a trigonometric identity,[4] we obtain

$$u_1 + u_2 = 2U \cos \tfrac{1}{2}(z\delta k - t\delta p) \cos \left[\left(k + \frac{\delta k}{2}\right)z - \left(p + \frac{\delta p}{2}\right)t\right]. \tag{I-4.40}$$

Since we consider $\delta k \ll k$ and $\delta p \ll p$, expression (I-4.40) describes a wave oscillating at a frequency which differs negligibly from $p/2\pi$ and travels with a phase velocity essentially equal to p/k, while its effective amplitude $2U \cos 1/2\,(z\delta k - t\delta p)$ varies slowly between the sum of the amplitudes of the component waves and zero. As a result of this constructive and destructive interference, the distribution with respect to both time and space appears as a series of periodically repeated "beats" or "groups," as shown in Fig. I.21. This situation is occasionally called a *modulation* impressed on a *carrier* wave.

As before with phase velocity, the surfaces over which the beat amplitude $2U \cos 1/2\,(z\delta k - t\delta p)$ is constant are defined by

$$z\delta k - t\delta p = \text{constant}, \tag{I-4.41}$$

and these surfaces are propagated with the (group) velocity

$$c_g = \frac{dz}{dt} = \frac{\delta p}{\delta k}. \tag{I-4.42}$$

Hence, c_g is determined by the ratio of the difference of angular frequency to the difference of wave number.

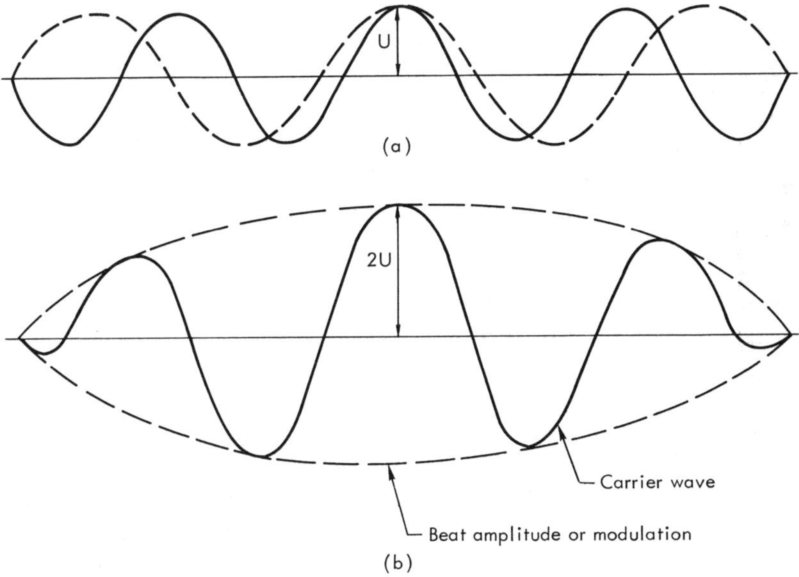

Fig. I.21. Superposition of two nearly identical waves. (a) Waves given by expressions (I-4.39). (b) Resultant, showing beats.

Several equivalent expressions for the group velocity may be derived which are helpful in understanding and applying this concept. As an extension to Eq. (I-4.42), we write

$$c_g = \frac{dp}{dk}\bigg|_{k_0} = \frac{d}{dk}(c_p k) = c_p + k\frac{dc_p}{dk} = c_p + \frac{2\pi}{\Lambda}\frac{\partial c_p}{\partial \Lambda}\frac{d\Lambda}{dk} = c_p - \Lambda\frac{\partial c_p}{\partial \Lambda} \qquad \text{(I-4.43)}$$

Thus, it is seen that if the medium is nondispersive and the propagation velocity is not a function of Λ, then c_g coincides with c_p. In a dispersive medium, their values are distinct.

The transport of mechanical energy is associated with the physical motion of particles. Hence the velocity of propagation of the beam amplitude c_g represents in general,[5] the velocity of propagation of energy. It is apparent, therefore, why the group velocity is of such importance in dispersive media.

As the interval δk is increased, the spread in phase velocity of the harmonic components in a dispersive medium becomes larger; the *wave packet* (a term also used in the subject of quantum mechanics) deforms rapidly, and the group velocity loses its physical significance. The concept is precise only when the wave packet is composed of waves lying within an infinitely narrow region of the spectrum, *i.e.*, at k_0. Hence, for practical purposes, we must interpret the group velocity as an approximation to the velocity with which the energy is propagated, the degree of such approximation depending upon the above considerations.

As an example of using expression (I-4.43), the reader may verify the group velocity for deep water waves for a small range of frequency to be

$$c_g = \frac{c_p}{2} = \frac{1}{2}(g\Lambda/2\pi)^{1/2}, \qquad \text{(I-4.44)}$$

where g is the acceleration due to gravity.

It can be shown mathematically (see Brillouin [6] or Stratton [7]) that if we interpret the term "wave front" to mean the very first arrival of the elastic energy disturbance, the wave front velocity is always equal to c_ϱ, no matter what the medium or what its boundary conditions. It may also be shown, however, that at this first instant the process starts always at zero amplitude.[6] The disturbance then changes slowly, both in period and amplitude, until with a sudden rise of amplitude the main body, or steady-state principal part, arrives, traveling with the *signal* velocity c_S. Figure I.22 describes qualitatively the theoretical arrival of a disturbance.

Here, the term "signal" physically represents that portion of the wave which actuates a measuring device. Hence, a measurement should indicate a velocity of

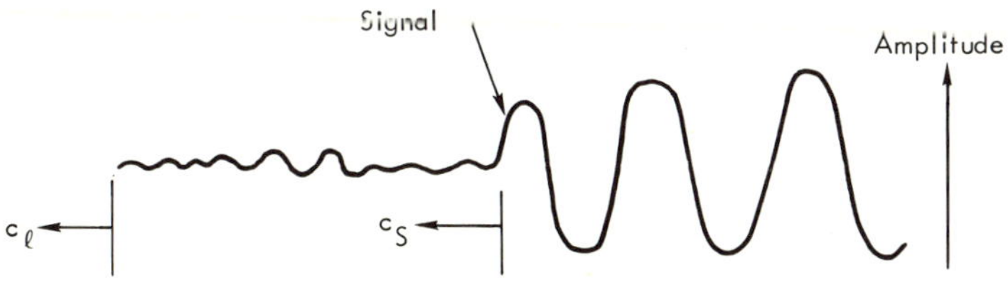

Fig. I.22. Theoretical movement of an elastic mechanical disturbance in a bounded medium.

propagation approximately equal to c_S which, under these conditions, coincides with c_g. It will be noted, however, that as the sensitivity of the detector is increased, the measured signal velocity also increases (theoretically), until in the limit of infinite sensitivity we should record the arrival of the front of the precursor which travels always with the velocity c_ϱ.

For purposes of plotting group velocity curves, it is helpful to rewrite relationship (I-4.43) in a nondimensional form. The rate of energy transmission in a normally dispersive medium c_g can be expressed as a ratio to the bar velocity c_0 as

$$\frac{c_g}{c_0} = \frac{c_p}{c_0} + \frac{R}{\Lambda}\frac{d(c_p/c_0)}{d(R/\Lambda)}. \tag{I-4.45}$$

[The important step in obtaining (I-4.45) from (I-4.43) is to recall $(-1)(d/d\Lambda)\Lambda^{-1} = \Lambda^{-2}$.]

Figure I.23 shows the group velocity curves corresponding to the fundamental and first two harmonics of longitudinal wave propagation for $\sigma = 0.25$. These are derived from Fig. I.20 using expression (I-4.45). The curves for the harmonics show that c_g, unlike c_p, does not exceed c_0. Note that in the fundamental c_g reaches a minimum at about 0.4 R/Λ while in the first harmonic c_g approaches a maximum at about the same value. Thus, when a pulse of longitudinal vibrations is transmitted in a glass cylinder ($\sigma = 0.25$), the Fourier components with wavelengths approximately $0.4\,R$ are found at the tail of the pulse and at its head in the first harmonic.

Returning to the question as to whether the Pochhammer-Chree analysis can predict transmission of energy with the longitudinal velocity c_ϱ, we may examine the harmonics in Fig. I.20. It is observed that these dispersion curves have "plateaus" in the region of c_ϱ, such plateaus becoming more pronounced (as a general rule[7]) as higher harmonics are reached. Hence in this region, the differential $dc_p/d\Lambda$ becomes small and $c_g \to c_p \approx c_\varrho$ according to expres-

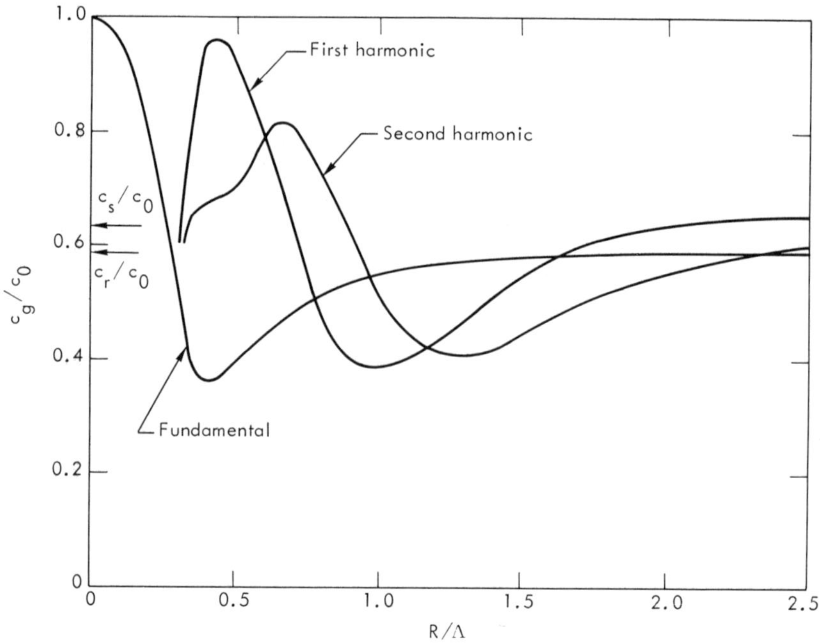

Fig. I.23. Group velocity of longitudinal waves in a solid circular cylinder of radius R and Poisson's ratio 0.25.

sion (I-4.45). Such a conclusion was originally advanced by this sort of plausibility argument, but recently proof (theoretical and experimental) has been given (see Zemanek [8]). Another type of plausibility argument that has been used occasionally can be stated as follows: If we consider a disturbance at a point in the interior of the cylinder, a spherical dilatation wave travels down the cylinder without suffering any reflections at the surface.[8] Although the effects of this disturbance are rapidly decreased, since expression (I-1.67) indicates the amplitude of this unreflected wave varies inversely with distance, some energy is propagated with the longitudinal velocity of the medium.

The general theory of wave propagation in a cylinder predicts there is non-uniformity of longitudinal stress and displacement over a cross section of the cylinder, that the radial displacement deviates from the simple linear law, and that the interior radial stress does not vanish (because of inertial effects). These results differ from the simple theory which is developed in some detail later. To determine these distributions of displacements and stresses across such a cross section corresponding to a given root of the dispersion equation, first Eqs. (I-4.31) and (I-4.32) and then (I-4.29) and (I-4.30) are used to find U_r and

U_z in terms of one constant; the constant can be calculated from the amplitude of vibration. Expressions (I-4.19) give u_r and u_z as functions of U_r and U_z, and the corresponding stress components can then be obtained from (I-4.18).

Illustrative results for a glass cylinder for two typical values of R/Λ are shown in Fig. I.24. In the main, and for purposes of clarity, we restrict ourselves here to longitudinal stresses and displacements arising from vibrations in the fundamental only. The particular harmonic depends upon the initial conditions, and it is found experimentally that it is the fundamental which is usually excited. More complete results are presented by Zemanek [8].

Some general comments with regard to Fig. I.24 are appropriate. First, it is clear that since the displacement u_z varies across the cross section, the disturbance is not generally planar.

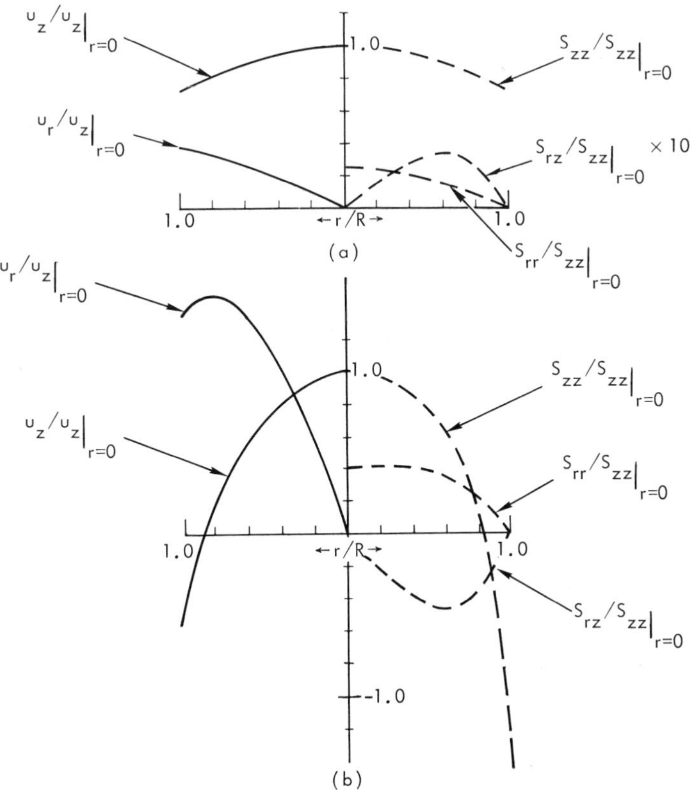

Fig. I.24. Variation of stresses and displacements for longitudinal fundamental waves over the cross section of a solid circular cylinder of radius R for two values of R/Λ and Poisson's ratio 0.25. (a) $R/\Lambda = 0.2$; (b) $R/\Lambda = 0.75$.

Second, at small values of R/Λ, one can infer that S_{zz} and u_z approach uniform distributions across a cross section, and S_{rr}, S_{zz}, and u_r approach zero. Hence with waves of long wavelengths, one approaches the results that are obtained when using the simple analysis (see Section II-B in this chapter).

Third, a comparison of Figs. I.24 (a) and (b) shows that a *nodal cylinder* develops over which the longitudinal displacement (but not the longitudinal stress) vanishes. Higher harmonics, if excited, involve higher numbers of nodal cylinders and are discussed briefly after the section on flexural waves.

Finally, at large R/Λ the longitudinal displacement is found to be very large at the surface of the cylinder and diminish rapidly with depth. From Fig. I.20, the phase velocity in the fundamental is observed to approach the velocity of Rayleigh surface waves. The analogy here is obvious.

The reader will recall we are dealing with a cylinder of infinite length. However, the preceding results are also reasonably appropriate to a cylinder of finite length (except near the ends) providing its length is large compared to its diameter, say $\ell/d > 10$.

    <u>b. Torsional Mode</u>. Torsional motion of a cylinder is characterized by each transverse cross section rotating about its center with the requirements that: (1) the section remain in its initial plane, (2) there are no lateral displacements, (3) the axis remains undisturbed, and (4) motion about this axis stay symmetrical. From these requirements, it follows that u_r and u_z must both vanish and u_θ must be independent of θ.

Such arguments introduce considerable simplicity into the analysis. Substituting these conditions into the expression for dilatation (I-4.12), we find that $\Delta = 0$. The components of rotation (I-4.14) become

$$\omega_r = -\frac{1}{2}\frac{\partial u_\theta}{\partial z},$$

$$\omega_\theta = 0, \tag{I-4.46}$$

and

$$\omega_z = \frac{1}{2}\left(\frac{u_\theta}{r} + \frac{\partial u_\theta}{\partial r}\right).$$

By methods analogous to that previously presented for longitudinal vibrations, the pertinent differential equations from the more general equations of motion (I-4.13) and assumed solutions (I-4.19) may be developed. Hence, we find that the first and third of (I-4.13) are satisfied identically, and with help of (I-4.46), the second becomes

$$\rho \frac{\partial^2 u_\theta}{\partial t^2} = \mu \left(\frac{\partial^2 u_\theta}{\partial z^2} + \frac{\partial^2 u_\theta}{\partial r^2} + \frac{1}{r} \frac{\partial u_\theta}{\partial r} - \frac{u_\theta}{r^2} \right) . \tag{I-4.47}$$

One can make direct substitution of the assumed solution for u_θ into the above expression (I-4.47) and obtain

$$\frac{\partial^2 U_\theta(r)}{\partial r^2} + \frac{1}{r} \frac{\partial U_\theta(r)}{\partial r} - \frac{U_\theta(r)}{r^2} + h^2 U_\theta = 0. \tag{I-4.48}$$

We can then make a suitable change in independent variables as we did in expressions (I-4.23) and (I-4.25) and write (I-4.48) in the form of Bessel's equation of order one. The solution that is finite at $r = 0$ is

$$U_\theta(r) = P J_1(hr), \tag{I-4.49}$$

where P is a constant.

The solution u_θ, using expression (I-4.49), is now introduced into the appropriate boundary condition, namely, the second of (I-4.18), and the dispersion equation is written

$$S_{r\theta} \big|_{r=R} = \mu \left[R \frac{\partial}{\partial R} \left(\frac{u_\theta}{R} \right) \right] = 0$$

or   (I-4.50)

$$\frac{\partial}{\partial R} \left[\frac{J_1(hR)}{R} \right] = 0.$$

The dispersion equation has multiple roots, the lowest of which is $h = 0$. However, we cannot immediately substitute this particular value into (I-4.49). Expression (I-4.48) can only be treated as a Bessel equation when $h \neq 0$,[9] and thus solution (I-4.49) is not valid under such a condition. Hence, Eq. (I-4.48) must be considered directly; we obtain as a solution

$$U_\theta(r) = P'r \tag{I-4.51}$$

where P' is a constant. From the second of expressions (I-4.18), we see that u_θ, now using (I-4.51), makes $S_{r\theta}$ vanish at $r = R$, *i.e.*, the boundary conditions are satisfied.

Since $h^2 = \rho p^2/\mu - k^2$, the phase velocity along the cylinder corresponding to $h = 0$ is given by

$$(c_p)_t = \frac{p}{k} = \left(\frac{\mu}{\rho}\right)^{1/2}. \tag{I-4.52}$$

Therefore, no dispersion occurs for waves propagating in the fundamental, and the phase velocity and group velocity are both equal to the velocity of distortion waves c_s in an extended medium.

To illustrate the dispersion characteristics of the higher harmonics of torsional vibrations, phase and group velocity curves are developed as before. In this situation, the treatment is considerably simpler than for longitudinal vibrations. We differentiate the dispersion equation (I-4.50) and use recurrence formulas for Bessel functions to obtain an expression from which we can readily extract the roots (see Kolsky [9]). Again we use a nondimensional form and

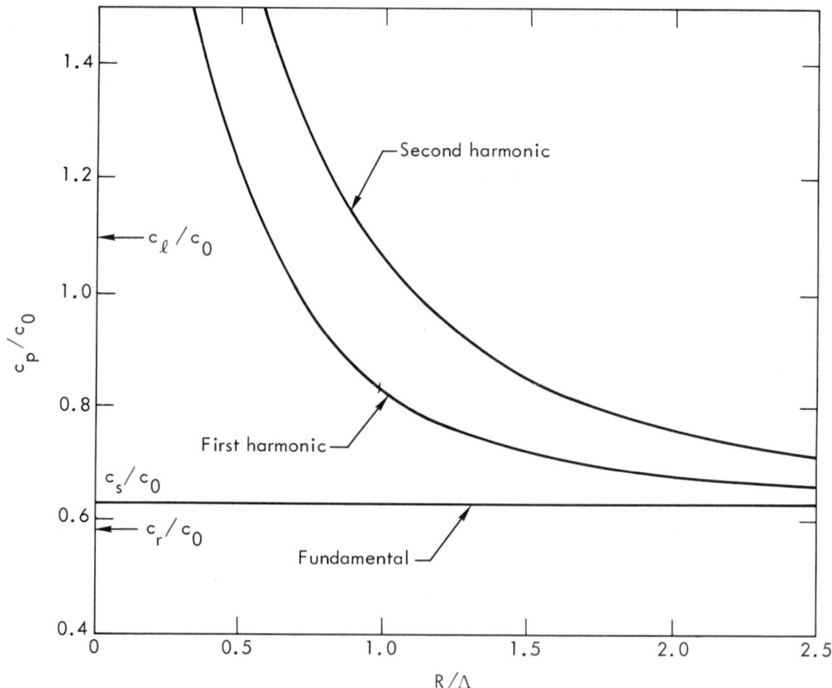

Fig. I.25. Phase velocity of torsional waves in a solid circular cylinder of radius R and Poisson's ratio 0.25.

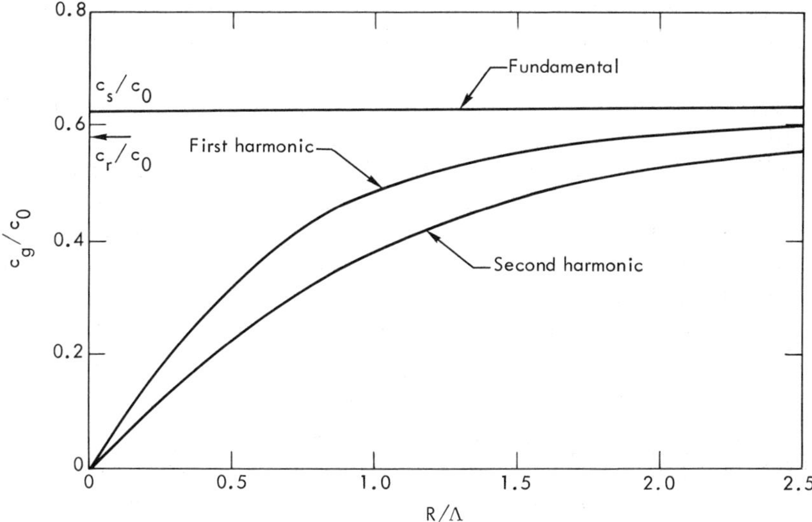

Fig. I.26. Group velocity of torsional waves in a solid circular cylinder of radius R and Poisson's ratio 0.25.

reduce the number of parameters; the dispersion diagram, c_p/c_0 versus R/Λ, is plotted for a given Poisson's ratio. Figure I.25 plots the fundamental and the first two harmonics over the most interesting range of wavelengths for $\sigma = 0.25$. The symbols used in the figure have their usual meanings. Note that the phase velocity for the two harmonics tends to infinity as R/Λ becomes infinite.

The group velocities are determined using Eq. (I-4.43) and the dispersion curves in Fig. I.25. Figure I.26 plots c_g/c_0 as a function of R/Λ for the first three branches in the dispersion diagram for torsional vibrations. It is seen here that c_g is always less than c_0; it is, in fact, always less than or equal to c_s.

We may proceed as described before to determine the distribution of (shearing) stress and displacement over a cross section of a cylinder subjected to such torsional vibrations. A brief discussion of this procedure is given by Mason [4]. From expressions (I-4.19) and (I-4.49), the relationship for the displacement with respect to any mode of vibration is given by

$$u_\theta = P J_1(hr) \exp\left[i(kz - pt)\right]. \tag{I-4.53}$$

For the fundamental, use of (I-4.51) reduces (I-4.53) to the equation

$$u_\theta = P'r \exp\left[i(kz - pt)\right]. \tag{I-4.54}$$

It can be seen from (I-4.54) that u_θ is directly proportional to r; hence, the motion in the lowest branch is a rotation of each transverse section of the cylinder as a whole about its center.

It is also interesting to note that torsional waves yield a solution that requires transverse sections to remain plane and parallel to each other during vibration; thus, when considering a cylinder of finite length, the boundary conditions on the ends are considerably simplified. A situation of somewhat lesser physical interest arises when $\partial u_\theta / \partial z$ vanishes. The traction across a normal section, $z = $ constant, then vanishes, and we have free torsional vibrations of a circular cylinder (of any length).

Since no dispersion occurs for the fundamental, it has been suggested to make use of a torsional pressure bar for the measurement of transient pressures and to use cylinders vibrating torsionally as delay lines.

c. <u>Transverse (Flexural) Modes</u>. Transverse motion of a cylinder is characterized by the periodic bending and straightening of its elements, together with lateral displacement of its neutral axis. We are not so fortunate here as we were in the situation of longitudinal or torsional waves in a cylinder, since motion is no longer symmetrical about the axis. Hence, all three components of the displacement must be considered in the analysis, and all three involve θ. The treatment in terms of the Pochhammer-Chree equations consequently becomes quite difficult, and is not given in detail.

We make the assumption that vibrations take place in the cylinder with respect to the plane of reference formed by the undisturbed axis of the cylinder and the line from which θ is measured. It is then reasonable to try solutions in which the displacements are proportional to trigonometric functions of θ; let us take u_r and u_z to be proportional to $\cos \theta$ and u_θ to $\sin \theta$. Hence, we may write instead of expressions (I-4.19)

$$u_r = U_r(r) \cos \theta \, \exp \left[i(kz - pt) \right],$$

$$u_\theta = U_\theta(r) \sin \theta \, \exp \left[i(kz - pt) \right], \qquad\qquad \text{(I-4.55)}$$

and

$$u_z = U_z(r) \cos \theta \, \exp \left[i(kz - pt) \right],$$

where U_r, U_θ, and U_z are again factors related to the amplitude of the waves, but are not to be taken equal to Eqs. (I-4.29) and (I-4.30).

For clarity and from convention, we define the plane from which θ is measured "vertical" and the direction of the z axis "horizontal." Several conclusions follow therefrom. The angle θ is zero for points in the vertical section of the bar which contains the z axis. Since the second of Eqs. (I-4.55) requires that for points on this plane $u_\theta = 0$, these points will remain in the plane during the vibration. For points in the horizontal plane containing the z axis, *i.e.*, the neutral axis, $\theta = \pi/2$, and from the first and third equations of (I-4.55), $u_r = u_z = 0$. Points in this plane will consequently perform only vertical oscillations. Thus the assumed expressions (I-4.55) appear to correspond to the desired flexural motion, and we are now able to determine whether expressions can be found for U_r, U_θ, and U_z which will satisfy the governing equations of motion and the boundary conditions. By methods analogous to that presented in developing the two differential equations (I-4.21) and (I-4.22) from the equations (I-4.20), we may derive from expressions (I-4.12)-(I-4.14) and (I-4.55), three equations:

$$\frac{\partial^2 \Delta}{\partial r^2} + \frac{1}{r}\frac{\partial \Delta}{\partial r} - \frac{\Delta}{r^2} + g^2 \Delta = 0,$$

$$\frac{\partial^2 \omega_z}{\partial r^2} + \frac{1}{r}\frac{\partial \omega_z}{\partial r} - \frac{\omega_z}{r^2} + h^2 \omega_z = 0, \tag{I-4.56}$$

and

$$\frac{1}{r^2}\frac{\partial}{\partial r}\left[r\frac{\partial}{\partial r}(r\omega_r)\right] - \frac{\omega_r}{r^2} + h^2 \omega_r + i\frac{2}{r}k\omega_z = 0,$$

where g and h have been defined in connection with (I-4.23) and (I-4.25).

Again, following the procedure for longitudinal vibrations, we can make suitable changes in variables in (I-4.56) to obtain Bessel's equations of order one and determine the following respective solutions:

$$\Delta = K(z,t) J_1(gr),$$

$$\omega_z = L(z,t) J_1(hr), \tag{I-4.57}$$

and

$$\omega_r = iM(z,t)\frac{\partial J_1(hr)}{\partial r} + iN(z,t)\frac{J_1(hr)}{r},$$

where the last expression is obtained by using the equation for ω_z immediately preceding.

Next we combine the assumed solutions (I-4.55) with expressions (I-4.12) and (I-4.14) and write

$$\Delta = \left[\frac{\partial U_r(r)}{\partial r} + \frac{U_r(r)}{r} + \frac{U_\theta(r)}{r} + ikU_z(r) \right] \cos \theta \, \exp \left[i(kz - pt) \right],$$

$$\omega_r = -\frac{1}{2} \left[\frac{U_z(r)}{r} + ikU_\theta(r) \right] \sin \theta \, \exp \left[i(kz - pt) \right],$$

$$\omega_\theta = \frac{1}{2} \left[ikU_r(r) - \frac{\partial U_z(r)}{\partial r} \right] \cos \theta \, \exp \left[i(kz - pt) \right], \tag{I-4.58}$$

and

$$\omega_z = \frac{1}{2} \left[\frac{\partial U_\theta(r)}{\partial r} + \frac{U_\theta(r)}{r} + \frac{U_r(r)}{r} \right] \sin \theta \, \exp \left[i(kz - pt) \right].$$

To satisfy all the relationships (I-4.57) and (I-4.58), U_r, U_θ and U_z can be written in the form (after some manipulation):

$$U_r(r) = A \frac{\partial J_1(gr)}{\partial r} + Bk \frac{\partial J_1(hr)}{\partial r} + \frac{C}{r} J_1(hr),$$

$$U_\theta(r) = -\frac{A}{r} J_1(gr) - \frac{Bk}{r} J_1(hr) - C \frac{\partial J_1(hr)}{\partial r}, \tag{I-4.59}$$

and

$$U_z(r) + iAkJ_1(gr) - iBh^2 J_1(hr),$$

where A, B and C are constants related to the amplitude of vibration, but are not the same as in (I-4.29) and (I-4.30).

Expressions (I-4.55) and (I-4.59) may then be introduced in the boundary conditions on the cylinder surface, and we obtain homogeneous algebraic equations containing the constants A, B and C as unknowns. In order that solutions (other than the trivial solution $A = B = C = 0$) of this set of equations exist, it is necessary that the determinant of the coefficients vanish. Expansion of the resulting determinant yields a dispersion equation similar to, although much more complicated than, expression (I-4.35). Both the determinant and the frequency equations are omitted here because of their length; the reader may consult Abramson [10].

To illustrate the dispersion characteristics of flexural waves, we may proceed in a manner similar to that presented in the study of longitudinal vibrations. Figure I.27 plots the fundamental and the first two harmonics[10] over the most interesting range of wavelengths for Poisson's ratio again equal to 0.25. The

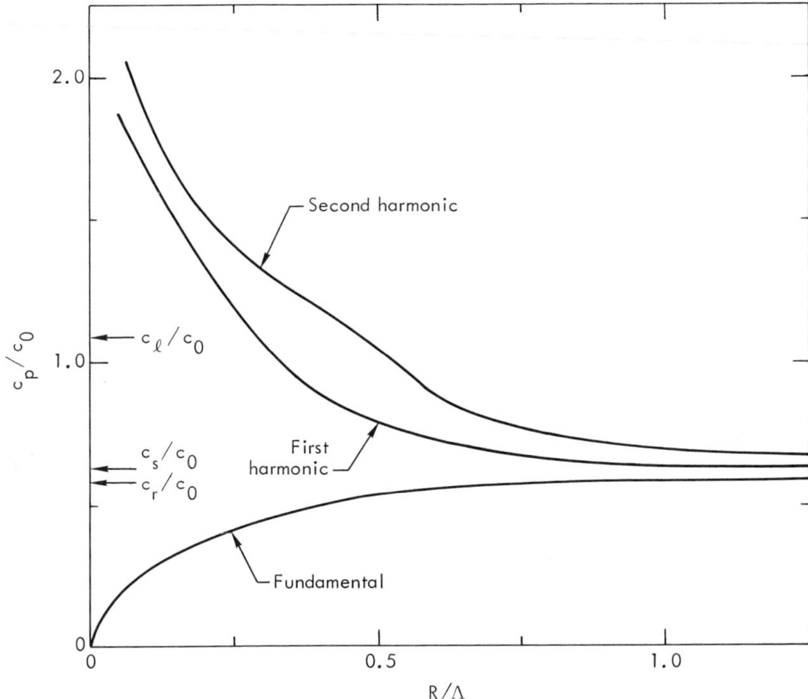

Fig. I.27. Phase velocity of transverse waves in a solid cylinder of radius R and Poisson's ratio 0.25.

symbols used in the figure have their usual meanings. Note that the fundamental approaches c_r/c_0 for large R/Λ. The effect of the variation of Poisson's ratio upon the phase velocity of flexural waves in the fundamental is much less than evidenced for the same mode of extensional waves.

The group velocities are determined using (I-4.45) and the dispersion curves in Fig. I.27. Figure I.28 plots c_g/c_0 as a function of R/Λ for the first two branches in the dispersion diagram for transverse wave propagation. It is noted again that although the phase velocities can exceed the dilatational velocity of an infinite medium the group velocities do not. The effect of the variation of Poisson's ratio is very pronounced in the second harmonic; for example, for $\sigma = 0.33$ (appropriate for aluminum), the initial peak of c_g/c_0 occurring at $R/\Lambda \approx 0.15$ is less than c_s/c_0.

To determine the distribution of stress and displacement over a cross section of a cylinder subjected to such vibrations, we can proceed in exactly the same manner as presented in the discussion of longitudinal waves in cylinders.

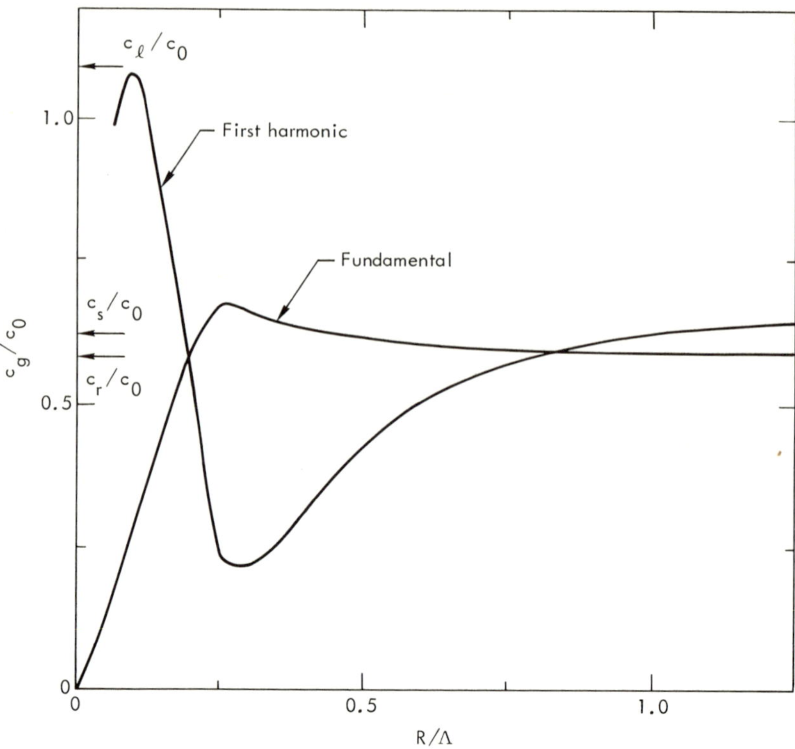

Fig. I.28. Group velocity of transverse waves in a solid cylinder of radius R and Poisson's ratio 0.25.

Again, the preceding results are reasonably appropriate to the analysis of a circular cylinder of finite length (except near the ends) providing its length is large compared to its diameter.

d. High Mode and High Harmonic Solutions. It is helpful to generalize the discussion and treatment pertaining to the dispersion equations for a circular cylinder. Briefly, it is found that in this general treatment two infinite sets of frequency equations are obtained; these are numbered ($n = 0, 1, 2, \ldots$, and $m = 0, 1, 2, \ldots$) in a manner which depends upon the particular coordinate system selected. Examination of these results, however, shows that the motions in both sets of equations, for the same value $n = m \geqslant 1$, are in fact identical, the displacements corresponding to the other by a rotation of coordinate axes. Therefore, these two infinite sets reduce to one infinite set containing two equations for $n = m = 0$ and one equation for each and all of the other values of

$n = m$. Of these dispersion equations, the first three ($n = 0, 0, 1$) correspond to the most important modes of vibration studies above, namely, longitudinal, torsional, and flexural.

In examining the dispersion curves for high-frequency waves in a circular cylinder, it is important to distinguish between the higher harmonics of the simple modes and the lower order harmonics of the higher modes. Both consist of high-frequency vibrations, but the resultant motions are physically very different.

With respect to the understanding of the higher modes of vibration, we can examine the nodal planes passing through the axis of the cylinder. A circular cylinder vibrating in the nth mode may be considered as consisting of n pairs of diametrically opposite sectors of a circle. The n pairs of sectors may be chosen in two ways, either bounded by nodal planes of radial motion or nodal planes of angular motion. These planes are separated by an angle $\pi/2n$. Figure I.29 illustrates schematically the motion across cross sections of cylinders vibrating in the modes $n = 0, 0, 1, 2,$ and 3 (the fundamental is considered for each mode). It is noted that $n = 0, 0,$ and 1 are "special," *i.e.*, the cylinder vibrates as a whole, and these modes do not conveniently lend themselves to generalizations; they are included for completeness.

From Fig. I.29 several observations and deductions can be advanced. The motion in the sectors bounded by nodal planes of radial displacement is more dilatational in character while the motion in the sectors bounded by nodal planes of angular displacement is more distortional. In the limit as n becomes very large, the sector pairs essentially reduce to thin planes of infinite length and width $2R$.

With respect to the understanding of the higher harmonics of vibration, we examine the number of amplitude nodal cylinders concentric with the axis of the cylinder. The formation of nodal cylinders is dependent upon the manner in which reflections occur at the boundary of the vibrating cylinder.

A simple generalization, which unfortunately is not completely correct in all situations but is convenient to remember, relates on a one-to-one basis a particular harmonic to the number of nodal cylinders that are formed. Thus, the fundamental corresponds to waves traveling parallel to the axis of the cylinder and which are never reflected at the boundary. The first harmonic is the simplest vibration for which reflection occurs, the reflected waves combining to produce one nodal cylinder. Higher harmonics would be of increasing complexities, the waves combining to give two, three, etc., nodal cylinders.

There are exceptions to this simple picture because the angle between the resultant wave front and the axis is a function of the wavelength (or frequency). A case in point is the fundamental of the longitudinal mode in which longitudinal vibration can occur without a nodal cylinder or with one nodal cylinder, depending upon the value of R/Λ.

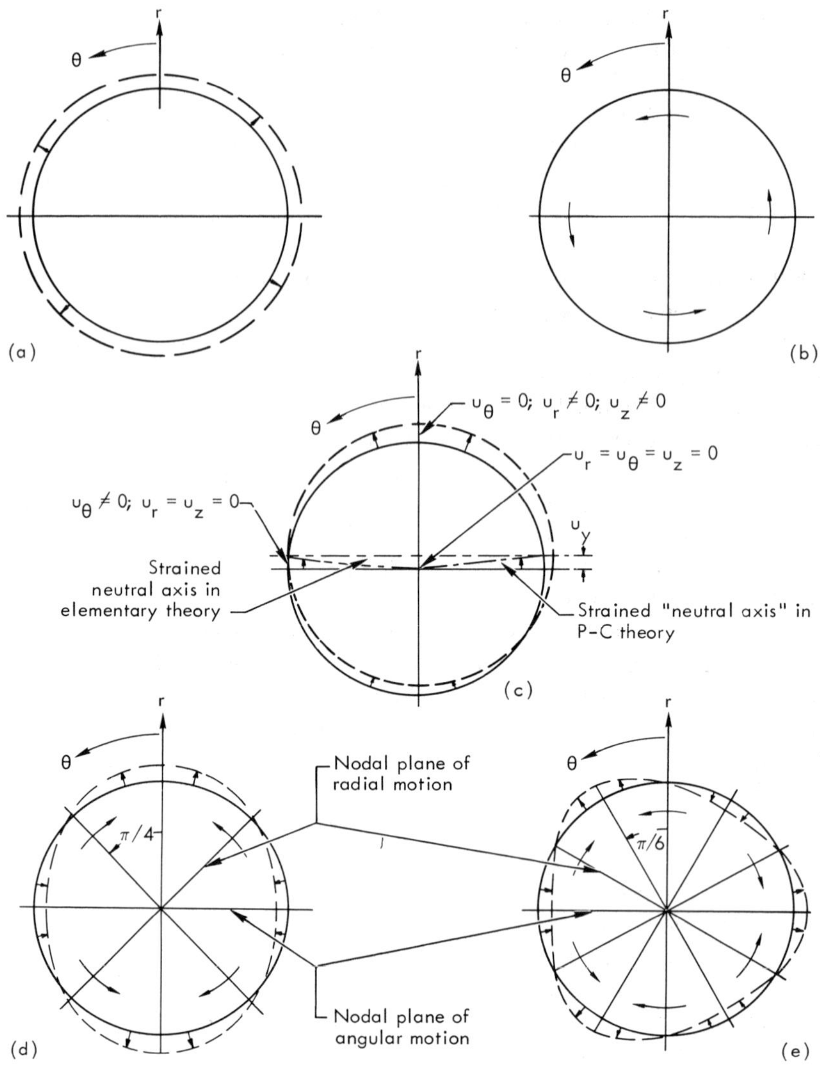

Fig. I.29. Motion across cross sections of cylinders vibrating in the fundamentals in the first five modes (schematic).

  (a) Longitudinal mode; $n = 0$; $u_\theta = 0$; $u_r \neq 0$; $u_z \neq 0$.
  (b) Torsional mode; $n = 0$; $u_r = u_z = 0$; $u_\theta \neq 0$.
  (c) Transverse mode; $n = 1$.
  (d) Screw mode; $n = 2$.
  (e) No designation; $n = 3$.

In addition to the infinity of roots of a given dispersion relation for real propagation constants, there also exists an infinite number of roots for complex propagation constants. These complex dispersion curves are extensions of the higher order harmonics below their cut-off frequencies.[11] Complex wave numbers are not important in the theory of wave propagation in infinitely long cylinders since the imaginary part can be interpreted as representing spatial attenuation of the wave, and consequently such vibrations eventually die out. On the other hand, they are important for problems in wave propagation in such bodies as semi-infinite or finite cylinders and plates, particularly near the ends.

The attenuation of the wave does not represent energy dissipation (fortunately so, since we are treating a perfectly elastic medium). It turns out that these waves are always generated in pairs having propagation constants which are negative complex conjugates of each other; it can be shown (see Zemanek [8]) that these two traveling waves form standing waves which decrease in amplitude and do not represent a transport of enengy.

B. Elementary and Approximate Methods

Because of the complexity in understanding and application of the exact analyses and the restrictions inherent therein, a number of elementary and approximate methods have been developed for treating elastic wave propagation in circular cylinders. Besides offering straightforward and relatively simple means of treating this problem, such methods enable a logical extension into the less-developed study of wave and pulse propagation in noncircular cylinders. In addition, less rigorous theories often assist in the examination of the dynamics of motion of other shapes and geometries, *e.g.*, cylinders of finite length, shells, beams, *etc*.

In the elementary approach, the equations of motion are derived directly, using certain simplifying assumptions. No allowance is made for the dispersion characteristics except in the instance of flexural wave transmission.

The approximate methods may be conveniently divided into two groups according to the method of approach (see Green [5]). The first group includes treatments that involve extensions and improvements of the above elementary methods so as to obtain the approximate dispersion curves. The method of approach of the second group is different in that approximate solutions are obtained to the exact dispersion equations – usually using power series techniques. Approximate methods of the first type are generally found to be the more satisfactory and more likely to give a working result for the engineer or physicist, although care must be exercised in application to problems treating waves of short length where higher modes carry a significant fraction of the total wave energy or where impact problems are considered in which pulses are generated.

Only discussion of the main features of certain techniques is given, and no attempt is made to be exhaustive in selection or complete in detail.

1. Longitudinal Mode

a. Elementary Method. Let us consider a uniform cylinder of cross-sectional area A and examine an infinitesimal element of that cylinder of length dz. If the stress on face P is S_{zz}, the stress on face Q is given by $S_{zz} + (\partial S_{zz}/\partial z)dz$, and if the displacement of the face P is u_z, the displacement of face Q is $u_z + (\partial u_z/\partial z)dz$. Figure I.30 illustrates these conditions.

Let us assume that each plane cross section remains plane during the motion, that the longitudinal stress over each plane is uniform (a one-dimensional stress condition), and that radial inertia is negligible. It follows from these assumptions that the longitudinal displacement is uniform, that the radial stress is everywhere negligible, and that the radial displacement at a distance r from the axis is $\sigma S_{zz}r/E$ (from Hooke's law), where E is Young's modulus. The equation of motion may now be obtained easily and directly. From Newton's second law of motion, we may write

$$(\rho A \; dz)\frac{\partial^2 u_z}{\partial r^2} = A\left(S_{zz} + \frac{\partial S_{zz}}{\partial z}\,dz - S_{zz}\right) \tag{I-4.60}$$

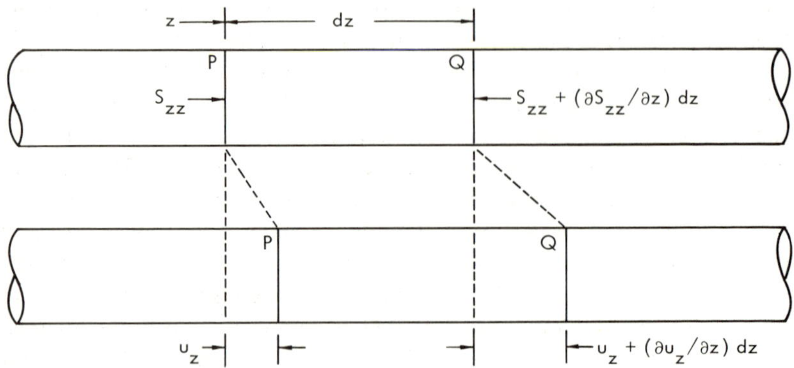

Fig. I.30. Forces acting on element of cylinder in (elementary) longitudinal motion.

where, as before, ρ is the density of the cylinder. We know from Eq. (I-1.34) that

$$S_{zz} = E \frac{\partial u_z}{\partial z}.$$
$$(\text{I-4.61})$$

Thus,

$$\frac{\partial^2 u_z}{\partial t^2} = \frac{E}{\rho} \frac{\partial^2 u_z}{\partial z^2}.$$
$$(\text{I-4.62})$$

Expression (I-4.63) is in the form of the equation of wave motion [see Eq. (I-1.59)], the general solution of which in one-dimension [see Eq. (I-2.63)] is

$$u_z = f\left(z - \sqrt{\frac{E}{\rho}}\, t\right) + g\left(z + \sqrt{\frac{E}{\rho}}\, t\right)$$
$$(\text{I-4.63})$$

where f and g are arbitrary functions that depend upon the initial conditions. The simple physical interpretation of solution (I-4.63) given in Chapter I-1, Section IV-C shows that this expression represents nondispersive disturbances propagating at velocity $(E/\rho)^{1/2}$. Therefore, it follows that the elementary theory predicts the phase velocity (and group velocity) to be given by the relation

$$c_p^2 = c_g^2 = c_0^2 = \frac{E}{\rho}$$
$$(\text{I-4.64})$$

and that an arbitrary disturbance will not suffer distortion as it is propagated along the cylinder. The constant velocity designated in Eq. (I-4.64) is often called the *rod* or *bar* velocity.

By suitable differentiation of expression (I-4.63) according to the wave equation (I-4.62), we can obtain the useful, well known result

$$S_{zz} = \rho c_0 \frac{\partial u_z}{\partial t}$$
$$(\text{I-4.65})$$

which shows that there is a linear relation between the stress at any point on a cross section and the particle velocity on that section, the ratio between them being the characteristic acoustic impedance.

A situation of considerable practical importance occurs when such an idealized disturbance encounters discontinuities in media and area, such as is illustrated in Fig. I.31. The conditions to be satisfied at the interface are equality of force and particle velocity,

$$A_1\left(S_{zz_i} + S_{zz_r}\right) = A_2 S_{zz_t},$$

and

$$\left(\frac{\partial u_z}{\partial t}\right)_i - \left(\frac{\partial u_z}{\partial t}\right)_r = \left(\frac{\partial u_z}{\partial t}\right)_t$$

(I-4.66)

where subscripts i, r and t refer to the incident, reflected and transmitted disturbances, and A_1 and A_2 are the areas of the cross sections, respectively. Equation (I-4.65) and boundary conditions (I-4.66) will be found to be of central importance in the discussion of uniaxial stress experimental techniques discussed in Chapter II-2. Combination of expressions (I-4.65) and (I-4.66) yields the ratios

$$\frac{S_{zz_r}}{S_{zz_i}} = \frac{A_2 \rho_2 c_{0_2} - A_1 \rho_1 c_{0_1}}{A_1 \rho_1 c_{0_1} + A_2 \rho_2 c_{0_2}}$$

and

(I-4.67)

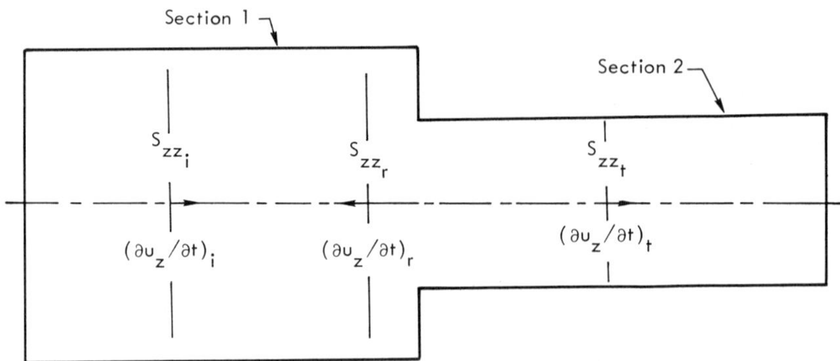

Fig. I.31. Reflection and transmission of a one-dimensional stress wave at an interface.

$$\frac{S_{zz_t}}{S_{zz_i}} = \frac{2A_1\rho_2 c_{0_2}}{A_1\rho_1 c_{0_1} + A_2\rho_2 c_{0_2}}.$$

Note the similarity between equation set (I-4.67) and equation set (I-3.46).

In this simple theory, no account is taken of the obvious stress and particle velocity concentrations occurring at the interface. However, providing the length of the actual elastic disturbance is long compared with the diameter of the bar and its amplitude varies relatively slowly with time, such as that described by a sinusoidal function, and providing observations are made of the wave profile several bar diameters away from the discontinuity in cross section, the elementary theory enables satisfactory prediction and analysis of results (see Habberstad and Hoge [11]). If the above conditions are not met, it may be necessary to use more sophisticated techniques or treatment, including perhaps numerical methods (see Habberstad [12] and Habberstad *et al.* [13]).

If we treat a cylinder of finite length and consider a free end, we may take $A_2 = 0$, and we find from the first of (I-4.66) or of (I-4.67) that

$$S_{zz_i} = -S_{zz_r} \tag{I-4.68}$$

which is the same result illustrated in Fig. I.11. On the other hand, if the end of the finite cylinder is fixed, we obtain

$$\left(\frac{\partial u_z}{\partial t}\right)_i = \left(\frac{\partial u_z}{\partial t}\right)_r \tag{I-4.69}$$

which states that the total particle velocity vanishes, as it must.

b. Approximate Methods of the First Kind. One of the earliest successful efforts to improve the results obtained with the elementary analysis was developed by Love [14]. The treatment takes into account the inertia of the lateral contraction. The assumptions are retained that plane sections remain plane and longitudinal stress over a cross section is uniform.

The derivation of the governing differential equation is most often made using a variational approach; more specifically, using Hamilton's principle. Hamilton's principle states that within an arbitrary interval of time, the true motion of a system can be characterized by the fact that the increment of the integral, $\int_{t_1}^{t_2} L\, dt$, vanishes for any continuously varying virtual displacement, provided this displacement vanishes at the limits of the arbitrary interval. The function L represents the difference of kinetic and potential energies of the system, and is often called the *Langrangian function* or the *kinetic potential*. This principle simply means that for the actual motion of a conservative system (*i.e.*, a system in which the total mechanical energy is constant in time), the system moves so that the time average of the difference between the kinetic and potential energies is an extremum, most often a minimum. Hamilton's principle can be shown to yield, in simple invariant form (since the kinetic and potential energies are scalar invariants to coordinate transformations), all the equations of classical dynamics. The mathematical statement of the principle is often referred to as *the* problem of the calculus of variations.

By application of Hamilton's principle, Love obtained

$$\frac{\partial^2 u_z}{\partial t^2} = c_0^2 \frac{\partial^2 u_z}{\partial z^2} = \sigma^2 K^2 \frac{\partial^4 u_z}{\partial z^2 \partial t^2} \tag{I-4.70}$$

where K is the radius of gyration of the cross section about the axis.[12] The last term represents the radial inertia [compare with Eq. (I-4.62)]. Assuming the propagation of sinusoidal waves similar in form to relationships (I-4.19), the approximate phase velocity dispersion equation for the fundamental obtained from the expression (I-4.70) is

$$\frac{c_p}{c_0} = (1 + 4\pi^2 \sigma^2 K^2/\Lambda^2)^{-1/2} = [1 + 2\pi^2 \sigma^2 (R/\Lambda)^2]^{-1/2}. \tag{I-4.71}$$

Using Eq. (I-4.43), one can obtain an expression analogous to (I-4.71) for the group velocity.

A significant improvement upon Love's method is an approach first presented by Mindlin and Herrmann [15]. This theory considers radial shear effects in addition to those of radial inertia, and gives rise to two dispersion curves in reasonable agreement with the fundamental and first harmonic, particularly so for the former. Plane sections are again assumed to remain plane.

Initial assumptions are made as to the form of the displacements, namely,

$$\bar{u}_r = \frac{r}{R} u_r(z,t)\, \bar{u}_\theta = 0, \text{ and } \bar{u}_z = u_z(z,t). \tag{I-4.72}$$

The appropriate stress equations of motion, written in cylindrical coordinates, are integrated over a cross section using definitions for point stresses in terms of average stresses that are consistent with the assumed displacements. The boundary conditions are obtained from energy considerations that utilize the following theorem (applicable in a conservative system): The rate of increase of energy in a given elastic medium (a unit length of cylinder in this situation) is equal to the rate at which work is done on its surfaces. To avoid certain mathematical difficulties, the results are then converted to displacement equations of motion through Hooke's law for isotropic materials. There result two coupled differential equations in u_r and u_z which can be solved (for an infinite train of waves in an infinite cylinder) using solutions of the form (I-4.19).

To make the wave velocities in the dispersion curves in the approximate theory agree more closely with the exact solution, two adjustable constants are introduced into the above stress-displacement relations. In particular, the fundamental dispersion curve is normalized to the exact curve by requiring $c_p/c_0 \to c_r/c_0$ for $R/\Lambda \to \infty$.

c. Approximate Method of the Second Kind. The most popular approach in this category is a method attributed first to both Pochhammer and Chree, working independently (see Love [14]). The restriction is placed upon analysis that the diameter of the cylinder is small compared to the wavelength of the vibrations.

Expansion of $J_0(gR)$ and $J_1(hR)$ in power series yields the results

$$J_0(gR) = 1 - \frac{1}{4}(gR)^2 + \frac{1}{64}(gR)^4 \cdots$$

$$\tag{I-4.73}$$

$$J_1(hR) = \frac{1}{2}(hR) - \frac{1}{16}(hR)^3 + \cdots .$$

If we substitute expressions (I-4.73) into the dispersion equation for longitudinal vibrations in a circular cylinder (I-4.35), we obtain the approximate equation for the fundamental,

$$\frac{c_p}{c_0} = 1 - \sigma^2 \pi^2 (R/\Lambda)^2. \tag{I-4.74}$$

Note that Eq. (I-4.74) also follows directly from (I-4.71) for small values of (R/Λ). Additional terms in the two series (I-4.73) can be retained, but the resulting approximate frequency equation rapidly becomes intolerably cumbersome.

d. Comparison with Exact Theory. The adequacy of these approximate methods is judged by comparison of their phase velocity spectra with that of the Pochhammer-Chree analysis. Only the fundamental is considered here.

Figure I.32 presents the desired comparison for Poisson's ratio equal to 0.25 as before. The curves are designated in the following manner: (1) Pochhammer-Chree (exact), (2) elementary, (3) Love, (4) Mindlin and Herrmann, and (5) Pochhammer-Chree (approximate). The figure illustrates that the range of the elementary theory is indeed only for small R/Λ. Note also that only the method of Mindlin and Herrmann gives clearly the same curvature as the exact solution.

2. Torsional Mode

a. Elementary Method. The equation of motion may be obtained by considering the forces (*i.e.*, moments) acting on an element of infinitesimal length dz
such as shown in Fig. I.33. In the derivation, each transverse section is required
to remain in its own plane and to rotate as a whole about its center.

Let the twisting couple acting on the section through P be T; the couple
acting on the section through Q is then $T + (\partial T/\partial z)dz$. If ϕdz is the relative
angular displacement of the two cross sections P and Q, we can consider $1/2\phi$ as
the mean angle through which the element rotates about its center. Note that ϕ
is the relative angular displacement of two cross sections unit length apart taken

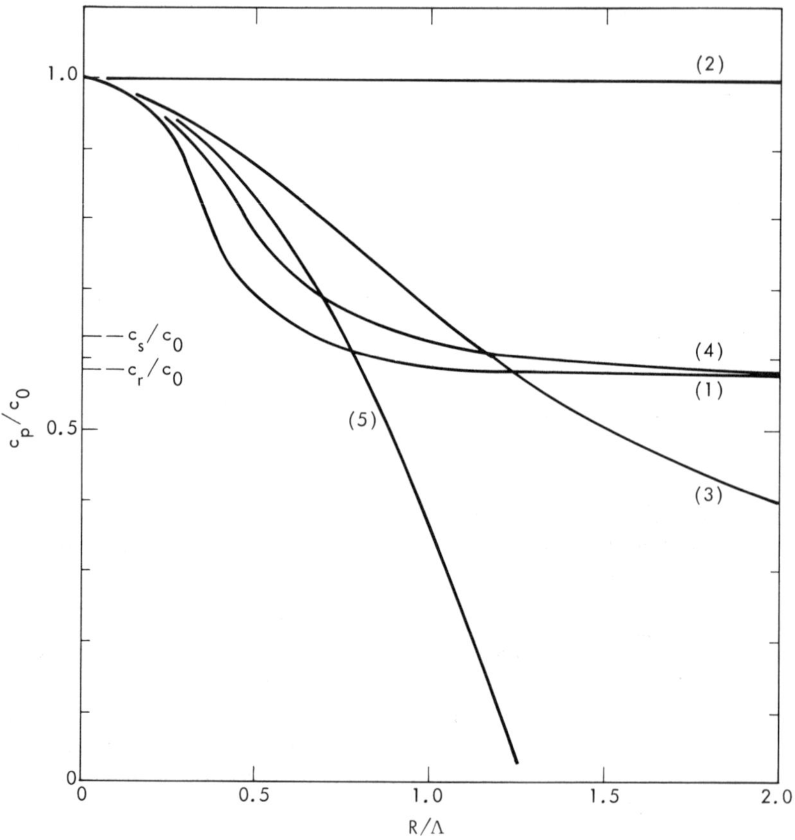

Fig. I.32. Comparison of approximate methods of calculating phase velocity of longitudinal waves (fundamental) in a solid cylinder of radius R and Poisson's ratio 0.25.

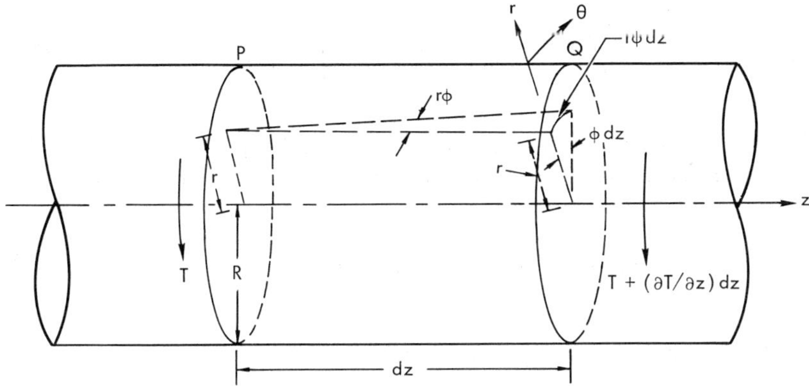

Fig. I.33. Couples acting on element of cylinder in (elementary) torsional motion.

in the same "direction" as θ, *i.e.*, the twist per unit length. Note also, it follows from Fig. I.33 that we can write $\phi = d\phi$ since $\phi_1 - \phi_2 = d\phi = \phi$ if ϕ_2 is taken equal to zero.

The equation of motion can be written

$$T + \frac{\partial T}{\partial z} dz - T = \frac{1}{2} I_m \, \ddot{\phi} \, dz$$

or

$$\frac{\partial T}{\partial z} = \frac{1}{2} I_m \, \ddot{\phi}. \tag{I-4.75}$$

The quantity I_m is the (mass) moment of inertia of the infinitesimal element PQ about its axis, *i.e.*, the z axis, and is given by

$$I_m = \rho \int r^2 \, dV = \rho \, dz \int_0^{2\pi} \int_0^R r^2 (r, d\theta) \, dr = \frac{1}{2} \pi \rho \, R^4 dz. \tag{I-4.76}$$

This result also follows directly from the definition of the radius of gyration of a mass m about an axis, $I_m = K^2 m = 1/2 \, R^2 m$ (see also Note 12).

Using Hooke's law, the definition of shearing strain, and Fig. I.33, the twisting couple can be related to the relative angular displacement of the two cross sections. Thus, the shearing stress at a point on section Q is

$$S_{\theta z}\big|_Q = S_{\theta z}\big|_P + \mu \frac{\partial u_\theta}{\partial z} = S_{\theta z}\big|_P + \mu \frac{r\phi dz}{dz} = S_{\theta z}\big|_P + \mu r\phi \qquad \text{(I-4.77)}$$

and the shearing stress at a point on a section midway between sections P and Q, *i.e.*, the mean stress to which the element is subjected, can be considered as

$$S_{\theta z}\big|_{\text{mean}} = S_{\theta z}\big|_P + \frac{1}{2}\mu r\phi = S_{\theta z}\big|_P + \frac{1}{2}\mu r\, d\phi. \qquad \text{(I-4.78)}$$

Then the resultant mean torque can be calculated,

$$T_{\text{mean}} = \int_A S_{\theta z}\big|_P\, r\, dA + \int_0^R \frac{1}{2}\mu d\phi\, 2\pi r^3\, dr$$

$$= T_P + \frac{1}{4}\pi\mu R^4 d\phi = T_P + \frac{1}{4}\pi\mu R^4 \frac{d\phi}{dz}\, dz. \qquad \text{(I-4.79)}$$

Combining Eqs. (I-4.75), (I-4.76), and (I-4.79), we obtain

$$\frac{1}{4}\pi\mu R^4 \frac{\partial^2\phi}{\partial z^2}\, dz = \frac{1}{2}\left(\frac{1}{2}\pi\rho R^4 dz\right)\frac{\partial^2\phi}{\partial t^2},$$

or

$$\frac{\partial^2\phi}{\partial t^2} = \frac{\mu}{\rho}\frac{\partial^2\phi}{\partial z^2}. \qquad \text{(I-4.80)}$$

By a similar treatment, we can derive the appropriate equation of motion for a cylinder of more general (although uniform) cross section, namely,

$$\frac{\partial^2\phi}{\partial t^2} = \frac{C}{\rho I}\frac{\partial^2\phi}{\partial z^2}, \qquad \text{(I-4.81)}$$

where I is the areal moment of inertia of the cross section about its axis (the polar moment of inertia) and C is the *torsional rigidity* (see Timoshenko and Goodier [16]), defined as the applied torque at a section divided by the twist per unit length at that section. It should be emphasized that we deal here with the areal moment of inertia rather than with the mass moment of inertia I_m as in Eq. (I-4.76). For our situation of an infinitesimal segment of a right circular cylinder, $I_m = \rho I\, dz$; or if the segment is of unit length and we use the torsional rigidity C, then $I_m = \rho I$, and Eq. (I-4.81) follows. The use of I_m in (I-4.76) is simply more convenient, but the use of areal moments of inertia is more conventional.

Relationships (I-4.80) and (I-4.81) are written in the form of the wave equation; hence, the phase velocity of torsional disturbances[13] is given by

$$c_p^2 = \frac{\mu}{\rho} \tag{I-4.82}$$

for cylinders of circular cross section, and by

$$c_p^2 = \frac{C}{\rho I} \tag{I-4.83}$$

for cylinders of general cross section. Equation (I-4.82) agrees with the result predicted in the exact theory [Eq. (I-4.52)] for the phase velocity of the fundamental.

With respect to approximate methods of the first kind, several techniques have been developed which, when applied to circular cylinders, are essentially useless because they all predict the same constant result as above. Love has taken into account the warping of the cross section for noncircular cylinders. He derives a result using energy methods similar to that given by Eq. (I-4.81), but includes another term involving the shape of the section. This additional term disappears as the section becomes circular.

With respect to approximate methods of the second kind, apparently relatively little work has been done here, with only one approach based upon perturbation theory being available (see Green [5]). The results obtained are similar to those obtained by Love's technique above.

Since we are primarily interested in propagation in circular cylinders, comparisons are not significant among the various treatments because all give the same result for the fundamental. In addition, there has been essentially no work done on approximate approaches for the harmonics, and no comparisons are possible at this time.

3. Transverse (Flexural) Mode

a. Elementary Method. As indicated earlier, the exact theory of flexural vibrations of (uniform) cylinders is more difficult than that of either longitudinal or torsional vibrations. This is true also in the elementary theory.

We consider as before that each plane cross section of the infinitely long cylinder remains plane during the motion. Also, we take the radius of the bar to be "small" and consider that all the motion in each element of the cylinder is normal to its length.

Figure I.34 shows the forces and moments acting on such an element of length dz. The bending moment varies along the cylinder, and is taken to have value M at section P and $M + (\partial M/\partial z)dz$ at section Q. The bending moment will be resisted by shearing forces acting "parallel" with the y axis. Through section P we take the shearing force to be F, and through Q, $F + (\partial F/\partial z)dz$. We label the lateral displacement of the cylinder u_y and write the equation of motion of the element in the y direction

$$(\rho A\ dz)\frac{\partial^2 u_y}{\partial t^2} = F + \left(\frac{\partial F}{\partial z}\right)dz - F, \tag{I-4.84}$$

where ρ and A have the usual meanings.

Note that $u_y \neq u_r$, although both are displacements in the Eulerian sense. The former represents the displacement of the whole section perpendicular to the unstrained neutral axis. The latter represents the displacement of a point in the cylinder with reference to the cylinder's axis, which axis is taken in the undisturbed state and labeled the z axis. The crucial difference here is that in the

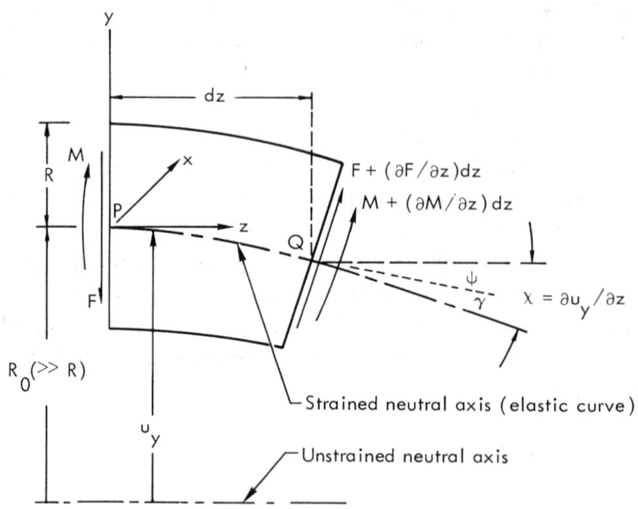

Fig. I.34. Forces and moments acting on element of cylinder in (elementary) fluxural motion.

latter situation the z axis is not permitted to move, *i.e.*, $u_r = u_\theta = u_z = 0$ at $r = 0$, even though we allow flexural vibrations to occur. Figure I.29 shows the manner by which u_y and u_r differ and how they represent these two physically different displacements.

To proceed, we must first relate F to the moment M about an axis in the x direction through the center of the element as shown. Thus, the equation of motion about such an axis is

$$-M + \frac{\partial M}{\partial z}\, dz + M + \left(F + F\frac{\partial F}{\partial z}\, dz\right)\frac{dz}{2} = \rho I_x \ddot{\chi}_x\, dz = 0, \qquad \text{(I-4.85)}$$

where I_x and $\ddot{\chi}_x$ are the appropriate areal moment of inertia[14] and angular acceleration of the element about the x axis, respectively. That $\rho I_x \ddot{\chi}_x\, dz = 0$ follows from the assumption that we are neglecting any rotary inertia (see expression (I-4.94)). Therefore,

$$F = -\frac{\partial M}{\partial z}. \qquad \text{(I-4.86)}$$

Note that we have considered in this most elementary treatment only the effects of strain energy of bending and of transverse inertia.

Now the relation among the moment M, the modulus of elasticity E, the radius of curvature of the section R_0, the moment of inertia I_x, and the displacement u_y can be readily obtained using methods in elementary mechanics of materials and some simple calculus. It is found that the *equation of the elastic curve* can be written for small deflections as

$$\frac{M}{EI_x} = \frac{1}{R_0} = \frac{\partial^2 u_y}{\partial z^2}. \qquad \text{(I-4.87)}$$

Therefore, Eqs. (I-4.84), (I-4.86), and (I-4.87) become

$$\rho A \frac{\partial^2 u_y}{\partial t^2} = \frac{\partial F}{\partial z} = -\frac{\partial^2 M}{\partial z^2} = -EI_x \frac{\partial^4 u_y}{\partial z^4} \qquad \text{(I-4.88)}$$

or

$$\frac{\partial^2 u_y}{\partial t^2} = -c_0^2 K_x^2 \frac{\partial^4 u_y}{\partial z^4} = -\frac{c_0^2 R}{4} \frac{\partial^4 u_y}{\partial z^4}. \qquad \text{(I-4.89)}$$

Expression (I-4.88) or (I-4.89) is the fourth-order differential equation for flexural wave transmission that is to be solved.

As before with our approximate solutions for longitudinal vibrations, we are interested primarily in the phase velocity associated with propagating disturbances and not the distribution of stress and displacement over a particular cross section. Nevertheless, a complete solution is often important, and an example can be found in Mason [4]. Recall that since a fourth order differential equation is being considered, there are four boundary conditions to satisfy.

If we assume propagation of a disturbance described by an expression similar to (I-4.63), we find that, in general, Eqs. (I-4.88) or (I-4.89) will not be satisfied. Thus a flexural disturbance of arbitrary shape will not be propagated along a cylinder without dispersion. If we assume propagation of a disturbance described by an expression similar to Eqs. (I-4.19), namely

$$u_y = U_y \exp[i(kz - pt)] \qquad \text{(I-4.90)}$$

where, as before, $p = 2\pi c_p/\Lambda$ and U_y is the amplitude, we find upon substitution into expression (I-4.89) that the phase velocity is given by the equation

$$c_p = kK_x c_0 = \frac{2\pi}{\Lambda} K_x c_0. \qquad \text{(I-4.91)}$$

For circular cylinders we obtain

$$\frac{c_p}{c_0} = \frac{\pi R}{\Lambda}, \qquad \text{(I-4.92)}$$

and from expression (I-4.43)

$$\frac{c_g}{c_0} = \frac{2\pi R}{\Lambda} = \frac{2c_p}{c_0}. \qquad \text{(I-4.93)}$$

Hence, the elementary theory leads to results that are physically unrealistic, namely, that a wave packet consisting of infinitely short wavelengths is propagated with an infinite velocity. Figure I.35 in Section 3-c will illustrate that the range of application is limited to small R/Λ.

There are several reasons why the above treatment becomes unsatisfactory when the wavelength is comparable to the lateral dimensions of the cylinder; specifically, when $R/\Lambda > 0.1$. First, the assumption that the motion is purely one of translation in the y direction becomes unjustifiable, and rotary inertia of the elements must be considered. Second, the effects of transverse shear contribute to the angle through which the element has been rotated, such effects becoming more pronounced for shorter wavelengths. The warping of the cross section, which is related to the second reason, also requires a correction. Finally, and apparently of least importance, the effects of lateral contraction must be included; the improvement from its inclusion, however, is relatively slight.

b. Approximate Methods of the First Kind. One of the earliest improvements on the elementary theory was that proposed by Lord Rayleigh (see Love [14]). In this approach, a term is added to Eqs. (I-4.88) or (I-4.89) to account for the rotary inertia of the element. Thus, in taking moments to relate the shear force F to the moment M, the right-hand side of expression (I-4.85) can no longer be set to zero. We must write instead

$$F + \frac{\partial M}{\partial z} = \rho I_x \ddot{\chi}_x. \tag{I-4.94}$$

For small deformations, χ_x will be given by $\partial u_y / \partial z$, so that expression (I-4.94) can be written

$$F = -\frac{\partial M}{\partial z} + \rho I_x \frac{\partial^2 u_y}{\partial z \partial t^2} \tag{I-4.95}$$

and

$$\frac{\partial F}{\partial z} = -\frac{\partial^2 M}{\partial z^2} + \rho I_x \frac{\partial^4 u_y}{\partial z^2 \partial t^2}. \tag{I-4.96}$$

Substituting the value for M from Eq. (I-4.87) into (I-4.96), and the resulting value for $\partial F/\partial z$ from (I-4.96) into Eq. (I-4.84), we obtain

$$\rho A \frac{\partial^2 u_y}{\partial t^2} = -EI_x \frac{\partial^4 u_y}{\partial z^4} + \rho I_x \frac{\partial^4 u_y}{\partial z^2 \partial t^2} \tag{I-4.97}$$

or

$$\frac{\partial^2 u_y}{\partial t^2} = -c_0^2 K_x^2 \frac{\partial^4 u_y}{\partial z^4} + K_x^2 \frac{\partial^4 u_y}{\partial z^2 \partial t^2}. \qquad \text{(I-4.98)}$$

By comparing expressions (I-4.97) and (I-4.98) with (I-4.88) and (I-4.89), it is apparent which term represents the correction for rotary inertia.

If we assume a solution of the form given in (I-4.90), we find the phase velocity to be

$$c_p = c_0 \left(1 + \frac{\Lambda^2}{4\pi^2 K_x^2}\right)^{-1/2} \qquad \text{(I-4.99)}$$

and for circular cylinders, we obtain

$$\frac{c_p}{c_0} = \left(1 + \frac{\Lambda^2}{\pi^2 R^2}\right)^{-1/2}. \qquad \text{(I-4.100)}$$

The group velocity can be readily computed using Eq. (I-4.43) if desired.

Hence, this corrected theory leads to results that are physically much more satisfactory than in the elementary theory in that it does not give an infinite value for phase (and group) velocity as $\Lambda \to 0$. Rather, the limiting phase velocity is c_0. Nevertheless, the range of application is still restricted to relatively small R/Λ.

Another method closely related to that given above was developed by Love, who made appropriate corrections for both rotary inertia and for the distortion of the cross section during bending. Suitable expressions for displacements are assumed, the various terms for potential and kinetic energies are formed, Hamilton's principle is invoked, and the equation of motion is obtained. This equation is not written here since it becomes clear upon examination of the results that the distortion of the cross section is of relatively minor importance unless the section is of peculiar shape, namely, such that K_y is much different (either more or less) than K_x.

The next approximate treatment for flexural vibrations is one which appears to be the most satisfactory (when compared with the exact fundamental dispersion diagram) over the entire range of frequencies and, indeed, is the one most widely in use. This approach, first derived by S. Timoshenko (see Kolsky [9] or Goldsmith [17]), includes corrective terms for both rotary inertia and transverse shear.

Whereas Rayleigh and Love both based their methods on a variational principle, Timoshenko examined the problem directly. Figure I.34 illustrates that the total slope $\partial u_y/\partial z$ can be considered to consist of an angle ψ due to bending and an angle γ due to shear. Hence, we may write for small deflections

$$\frac{\partial u_y}{\partial z} = \chi = \psi + \gamma. \tag{I-4.101}$$

We rewrite the rotational and translational equations of motion, (I-4.94) and (I-4.84), respectively, in the forms

$$\rho I_x \frac{\partial^2 \psi}{\partial t^2} = F + \frac{\partial M}{\partial z} \tag{I-4.102}$$

and

$$\rho A \frac{\partial^2 u_y}{\partial t^2} = \frac{\partial F}{\partial z}. \tag{I-4.103}$$

Note the angular acceleration in expression (I-4.102) considers only the bending angle ψ; the shear angle γ (or, equivalently, the shear strain) is assumed to act independently during the bending process, entering into the above equations only through F. There is some influence of one angle upon the other, but for these small motions such influence is slight. Furthermore, there are two important practical reasons for this assumption: (1) the sets of equations (I-4.102) through (I-4.105) enable a tractable analysis, and (2) regardless of the theoretical arguments advanced pro or con, the results obtained are very satisfactory.

The equation of the elastic curve for a uniform cylinder due to the bending moment can be expressed in a fashion similar to the relationship (I-4.87), namely

$$\frac{M}{EI_x} = \frac{\partial \psi}{\partial z}. \tag{I-4.104}$$

Using Hooke's law with respect to shearing stresses, we may write

$$S_{yz} = \frac{F}{A} = \mu \eta \gamma = \mu \eta \left(\frac{\partial u_y}{\partial z} - \psi \right) \tag{I-4.105}$$

where μ is the modulus of rigidity defined in Chapter I-1, and η is a parameter accounting for the nonuniform shear distribution across the cross section. Numerical values for this factor are 2/3 and 3/4 for a rectangular and a circular cross section, respectively. These values represent the ratio of average shear stress to shear stress at the neutral axis for static loading conditions.

This above set of four equations can be readily reduced to one fourth-order differential equation by the elimination of ψ; this expression is known as the *Timoshenko beam equation*:

$$\frac{\partial^2 u_y}{\partial t^2} = -c_0^2 K_x^2 \frac{\partial^4 u_y}{\partial z^4} + K_x^2 \left(1 + \frac{E}{\eta\mu}\right)\frac{\partial^4 u_y}{\partial z^2 \partial t^2} + \frac{EK_x^2}{c_0^2 \eta\mu}\frac{\partial^4 u_y}{\partial t^4}. \tag{I-4.106}$$

It is interesting and instructive to compare expressions (I-4.89), (I-4,98), and (I-4.106), observing the form in which the various corrective terms enter.

For a simple harmonic wave train of the form (I-4.90) propagating in a circular bar of radius R, the phase velocity dispersion equation obtained from Eq. (I-4.106) is

$$\left(\frac{c_0}{c_p}\right)^2 + \frac{E}{\eta\mu}\left(\frac{c_p}{c_0}\right)^2 = 1 + \frac{\Lambda^2}{\pi^2 R^2} + \frac{E}{\eta\mu}. \tag{I-4.107}$$

Expression (I-4.107) has two real roots which correspond to the fundamental and the first harmonic. The fundamental will be compared with the exact results in Fig. I.28. Note that from Eq. (I-1.41), $E/\mu = 2(1 + \sigma)$. Again, the group velocity can be computed from relation (I-4.43).

Apparently, the only approach in the category of approximate methods of the second kind is one originally developed by Lord Rayleigh (see Love [14]) where Bessel's functions are expanded in series [see expressions (I-4.73)]. These series solutions can be substituted into the dispersion equation for transverse vibrations. The procedure is quite complicated; to date, only enough terms have been retained in the series upon substitution to derive the same result that was obtained in the elementary method, namely Eq. (I-4.92).

c. Comparison with Exact Theory. As before, we compare the results from these approximate methods and those of the Pochhammer-Chree analysis with respect to the phase velocity spectra in the fundamental. Figure I.35 presents the desired comparison for Poisson's ratio equal to 0.25. The curves are designated in the following manner: (1) Pochhammer-Chree (exact); (2) elementary; (3) Rayleigh; and (4) Timoshenko.

As indicated, Timoshenko's treatment is the most satisfactory over the whole range of wavelengths. It is probable that this method is of the most engineering

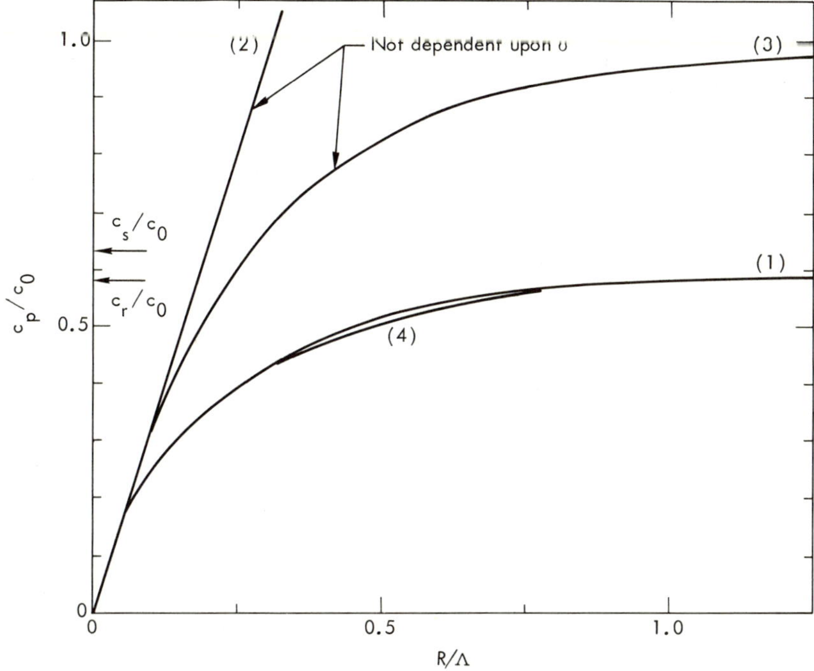

Fig. I.35. Comparison of approximate methods of calculating phase velocity of transverse waves (fundamental) in a solid cylinder of radius R and Poisson's ratio 0.25.

value when considering the propagation of flexural pulses in problems of impact or in the transmission of transverse disturbances in uniform cylinders with non-circular cross sections. However, it should be remembered that for $R/\Lambda < 0.1$, the elementary treatment also gives reasonable results and has the advantage of being simpler.

III. Associated Treatments

The foregoing analyses are only directly appropriate for the study of elastic stress wave propagation in bounded geometries with simple circular (or near circular) cross sections. Moreover, it is apparent that we have limited our consideration to a rather idealized mathematical form of such wave propagation. Although distinct similarities exist in concept when problems outside the above rather specific areas are examined, usually some extension – or even additional analysis – is required. The following comments give some examples.

Analysis of the transmission of waves in a uniform circular cylinder of finite length is made difficult because of the introduction of the end conditions on the bar. The exact solutions derived previously for the longitudinal and transverse modes give results close to reality (except near the ends) so long as the length of the cylinder is large compared with its diameter. Because of the simplified conditions on transverse sections, the solution for the torsional mode has greater applicability than the other two to elastic wave propagation in cylinders of finite length.

A direct treatment of the problem was not achieved until 1957 when Skalak [18] analyzed the elastic impact of two isotropic semi-infinite bars using a superposition technique. The problem of a pressure wave applied to a radially constrained end of a semi-infinite bar was studied by Folk *et al.* [19] and Jones and Norwood [20]. The same problem was solved by Kennedy and Jones [21], but they allowed the pressure to vary as a function of the radial position on the end of the bar. However, in all these analytical treatments, the resulting integrals are so complicated that at present only asymptotic solution is available, correct at large distances from the end of the bar. Recent two-dimensional numerical solutions (see for example, Habberstad [12] and Habberstad *et al.* [13]) have provided significant insight.

A useful and practical approach to the problem involves a manipulation on some kind of an elementary or approximate solution. A specific example of an elementary approach (with longitudinal waves) is given by Kolsky [9]; examples of approximate approaches (again with longitudinal waves) are given by Miklowitz [22] and Miklowitz and Nisewanger [23]. Some relevant comments are reported by Lindholm and Doshi [24], although their work involves principally the propagation of a pulse in an inhomogeneous bar.

With respect to the problem of propagation of waves in uniform noncircular cylinders of infinite length, some exact analytical solutions have been obtained of specific problems. For example, Morse [25,26] has considered the propagation of infinite trains of longitudinal waves along bars of rectangular cross section in terms of the exact equations and has obtained solutions that are in reasonable agreement with his experimental data. Similar investigations have been made by Mindlin and Fox [27].

At present, elementary and approximate theories offer the best practical means of analytically investigating disturbance propagation in cylinders of rectangular cross section because the addition of an additional characteristic length makes the exact solutions exceedingly complex (see Medick [28]). The application of elementary theories is straightforward, so long as the noncircular cylinder is of uniform section and $\Lambda \gg R$. According to these treatments, longitudinal waves are propagated with the constant rod velocity $c_0 = (E/\rho)^{1/2}$, the velocity of torsional waves is a constant for any given shape of cross section [Eq. (I-4.83)], and transverse (flexural) simple harmonic waves are dispersive with their velocity being a function of K_x/Λ [Eq. (I-4.91)]. Studies of the velocities of propagation of flexural and extensional waves in rectangular sections using approximate equations of motion have been reported by Gazis and Mindlin [29].

The associated subject of the propagation of elastic stress disturbances in anisotropic circular cylinders has not received much attention. Some thoughts and references are given by Morse [30].

An extension of the above problem is the propagation of waves in nonuniform cylinders of infinite length. If the cross section is not uniform along the length of an isotropic cylinder, the shape and amplitude of a disturbance changes during passage, such change being in addition to that caused by dispersion. Most of the work done to date concerns the elementary analysis of the propagation of longitudinal waves along conical circular rods (see Kolsky [9]). For a disturbance where the wavelength is great compared with the local diameter of the cylinder, this treatment applies. It can be shown that for a disturbance traveling toward the apex of the cone, a pulse must consist of two parts in order to conserve momentum — a compression wave followed by a wave of tension. As this disturbance approaches the apex, the compression region becomes shorter and shorter; when the disturbance arrives at the apex, the conical cylinder is entirely under tension and spall of the tip can occur. Goldsmith and Kenner [31] give a recent examination of this subject.

A very important situation is the propagation of waves in flat plates of infinite extent. This problem has been treated exactly as a two-dimensional investigation wherein longitudinal waves are propagated in one of the directions of infinite extent and considered only as functions of that direction and the thickness of the plate. The analysis can also be considered as an extension of the problem of waves in a long bar of rectangular cross section where one lateral dimension is much larger than the other. One can conclude from the analysis that longitudinal waves in an infinite plate travel with the constant velocity

$$c_{plate} = \left[\frac{E}{\rho(1-\sigma^2)}\right]^{1/2} \tag{I-4.108}$$

when the wavelength is very large compared with the thickness of the plate, and with the velocity of Rayleigh surface waves when the wavelength is very small compared with the thickness. For wavelengths comparable with the thickness dispersion occurs, and the phase velocity depends upon the ratio of wavelength to plate thickness. Expression (I-4.108) can be derived using both exact methods (above) and elementary considerations (assuming the stress is uniform over a cross section of the plate perpendicular to the direction of wave travel).

Considerable progress has been made, and much has been written concerning this topic. Most of the work has appeared since 1950, although an initial main contribution was made by Lamb in 1917 [32]. Examples of the available literature are efforts by Holden [33], Tolstoy and Usdin [34], Sherwood [35], Meeker and Meitzler [36], and the text by Ewing *et al.* [37]. Mindlin has been particularly active in this field, especially in the study of plate vibrations. Some references are Mindlin [38,39], Kane and Mindlin [40], Mindlin and Medick [41], and Gazis and Mindlin [42]. Some analysis has also been given to the propagation of elastic waves in anisotropic, heterogeneous plates by Yang *et al.* [43]. Extension to the study of waves in shells has been made by Naghdi and Cooper [44] and Cooper and Naghdi [45]. Elastic waves in semi-infinite plates has been examined by Wu [46].

The topic of propagation of transient disturbances in bounded media is becoming increasingly important. The discussion of pulse propagation in Chapter I-2, Section V was specifically directed to nondispersive propagation. However, the implication was made therein, and later in the comments in connection with Fig. I.23, that Fourier analysis could be applied to a particular disturbance with velocity of propagation of each component found from the appropriate exact dispersion curve. Such a technique is generally possible but very often laborious, and usually some approximation is made, *e.g.*, reducing the number of components in the analysis.

Another method — actually a special treatment of the problem using Fourier analysis — in which longitudinal or flexural pulses are analyzed was developed by Lord Kelvin and extended by R. Davies (see Zemanek [8], Abramson [10]; and Davies [47,48]). This method essentially treats the propagation of an infinitely thin pulse of infinitely large amplitude. Such a stress pulse may be considered as a superposition of sinusoidal waves incorporating a spectrum of wavelengths. The disturbance may be expressed as a Fourier integral, and we thus have at our disposal a well-developed mathematical tool. Of course, the situation of a pulse of infinitely short duration obviously cannot be realized in practice, but by superposing the results of a number of such disturbances it is possible to proceed theoretically (although necessarily approximately) to the case of a disturbance of finite initial duration and determine the distortion that occurs. Davies also experimentally investigated the problem of longitudinal pulses, and has found the theory to agree with his observations.

Literature is becoming rather extensive on the subject. The references cited at the first of this section treating impact of two cylinders consider problems in pulse transmission. Other examples of similar articles are Miklowitz [49,50]; Fox and Curtis [51]; Jones and Ellis [52]; and Kaul and McCoy [53]. Examples of articles more specifically oriented to plate geometry are Huth and

Cole [54] ; Davids [55] ; Knopoff [56] ; Broberg [57] ; Medick [58] ; Goodier *et al.* [59] ; and Scott and Miklowitz [60] .

IV. Summary

The transmission of stress disturbances in circular cylindrical rods is examined. The complexities of analysis, compared with wave propagation in extended and semi-extended media, are greatly increased because of geometry considerations. An important analytical exact solution treats the problem of propagation of trains of progressive simple harmonic waves of infinite duration in isotropic solid circular cylinders of infinite length. This solution is called the Pochhammer-Chree analysis, and forms the basis for much of the work in this field.

The governing equations of small motion for a Hookean elastic isotropic medium are first written in a system of cylindrical coordinates. The solutions of these equations are found using expressions for sinusoidal waves of infinite duration propagating along an isotropic solid cylinder (*i.e.*, in the z direction) of infinite length and the boundary conditions that the three stress components, S_{rr}, $S_{r\theta}$, and S_{rz}, vanish at the surface of the cylinder. Expressions are derived involving the frequency and wavelength of the vibrations, the radius of the cylinder, the elastic constants of the material, and its density. These expressions are termed the dispersion equations.

Three particular types or modes of guided waves that are propagated in circular bars are investigated in detail, namely, longitudinal, torsional, and transverse (or flexural). Longitudinal motion of a cylinder is characterized by the periodic extension and contraction of its elements without lateral displacement of its axis. Torsional motion is characterized by each transverse cross section rotating about its center with the requirements that the section remain in its initial plane, that there are no lateral displacements, that the axis remain undisturbed, and that motion about this axis stays symmetrical. Transverse motion is characterized by the periodic bending and straightening of its elements, together with lateral displacement of its neutral axis.

Each of these modes is identified with a specific dispersion equation, each equation having an infinite number of roots. Each root corresponds to a particular deformation pattern that exists independently of any other. The first root is termed the fundamental, the second root the first harmonic, and so on. The disturbances are transmitted at velocities which depend upon both the mode and harmonic of transmission. For a given value of Poisson's ratio σ, the particular dispersion equation desired can be obtained involving only c_p/c_0 and R/Λ, where c_p is the phase velocity, c_0 is the rod velocity $\left[c_0 = (E/\rho)^{1/2}\right]$, R is the radius of the bar, and Λ is the wavelength. A numerical solution can then be computed, and the results for the various harmonics may be conveniently graphed.

The transport of mechanical energy is associated with the physical motion of particles, and can be related with the group velocity c_g. The latter is given by

$$c_g = c_p - \Lambda \frac{\partial c_p}{\partial \Lambda}. \tag{I-4.43}$$

It is seen that if the medium is nondispersive and the propagation velocity is not a function of Λ, then c_g coincides with c_p. In a dispersive medium, their values are distinct. Group velocity curves can also be readily plotted for each harmonic of the various modes.

The treatment can be generalized to higher mode and higher harmonic solutions. There is an infinite set of dispersion equations containing an infinite number of modes. Of these equations, the first three correspond to the most important modes of vibration studied in detail in the chapter.

Because of the complexity in understanding and application of the exact analyses, a number of elementary and approximate methods have been developed for treating elastic wave propagation.

In the elementary approach, the equations of motion are derived directly, using certain simplifying assumptions. No allowance is made for the dispersion characteristics except in the situation of flexural wave transmission. For example, in the longitudinal mode, the elementary approach predicts the phase velocity (and group velocity) to be given by the relation

$$c_p^2 = c_g^2 = c_0^2 = \frac{E}{\rho} \tag{I-4.64}$$

and that an arbitrary disturbance will not suffer distortion as it is propagated along the cylinder. This simple theory of longitudinal wave propagation, as opposed to the general (Pochhammer-Chree) theory, assumes there is uniformity of longitudinal stress and displacement over a cross section of the cylinder, that the radial displacement follows a simple linear law, and that the interior radial stress vanishes.

The approximate methods may be conveniently divided into two groups according to the method of approach. The first group includes treatments that involve extensions and improvements to the above elementary methods so as to obtain the approximate dispersion curves. The method of approach of the second group is different in that approximate solutions are obtained to the exact dispersion equations — usually using power series techniques. Approximate methods of the first type are generally found to be the more likely to give a working result for the engineer or physicist, although care must be exercised in application to problems treating waves of short length where higher modes carry a significant fraction of the total wave energy or where impact problems are considered in which pulses are generated.

The adequacy of the elementary and approximate methods considered are evaluated by comparing them with exact results for wave propagation in a circular cylinder.

Brief discussion is also given on theoretical analyses of elastic stress wave propagation in bounded geometries, other than for a simple circular (or near-circular) cylindrical shape. For example, the propagation of waves and transient disturbances in noncircular cylinders, nonuniform cylinders, and flat plates are examined. Representative references are given.

Notes

[1] The distinction between covariance and contravariance does not arise for Cartesian tensors.

[2] An example of the intermediate mathematics necessary to obtain (I-4.20) is:

$$\frac{\partial \Delta}{\partial z} = \frac{\partial}{\partial z}\left[\frac{1}{r}\frac{\partial(ru_r)}{\partial r} + \frac{1}{r}\frac{\partial u_\theta}{\partial \theta}^{0} + \frac{\partial u_z}{\partial z}\right]$$

$$= \frac{1}{r}\frac{\partial}{\partial r}\left[\frac{\partial(ru_r)}{\partial z}\right] + \frac{\partial}{\partial z}\left(\frac{\partial u_z}{\partial z}\right)$$

$$= \frac{1}{r}\frac{\partial}{\partial r}\left(r\frac{\partial u_r}{\partial z} + u_r\frac{\partial r}{\partial z}^{0}\right) + \frac{\partial}{\partial z}\left(\frac{\partial u_z}{\partial z}\right), \quad r \neq r(z)$$

$$= \frac{1}{r}\frac{\partial}{\partial r}\left[r(iku_r)\right] + \frac{\partial}{\partial z}(iku_z)$$

$$= ik\frac{1}{r}\left[\frac{\partial(ru_r)}{\partial r} + \frac{\partial u_z}{\partial z}\right]$$

$$= ik\Delta.$$

[3] As with the differential equations (I-4.20), some mathematical manipulation is required here to obtain conditions (I-4.29) and (I-4.30). The first of these expressions can be shown to be correct as follows:

$$\Delta = GJ_0 = \left[A\,\frac{\partial^2 J_0}{\partial r^2} + Bk\,\frac{\partial J_1}{\partial r} + \frac{A}{r}\,\frac{\partial J_0}{\partial r} + \frac{BK}{r}\,J_1 - Ak^2 J_0 - Bk\,\frac{\partial J_1}{\partial r} - \frac{Bk}{r}\,J_1 \right] \exp\,[i(kz - pt)]$$

$$= A\left[\frac{\partial^2 J_0}{\partial r^2} + \frac{1}{r}\,\frac{\partial J_0}{\partial r} - k^2 J_0 + J_0 - J_0 \right] \exp\,[i(kz - pt)]$$

$$= -A(1 + k)^2 \exp\,[i(kz - pt)]\,J_0$$

$$= GJ_0$$

where we have set $G(z,t)$ equal to $-A(1 + k^2) \exp\,[i(kz - pt)]$. Condition (I-4.30) can be shown to be correct in a similar manner.

[4] $\cos\alpha + \cos\beta = 2 \cos\left[\frac{1}{2}(\alpha - \beta)\right] \cos\left[\frac{1}{2}(\alpha + \beta)\right]$.

[5] It is important to point out that energy is actually transported at the signal velocity (discussed below), which differs from the group velocity in the neighborhood of a resonance. As we do not study driven (forced) vibrations, this distinction does not arise.

[6] The reader is reminded that we are considering here bounded media, in which steady-state waves of "practical" interest propagate at velocities less than c_{ϱ}. If extended media are considered, waves with *finite* amplitude can be transmitted with velocity c_{ϱ}.

[7] The second harmonic is an exception. However, the higher harmonics do indeed show the behavior indicated.

[8] If the cylinder is infinite in length, only one "ray" is unaffected by reflections.

[9] The relationship, $r^2\left[\partial^2 U_\theta(r)/\partial r^2\right] + r\left[\partial U_\theta(r)/\partial r\right] - U_\theta(r) = 0$, is not written in the form of Bessel's equation, nor can it be transformed into such a form.

[10] As an interesting mathematical aside, it has been stated only recently (by Abramson [10] in 1957) that an infinite number of branches exist in the dispersion diagram for flexural vibrations. All other literature preceding this date indicated only one branch. Yet the presence of Bessel functions in the determinant of the coefficients of Eqs. (I-4.59) (written in nondimensional form) indicates immediately that the determinant is equal to zero for infinitely many values of c_p/c_0 when R/Λ is held constant so that an infinity of branches in the dispersion diagram should have been expected.

[11] The cut-off frequency is related to the value of R/Λ at which the phase velocity approaches infinity (or equivalently, the wave number goes to zero). For example, See Figs. I.20, I.25 and I.27.

[12] For a cylinder of circular cross section, the "polar" radius of gyration (*i.e.*, with respect to the z axis) is given by $K^2 = 1/2\,R^2$.

[13] The twist per unit length, ϕ [the dependent variable in Eqs. (I-4.80) and (I-4.81)], can be thought of as a "displacement," *i.e.*, u_θ is proportional in some fashion to ϕ [see expression (I-4.77)].

[14] The moment of inertia of the cross section is taken about a diameter in the x direction through the unstrained neutral axis. Hence the radius of gyration here for this circular cylinder is given by $K_x^2 = I_x/A = 1/4\,R^2$.

References

[1] P. C. Chou and N. J. Pagano, *Elasticity: Tensor, Dyadic, and Engineering Approaches*, Chap. 8, D. Van Nostrand, Princeton, N. J., 1967.

[2] F. P. Hildebrand, *Advanced Calculus for Engineers*, 7th printing, Prentice-Hall, Englewood Cliffs, N. J., 1957, pp. 299-301.

[3] W. P. Mason, *Piezoelectric Crystals and Their Application to Ultrasonics*, 5th printing, D. Van Nostrand, Princeton, N. J., 1964, pp. 439, 486-489.

[4] W. P. Mason, *Physical Acoustics and the Properties of Solids*, D. van Nostrand, Princeton, N. J., 1958, pp. 35-39, 43-50.

[5] W. A. Green, "Dispersion Relations for Elastic Waves in Bars," in *Progress in Solid Mechanics*, Vol. I (I. N. Sneddon and R. Hill, eds.), North-Holland, Amsterdam, 1960, pp. 227-246, 257, 258.

[6] L. Brillouin, *Wave Propagation and Group Velocity*, 3rd printing, Chap. I, Academic, New York, 1964.

[7] J. A. Stratton, *Electromagnetic Theory*, McGraw-Hill, New York, 1941, pp. 330-340.

[8] J. Zemanek, *An Experimental and Theoretical Investigation of Elastic Wave Propagation in a Cylinder*, Technical Report No. XVIII, Dept. of Physics, University of California, Los Angeles, 1962.

[9] H. Kolsky, *Stress Waves in Solids*, Oxford University Press, London, 1953; Dover, New York, 1963, pp. 48-54, 68-69, 73-79.

[10] H. N. Abramson, *J. Acoustical Soc. Am.*, 29, 42, (1957).

[11] J. L. Habberstad and K. G. Hoge, *The Effects of a Discontinuity in Cross Section on an Elastic Pulse*, University of California, Lawrence Livermore Laboratory, Rept. UCRL-71681, 1969.

[12] J. L. Habberstad, *J. Appl. Mech.*, 38, 62, (1971).

[13] J. L. Habberstad, K. G. Hoge and J. E. Foster, *An Experimental and Numerical Study of Elastic Strain Waves on the Center Line of a 6061-T6 Aluminum Bar*, University of California, Lawrence Livermore Laboratory, Rept., UCRL-72717, 1970.

[14] A. E. H. Love, *A Treatise on the Mathematical Theory of Elasticity*, 4th ed., Cambridge University Press, 1927; Dover, New York, 1944, pp. 289-290, 292, 428-431.

[15] R. D. Mindlin, and G. Herrmann, "A One-Dimensional Theory of Compressional Waves in an Elastic Rod," in *Proc. First U. S. National Congress of Appl. Mech.*, ASME, Chicago, 1952.

[16] S. Timoshenko and J. N. Goodier, *Theory of Elasticity*, 2nd ed., McGraw-Hill, New York, 1951, p. 264.

[17] W. Goldsmith, *Impact*, Edward Arnold, London, 1960, pp. 36-38.

[18] R. Skalak, *J. Appl. Mech.*, 24, 59 (1957).

[19] R. Folk, G. Fox, C. Shock, and C. Curtis, *J. Acoustical Soc. Am.*, 30, 552 (1958).

[20] O. E. Jones and F. R. Norwood, *J. Appl. Mech.*, 34, 718 (1967).

[21] L. W. Kennedy and O. E. Jones, *J. Appl. Mech.*, 36, 470 (1969).

[22] J. Miklowitz, *J. Appl. Mech.*, 24, 231 (1957).

[23] J. Miklowitz and C. R. Nisewanger, *J. Appl. Mech.*, 24, 240 (1957).

[24] U. S. Lindholm and K. D. Doshi, *J. Appl. Mech.*, 32, 135 (1965).

[25] R. W. Morse, *J. Acoustical Soc. Am.*, 20, 833 (1948).

[26] R. W. Morse, *J. Acoustical Soc. Am.*, 22, 219 (1950).

[27] R. D. Mindlin and E. A. Fox, *J. Appl. Mech.*, 27, 152 (1960).

[28] M. A. Medick, "One-Dimensional Theories of Wave Propagation and Vibrations in

Elastic Bars of Rectangular Cross Section," in *Proc. Fifth U. S. National Congress of Appl. Mech.*, ASME, Minneapolis, 1966.

[29] D. C. Gazis and R. D. Mindlin, *J. Appl. Mech.*, **24**, 541 (1957).

[30] R. W. Morse, *J. Acoustical Soc. Am.*, **26**, 1018 (1954).

[31] W. Goldsmith and V. Kenner, "Wave Propagation in Conical Bars," in *Proc. Fifth U. S. National Congress of Appl. Mech.*, ASME, Minneapolis, 1966.

[32] H. Lamb, *Proc. Royal Soc. London*, **93A**, 114 (1917).

[33] A. Holden, *Bell System Tech. J.*, **30**, 956 (1951).

[34] I. Tolstoy and E. Usdin, *J. Acoustical Soc. Am.*, **29**, 37 (1957).

[35] J. W. C. Sherwood, *J. Acoustical Soc. Am.*, **30**, 979 (1958).

[36] T. R. Meeker and A. H. Meitzler, "Guided Wave Propagation in Elongated Cylinders and Plates," in *Physical Acoustics*, Section 2, W. P. Mason (ed.), Vol. 1, Part A, Academic, New York, 1964.

[37] W. M. Ewing, W. S. Jardetsky, and F. Press, *Elastic Waves in Layered Media*, McGraw-Hill, New York, 1957, Chap. 6.

[38] R. D. Mindlin, *J. Appl. Mech.*, **18**, 31 (1951).

[39] R. D. Mindlin, *An Introduction to the Mathematical Theory of Vibrations of Elastic Plates*, U. S. Army Signal Corps Engineering Laboratories, New Jersey, Project 142B, Contract DA-36-039 SC-56772, 1955.

[40] T. R. Kane and R. D. Mindlin, *J. Appl. Mech.*, **23**, 277 (1956).

[41] R. D. Mindlin and M. A. Medick, *J. Appl. Mech.*, **26**, 561 (1959).

[42] D. C. Gazis and R. D. Mindlin, *J. Appl. Mech.*, **27**, 541 (1960).

[43] P. C. Yang, C. H. Norris, and Y. Stavsky, *Intern. J. Solids Structures*, **2**, 665 (1966).

[44] P. M. Naghdi and R. M. Cooper, *J. Acoustical Soc. Am.*, **28**, 56 (1956).

[45] R. M. Cooper and P. M. Naghdi, *J. Acoustical Soc. Am.*, **29**, 1365 (1957).

[46] C. -H. Wu, "Elastic Waves in Semi-Infinite Plates," in *Proc. Fifth U. S. National Congress of Appl. Mech.*, ASME, Minneapolis, 1966.

[47] R. M. Davies, *Roy. Soc. Phil. Trans.*, **A240**, 325 (1948).

[48] R. M. Davies, "Stress Waves in Solids," in *Surveys in Mechanics, G. I. Taylor Edition* (G. K. Batchelor and R. M. Davies, eds.). Cambridge University Press, London, 1956, pp. 95, 102-107.

[49] J. Miklowitz, "On the Use of Approximate Theories of an Elastic Rod in Problems of Longitudinal Impact," in *Proc. Third U. S. National Congress of Appl. Mech.*, ASME, Providence, R. I., 1958.

[50] J. Miklowitz, *J. Geophys. Res.*, **68**, 1190 (1963).

[51] G. Fox and C. Curtis, *J. Acoustical Soc. Am.*, **30**, 559 (1958).

[52] O. E. Jones and A. T. Ellis, *J. Appl. Mech.*, **30**, 51 and 61 (1963).

[53] R. K. Kaul and J. J. McCoy, *J. Acoustical Soc. Am.*, **36**, 653 (1964).

[54] J. H. Huth and J. D. Cole, *J. Appl. Mech.*, **22**, 473 (1955).

[55] N. Davids, *J. Appl. Mech.*, **26**, 651 (1959).

[56] L. Knopoff, *J. Appl. Phys.*, **29**, 661 (1958).

[57] K. B. Broberg, *Trans. Roy. Inst. Technol., Stockholm*, 139 (1959).

[58] M. A. Medick, *J. Appl. Mech.*, **28**, 223 (1961).

[59] J. N. Goodier, I. K. McIvor, and H. E. Lindberg, *Elastic Motion and Plastic Buckling of Cylindrical Shells Under Nearly Uniform Radial Impulse*, Poulter Laboratories Tech. Rept. 001-63, Stanford Research Institute, Menlo Park, California 1963.

[60] R. A. Scott and J. Miklowitz, *J. Appl Mech.*, **34**, 104 (1967).

I-5

SELECTED APPLICATIONS OF
CONCEPTS OF ELASTICITY

I. General: Applications of Quasistatic Elasticity

As indicated in the Preface and in the beginning of Chapter I-1, the mathematical theory of elasticity in itself constitutes an important part of physical science and forms the basis for many practical analyses of the mechanical loading of structures and systems. Although the direct application of the principles of elasticity to such problems is of lesser importance for the purposes of this book, discussion of selected examples are given below to illustrate part of its utility. In keeping with the theme of the manuscript, emphasis is given to dynamic loading (Section II). In addition, there is a section which compares with experiment the exact Pochhammer-Chree Solution previously presented.

Many books have been written on the subject of the application of quasistatic elastic principles. Two of the most widely used texts for practical engineering analysis, even through both were written some years ago, are *Theory of Elasticity* by Timoshenko and Goodier and *Mathematical Theory of Elasticity* by Sokolnikoff. Both books provide numerous practical examples, although the latter tends to be considerably more mathematical. Examples therein involve elasticity problems in two and three dimensions in rectangular and curvilinear coordinates, including problems with bending and torsion of prismatical bars, thermal stress, and strain energy. The latter technique is particularly useful in situations involving the integration of complicated differential equations; such integration is replaced by the investigation of minimum conditions of certain integrals. An example of an energy approach is the Rayleigh-Ritz method, mentioned briefly in Chapter I-1.

Perhaps more in line with our limited interests in quasistatic applications is the study of crystallography, or crystal physics, commented upon in Chapter I-1 (and from a dynamic point of view, in Chapter I-2). A good grasp of the fundamentals of elasticity and of tensor analysis is required for such investigation.

187

Representative books are *Physical Properties of Crystals* by Nye and *Piezoelectric Crystals and Their Application to Ultrasonics* by Mason.

Two points should be made regarding the solution of engineering problems. First, in the treatment of specific situations, the method of direct determination of stresses and the use of the compatibility equations in terms of stress components is usually applied. We have mentioned the compatibility equations in terms of strain components in Chapter I-1, Section II-B.

Second, most practical quasistatic engineering design problems, *e.g.*, those involving the design of building, bridges, dams, and similar structures, are based upon elementary methods of *strength of materials*, in which approximations and simplifications of actual stress distributions are made that reduce considerably the complexity of analysis. This approach had its beginnings with B. de Saint-Venant in the middle of the 19th century with his suggestion that, for the most part, engineering analyses can neglect many local perturbations of stresses. It is now appropriately called *Saint-Venant's Principle*. However, the elementary theory is not sufficient to provide information regarding stress distribution in bodies in which all dimensions are of the same order, such as ball bearings; nor is it adequate to calculate local stress concentrations occurring at load points on beams or in regions of sharp variation in cross section, such as an abrupt change in diameter in a shaft. Recourse must then be made in these situations to the more powerful methods of the mathematical theory of elasticity.

II. Applications of Dynamic Elasticity

Several applications are discussed that illustrate the principles of elasticity under dynamic conditions. Again, as with quasistatic elasticity, many books have been written on the subject. However, they are usually more specific in nature, treating particular topics rather than the general subject. References are given where appropriate.

The wave equation (I-1.59) examined in Chapter I-1 forms the basis for all types of wave motion (and vibration phenomena) in conservative media or a vacuum in which the velocity of propagation is constant. Differentiation was made at that time between elastic traveling wave solutions and elastic standing wave (vibration) solutions to the wave equation. As indicated, we are obviously more concerned with the former, and the discussion below reflects this interest.

A. *Resonance and Pulse Ultrasonic Techniques*

Ultrasonic (elastic) waves can be used to measure certain properties and response of elastic materials. They are used in techniques that test materials in a *nondestructive* manner. Nondestructive testing is the technology by which quality of materials, reliability of components, and systems safety are monitored and maintained. Two basic ultrasonic methods of specific interest to us are available, one in which resonance conditions are obtained and one in which pulse transmission is employed. The first approach allows determination of the phase velocity as a function of mode and frequency of vibration; the second is used generally to determine the Lamé elastic constants, λ and μ, for isotropic, homogeneous unbounded media. (Recently, effort has also been directed toward the determination of elastic constants for several anisotropic (cubic and hexagonal) metals; (see Henneke and Green [1]). A recent contribution to the state of the art in research techniques in nondestructive testing is the book edited by Sharpe [2].

It is appropriate first to digress briefly to say what is meant by the term ultrasonics, and what can generally be accomplished using the principles of this subject, before we examine the above two methods. Phenomenologically, *ultrasound* is considered to consist of sound (elastic and/or acoustic) waves at frequencies above the upper range of human hearing. The study of the scientific principles governing ultrasound, of the apparatus used to generate it, and of its applications constitute the branch of physical acoustics called *ultrasonics*. Most ultrasonic applications occur in the frequency range of about 20 to 100 kHz, although often for work in solids frequencies up to and exceeding 5 MHz are used. Ultrasonic radiation of frequency 20,000 MHz has been produced (see Lindsay [3]) by driving a piezoelectric quartz rod at cryogenic temperatures.

The associated stresses and strains generated with ultrasonic methods are small, considerably smaller than those measured satisfactorily with the dynamic Hopkinson pressure bar (see Chapter II-2). For example, a moderately severe sound wave intensity is 60 dB (decibels), which is equal to 2×10^{-7} bar[1] (using a defined reference pressure level of 0.0002 dyne/cm^2). The strain levels with which we are dealing in this case are customarily in the range of tens of microinches per inch.

At present, the most active areas of applications in ultrasonics are cleaning by cavitation, processing of liquids (emulsification, *etc.*) and metals (drilling, welding, *etc.*), nondestructive testing, and underwater echo-ranging (*e.g.*, SONAR). Other increasingly active areas of application include instrumentation (delay lines, liquid-level control, *etc.*) and medical applications (ultrasonic surgery, dentistry, *etc.*). As stated, we are concerned here with nondestructive testing, in particular the determination of phase velocities in cylinders and the determination of the Lamé elastic constants. Other examples of work in the field of nondestructive testing using ultrasound are determination of thicknesses, imperfections, cracks, and voids (or inclusions) in solid bodies, of delamination in composites, of the existence of anisotropy (*e.g.*, the orientation of glass fibers in thermoplastics), and of moisture content.

Returning to the two ultrasonic methods we want to examine, we find the first technique, involving resonance, has a considerably longer history. Essentially, a cylinder (of finite length) of the material under investigation is vibrated at a certain resonant frequency so as to produce a steady-state harmonic standing wave pattern, after which the frequency and wavelength are measured to determine the phase velocity [see Eq. (I-1.70)]. With a given rod, a number of resonances can be observed corresponding to the fundamental frequency and a series of harmonics. A cylinder of intermediate length is usually chosen, say 10 ft, as a compromise between a very long cylinder with the desirable closely spaced resonant frequencies and a short cylinder with widely separated resonant frequencies. By this method, a cylinder of infinite length is experimentally "created," that is, the standing waves appear − via a reference frame translating with velocity c_p − as an infinite train of such waves in an infinite cylinder. Such a situation enables an experimental comparison with the exact theory in Section III.

The resonant frequencies are experimentally obtained by adjusting a suitable driving oscillator to positions where the amplitudes of vibration, *i.e.*, the voltage output, pass through maximums.[2] The wavelengths are then measured in some fashion, perhaps with the use of powder sprinkled on the cylinder (Kolsky [4]) or perhaps with some sort of electronic level indicator that can be passed along the rod and that is sensitive to the radial component of the displacement (Zemanek [5]). Actually, in practice, a particular harmonic of vibration is first selected to be investigated, *e.g.*, the fundamental, and the appropriate standing wave frequencies and wavelengths are then found in order to map the dispersion curve.

Care must be exercised to be certain that the proper harmonic (as well as the particular mode desired) is excited. Driving oscillators can be constructed to generate many reasonable combinations of mode and harmonic. The methods for obtaining such combinations are described in detail in the reference by Zemanek. Appropriate checks are generally made using "scanning" apparatus to determine the number of nodal planes (for the mode) and the number of nodal cylinders (for the harmonic) that are excited.

As an example, consider the situation of longitudinal vibrations (which is the easiest to investigate experimentally). The condition for resonance of the finite cylinder in this situation is that there exists an integral number of half-wavelengths along its length, with nodes occurring at or very near the ends. This supposes that end effects do not occur or are negligible, *i.e.*, the normal and shearing stress and displacement distributions are equal on both sides of the interfaces between the cylindrical specimen ends and the respective oscillator and detector. Theoretically, this is extremely difficult to achieve since the variation of stresses and displacements depends upon the particular frequency (see Fig. I.24); practically, however, an adequate experimental approximation can be made. This condition for resonance in our case can be expressed as

$$\Lambda = \frac{2L_0}{N}$$

or   (I-5.1)

$$\nu = \frac{c_p}{\Lambda} = \frac{c_p N}{2L_0}$$

where L_0 is the length of the cylinder, N is an integer greater than or equal to zero, and the other quantities have been previously defined. For $N = 0$, $\Lambda \to \infty$ and $\nu = 0$ and $c_p = c_0$ for longitudinal vibrations in the fundamental mode. Although theoretically, c_0 is undefined in our steady-state condition and one cannot assign a distinct value for the velocity of a specific element on the "wave," it can be approached closely enough by making L_0 large so that extrapolation to $c_p = c_0$ is relatively easy.

The second approach uses the information gained from the transmission of short, well-defined pulses through samples (usually circular cylinders) of the materials under study. It has become a very common technique, especially since the advent of ultrafast electronic equipment. The method is accurate, easy to use, and relatively inexpensive. Depending upon the situation or upon the user's preference, the method can be employed as either a pulse echo technique or through pulse transmission technique; the latter is the approach more often taken. The usual use of the method is in the determination of the Lamé constants for isotropic unbounded media. An electronically generated pulse of desired shape and repetition frequency is converted into a mechanical pulse using a transducer, generally made of barium titanate or quartz. The travel time of this pulse in the specimen is then accurately measured (in the case of the transmission technique, using a suitable receiving transducer).

Although it is not necessary to do so, both longitudinal and shear pulses are usually generated (in different experiments) to enable the calculation of λ and μ [see Eqs. (I-2.10) and (I-2.11)]. The principle of mode conversion at the specimen boundaries is sometimes used to obtain the distortional velocity c_s without the actual use of a transducer that generates shear waves. A longitudinal wave propagating nearly parallel to the axis of the cylindrical specimen is reflected, in part, as a distortion wave at some calculable angle. This shear wave, in turn, reflects from the opposite surface, and some of its energy is transmitted as a dilatation wave, again in the same direction as the originally propagating disturbance. The time delay caused from these reflections can be measured on an oscilloscope and related to c_ϱ and c_s [see Hughes *et al.* [6]; Hughes *et al.* [7]; Dunegan [8]; and the discussion about dispersion associated with Eq. (I-4.35)].

For determining the Lamé constants appropriate to unbounded media, the quantity R/Λ (assuming the specimen is a cylindrical rod) should be about 2.5 or greater, or else dispersion of the pulse will occur. This restriction is not difficult to maintain in practice. Absence of dispersion is usually checked by varying the geometry of the specimen rather than by changing the frequency of the forced transients.

It should be noted that the comments with regard to dispersion are relevant only to longitudinal vibrations, *i.e.*, when determining c_ϱ. When pulsed shear waves are used instead of compressional waves, no dispersion is found; strictly unidirectional distortional waves are associated only with an unbounded medium (see Tu *et al.* [9] and the comments about the propagation of torsional disturbances in Chapter I-4). Nor is the statement with respect to dispersion in conflict with the arguments given in Chapter I-4 in association with Fig. I.20 to the effect that some energy is propagated at the longitudinal velocity at certain values of $R/\Lambda < 2.5$. It can be recalled that such a condition was observed only with regard to the higher harmonics of longitudinal waves. Relatively small amounts of energy are propagated in these higher harmonics. Hence, most practical experimental testing will not be able to detect readily this higher harmonic energy transmission at these lower values of R/Λ.

Recently, considerable effort has been directed to the application of ultrasonics in determining the elastic Lamé constants as a function of temperature and pressure. Representative publications in this area are by Hughes *et al.* [7]; Dunegan [8]; and Fahey [10].

As a final comment relevant to the practical application and use of certain ultrasonic data, it is important to remember the comments in Chapter I-1 to the effect that if the material is perfectly elastic, the moduli are physical constants of the medium and are not functions of strain rate and strain amplitude. If such is the case, the moduli generated by relatively high frequency and low strain ultrasonic techniques can be used in quasistatic engineering design problems; otherwise, one must be assured that the design conditions are similar to those conditions under which the values were obtained. More will be said on this point in Chapter II-2, Sections I-B and I-C.

B. Other Applications and Examples Involving Traveling Waves

Two subjects of practical interest that involve the concepts of elastic waves are earthquake seismology and geophysical prospecting. They have as their basis the principles of surface waves in solids advanced by Lord Rayleigh in the 19th century. Rayleigh (surface) waves were treated in some detail in Chapter I-3. It can be shown using energy considerations that since elastic surface waves propagate (theoretically) only in two dimensions, the fall in amplitude per unit distance propagated (in real media) is less for these waves than for elastic body waves. This difference in rate of decrease in amplitude is one of the principal reasons why surface waves are of importance in seismic phenomena, becoming relatively more important (*e.g.*, more damaging) than other types of vibrations as distance increases. One of the best references on the subject is the text by Ewing, Jardetsky and Press, *Elastic Waves in Layered Media*.

It has been noted in seismographic records that the direction of vibration of the horizontal components is not necessarily parallel to the direction of propagation (as in Rayleigh waves), but may also involve components horizontally perpendicular to this direction (such that $u_2 \neq 0$; see Chapter I-3). It has been shown that these latter waves can be treated analytically by assuming that the elastic constants and density of the outer layer of earth in which these waves originally develop differs from that in the interior. These waves are termed *Love* waves after the investigator who first studied them. It is apparent that often both types of waves appear intermingled in a seismic record.

The more general problem of elastic waves at the surface of separation of two solid media (or a solid-liquid combination) has also been considered. These are termed *Stoneley* waves, again named after their original investigator. These disturbances will be propagated in both media, the amplitudes in the media being maximum at the surface of separation. Another extension of Rayleigh waves involves the consideration of surface waves in an elastic solid body bounded by two parallel planes; these are termed *Lamb* waves.

An associated subject was also treated in Chapter I-3, namely, the reflection and refraction of elastic waves at the boundary of solids. Such waves have considerable practical importance. For example, the topic of propagation of multiple reflections and refractions of elastic waves through layered media with plane interfaces follows from the treatment given in Chapter I-3, Section III-C. This analysis, as with Rayleigh wave analysis, can also be applied to the fields of seismology and geophysical prospecting, although it is perhaps more related to the latter. A recent article by Speilvogel [11] examines this problem in some mathematical generality, and also indicates how the results can also be useful in tsunami investigations.

Further application of this multiple reflection-refraction analysis can be made in the (idealized) inspection of solids for the presence of imperfections. Note that this provides a basis for a theoretical approach to supplement the experimental method using ultrasonics described in Section II-A. When oddly shaped defects such as cracks or blowholes are possible in the solid, the theoretical concepts of elastic wave scattering (Chapter I-3, Section III-D) may be more appropriate. The importance of the subject of mechanical wave scattering is also apparent when investigating practical design problems in construction such as a stress wave initiated by a nuclear explosion acting on an underground protective installation.

In addition, there are special properties of reflected and refracted waves that can be utilized. For instance, shear waves with their particle motions parallel to the reflecting surface are reflected, with respect to the normal plane, as exact images of the impinging shear waves; this property is useful in constructing delay systems and information storing devices. In a similar fashion, bounded media (Chapter I-4) that transmit guided waves can also be used as delay lines, as well as in such devices as tuning forks, strings on musical instruments, and mechanical filters to separate one region of the frequency range from others.

C. *Application Involving Standing Waves (Vibrations)*

The study of mechanical vibrations is a well-developed branch of engineering, and many useful techniques of analysis exist. Excellent texts covering the subject are in existence. Two of the more widely used are *Advanced Dynamics* by Timoshenko and Young and *Mechanical Vibrations* by Den Hartog, even though, as with the books listed in Section I, they were written some time ago.

A special method of treating vibration problems, more direct than the approach outlined in Chapter I-1, Section IV-C, that is of considerable practical use in engineering applications is the so-called *lumped parameter* or *lumped mass* technique,[3] mentioned in connection with matrix notation in the Introduction [see Eq. (9)]. The elastic system or structure to be analyzed is imagined as consisting of a number of concentrated point masses appropriately spaced and connected by suitable springs. The parameters selected depend to a great extent upon the analyst's experience and cleverness in producing a tractable idealization that satisfactorily duplicates the actual physical behavior of the structure. Newton's second law is then applied to the spring-mass system and the resulting set of coupled ordinary differential equations are solved (usually by matrix methods) to give the position of every point mass as a function of time. The number of allowed frequencies and normal modes of vibration in the solution depends upon the number of point masses selected. The connection of the lumped parameter solution and standing wave solution is simply one of generalization. If we increase the number of point masses selected above to an infinite number of infinitesimally spaced masses so as to gain equivalence with the actual continuous structure, the infinite set of equations obtained represent an infinite number of normal modes of vibration of the system and correspond to the formal solution to the boundary value problem given in Chapter I-1.

III. Comparison of Elastic Theory and Experiment

This section forms one of the more important parts of Part I of the text (even though it is rather brief), namely, how close to reality are the various elastic models presented. It is sufficient that we make comparisons only with the results of the Pochhammer-Chree analysis pertinent to elastic wave propagation in isotropic, homogeneous, bounded circular cylinders — first because discussion of both transmission in extended and semi-extended media also enters as a consequence of such comparison, and second because comparisons have previously been given between the Pochhammer-Chree and elementary and approximate theories.

Some of the earliest work involved experiments using ultrasonic standing wave resonance techniques, primarily because the required level of sophistication of electronic instrumentation was not great. Kolsky [4] plots phase velocity data obtained in the early 1940's for the fundamentals of longitudinal and flexural waves. Although the data do not extend over a large portion of the wavelength spectrum (about $R/\Lambda = 0.2$ to 0.5), agreement with the exact theory is excellent — within 1%.

One of the earliest (and non-ultrasonic) efforts that made direct use of some of the results calculated from the exact solutions was presented by Davies [12]. He concerned himself primarily with the examination of longitudinal pulse distortion in cylinders using his modification of the Hopkinson pressure bar[4] and made comparisons of theory and experiment using Fourier analysis techniques. Fundamental to the Hopkinson pressure bar method are the assumptions that the stress is uniaxial and that $c_p = c_0 = $ constant — and thus that the elementary approach discussed in Chapter I-4, Section II-B is adequate. Davies "pushed" these assumptions by his study of pulse propagation in rods so that he was no longer able to use results from the elementary theory. Thus, Davies confirmed dispersive effects indicated by the Pochhammer-Chree theory. He found that he could predict test results to within about 2.5%, provided that examination of sudden changes in stress occurring in less than about 5 μsec were not required. This approach is important because until the publication of Davies' work, the theory had been correlated only with steady-state experiments. However, for our considerations it does not come directly to the desired end, namely, a plot of c_p/c_0 versus R/Λ with both the exact results and experimental data indicated.

Other relatively recent contributions have given considerable confidence with respect to the reliability of the exact analysis. Tu *et al.* [9] investigated the velocities of longitudinal waves in the fundamental mode in cylindrical rods of various metals using both resonance and pulse transmission ultrasonic techniques. With the former method, they found that the dispersion curves of group velocity plotted using the experimental results were within about 3% of that predicted by the theory for $0 \leqslant R/\Lambda \leqslant 0.8$.

The pulse method was used for $R/\Lambda > 1.1$. It was observed that as R/Λ was increased, the pulse velocity also increased and approached the longitudinal velocity of the material as a limit as R/Λ approached values 2.5 and greater. This is a very important practical point. It is contrary to the theoretical expectation discussed in Chapter I-4 that the velocity should approach the Rayleigh surface wave velocity. The reason for this inconsistency, alluded to in Section II-A in the discussion of end effects in the standing wave technique, results because the exact solution requires a special particle velocity function to be impressed upon the end surface of the specimen, and this condition is not realized by the transducer in this range of R/Λ.

Some general observations are appropriate at this point. The standing wave technique appears to offer more reliable data pertinent to the investigation of dispersion of elastic waves in cylinders for R/Λ less than unity. The pulse transmission method appears more useful for values of R/Λ greater than unity, but it is primarily relevant to the determination of the Lamé constants. It is experimentally difficult to obtain measurements when the ratio R/Λ is in the region near unity with either technique. The principal reason for this difficulty is that the energy tends to divide among a number of modes causing the records (*i.e.*, either the standing wave pattern or the transmitted pulse) to become weak and blurred. Note that the group velocity c_g is obtained when the pulse method is used, whereas the phase velocity c_p is found when resonance is used. Conversion can be made from one to the other value with Eq. (I-4.43).

One of the most extensive works to date that examines the applicability of the Pochhammer-Chree theory over several modes and harmonics is the study reported by Zemanek [5]. Zemanek has obtained the experimental dispersion curves using standing wave techniques for the fundamental and first harmonic of the longitudinal, flexural, and screw modes. The wavelength spectrum investigated was from R/Λ essentially equal zero to R/Λ equal about 0.7. Agreement of the data with exact theory can be stated unequivocally to be remarkably good in the range of R/Λ measured. In fact, curves illustrating the discrepancies cannot be drawn on reasonably-sized plots since the average deviation is about 0.3%.

IV. Summary

The mathematical theory of elasticity in itself constitutes an important part of physical science and forms the basis for many practical analyses of the quasistatic and dynamic mechanical loading of structures and systems. Discussion of selected examples of application are given; emphasis is placed upon dynamic phenomena, and in particular, upon the use of ultrasonic techniques to measure certain properties of elastic materials.

Two basic methods using ultrasonic waves are available. The first approach allows determination of the phase velocity as a function of mode and frequency of vibration. Essentially, a cylinder of finite length of the material under investigation is vibrated with a suitable driving oscillator at a certain resonant frequency so as to produce a steady-state harmonic standing wave pattern, after which the frequency and wavelength are measured to determine the phase velocity. With a given rod, a number of resonances can be observed corresponding to the fundamental frequency and a series of harmonics.

The second approach uses the information gained from the transmission times of short, well-defined pulses through samples (usually circular cylinders) to determine (generally) the Lamé elastic constants, λ and μ, for isotropic, homogeneous, unbounded media. Although it is not necessary to do so, both longitudinal and shear pulses are usually generated (in different experiments).

The last section presents comparisons between the Pochhammer-Chree theory and elastic experimental data. One of the most extensive studies to date is that reported by Zemanek in 1962 [5]. Zemanek obtained the experimental dispersion curves using standing wave techniques for the fundamental and first harmonic of the longitudinal, flexural, and screw modes. The wavelength spectrum investigated was from R/Λ essentially equal zero to R/Λ equal about 0.7. Agreement with exact theory is found to be remarkably good in the range of R/Λ measured.

Notes

[1] 1 bar = 10^6 dyne/cm^2 = 0.9869 atm = 14,504 psi.

[2] The reason we do not obtain amplitudes (or voltages) that increase without limit when the specimen is driven at its resonant frequencies by an external forcing function is because of internal friction (Chapter II-1).

[3] In courses on simple mechanical dynamics, this method is also known more generally as the analysis of free small vibrations of coupled dynamical systems with a finite number of degrees of freedom.

[4] Complete avoidance of the mention of the Hopkinson pressure bar technique until Chapter II-2 is not possible if we wish to compare elastic theory and experiment at this point, since the initial experimental work with the pressure bar was done with the sample materials remaining elastic. If concern arises in the reader's mind here about the details of the Davies' modification, he can refer to Chapter II-2, Section I-B.

References

[1] E. G. Henneke and R. E. Green, *J. Appl. Phys.*, **40**, 3626 (1969).

[2] R. S. Sharpe (ed.), *Research Techniques in Nondestructive Testing*, Academic, New York, 1970.

[3] R. B. Lindsay, *Mechanical Radiation*, pp. 370-371, McGraw-Hill, New York, 1960.

[4] H. Kolsky, *Stress Waves in Solids*, Oxford University Press, London, 1953; Dover, New York, 1963, pp. 95-98.

[5] J. Zemanek, *An Experimental and Theoretical Investigation of Elastic Wave Propagation in a Cylinder*, Technical Report No. XVIII, Dept. of Physics, University of California, Los Angeles, 1962.

[6] D. S. Hughes, W. L. Pondron and R. L. Mims, *Phys. Rev.*, **75**, 1552 (1949).

[7] D. S. Hughes, E. B. Blankenship and R. L. Mims, *J. Appl. Phys.*, **21**, 294 (1950).

[8] H. L. Dunegan, *Determination of Dynamic Modulus at High Temperatures by the Use of Ultrasonics*, University of California Lawrence Livermore Laboratory, Report UCRL-7326, Rev. 1, 1963.

[9] L. Y. Tu, J. N. Brennan and J. A. Sauer, *J. Acoustical Soc. Am.*, **27**, 550 (1955).

[10] N. H. Fahey, *Ultrasonic Determination of Elastic Constants at Room and Low Temperatures*, Watertown Arsenal Laboratories, Watertown, Mass., Report TR 118.1/1, 1960.

[11] L. Speilvogel, *J. Appl. Phys.*, **42**, 3667 (1971).

[12] R. M. Davies, *Roy. Soc. Phil. Trans.*, **A240**, 375 (1948).

Part II
EXTENSION OF APPLICATIONS OF ELASTIC CONCEPTS

NONELASTIC MATERIAL BEHAVIOR

I. General

An obvious, yet helpful, division that can be given when characterizing mechanical behavior of materials under loading is that response is either elastic or that it is not elastic. As stated early in the text, and in Chapter I-5, the subject of elasticity constitutes only a part — although a very important part — of the total scheme of methods to represent mathematically material behavior. We take *nonelastic* behavior as any behavior that is not characterized as perfectly elastic; and in this chapter some of the main features of description of this type of response are qualitatively discussed.

From a theoretical standpoint, the subject falls naturally into two general parts which correspond to whether or not the medium can be considered to obey Hooke's law during loading. Obvious analytical difficulties develop when nonelastic deformations are treated. One of the major problems arises because of energy dissipation. When any real material is subjected to a stress cycle, some of the mechanical energy is lost (converted into heat) as a result of dissipative mechanisms in the material. These dissipative mechanisms are collectively termed *internal friction*. For liquids and gases the dissipative forces are due to viscosity and thermal conduction (and, generally to a lesser degree, chemical reaction, mass diffusion, and radiation); but in solids the behavior is more complex. A measure of the energy dissipated per cycle is the area of the closed hysteresis loop of the stress-strain diagram of the cyclically loaded specimen. The energy dissipated depends upon the material and usually upon the amplitude and speed of the cycle and on the past loading history of the specimen.

Other analytical difficulties are also evident. For example, superposition is no longer valid, and often nonelastic materials behave both in a nonlinear and in a time-dependent manner. In fact, obtaining a realistic description of a nonelastic response is a difficult undertaking, and no one expression or mathematical treatment universally representative for all conditions is possible (although some general progress is being made; see for example Lee [1]). Every theory is directed at a certain segment or portion of these phenomena and incorporates only their essential characteristics. It is often helpful or advantageous in this regard to use approximate analyses and/or analyses based upon linear or elastic theory. As with the subject of elasticity, the material is treated as a continuum in most situations.[1]

From an experimental point of view the techniques are generally similar, regardless of whether elastic or nonelastic solids are being investigated; but the purpose of carrying out experiments is usually quite different in the two types of investigation. In the former, experiments are often conducted to determine the appropriate elastic constants of the medium or to test the validity of approximate theories. Since exact analytical treatments of nonelastic behavior have very limited physical validity, experimental investigations are generally the only practical method of determining the real mechanical response.

Although nothing in the description of nonelasticity limits consideration to either a "slow" or "fast" rate of loading, it is nevertheless convenient to do so, just as was done in Chapter I-1 for elasticity. With reference to Fig. I.1, inertial (or wave) effects become important when strain rates (related to frequency) become greater than, say, 10^2 sec^{-1}. Also, nonlinear effects often become important as strain rates are reached of the order 10^3 - 10^6 sec^{-1}, depending upon the amplitude of the mechanical disturbance.

The history associated with the treatment of nonelastic phenomena is much shorter than that of its counterpart, elasticity. Although there were some prior notable developments, significant impetus in nonelastic effort came during the two world wars, particularly the second. For example, during WW II, the properties of materials at high rates of loading increased in importance for military reasons and led to the extension of the classical elastic theory to solids that were nonelastic and to the development of the theory of uniaxial stress plastic waves. Much of the present work in such waves is from engineers interested in studies related to the rapid loading of structures.

A tremendous amount of effort has also been accomplished during the past two decades in understanding the generation and effects of high intensity shock waves. The interest in such loading conditions resulted from the necessity of generating Hugoniot data and high pressure thermodynamic information for specific equation-of-state purposes. The work has been aided by significant improvements in electronic instrumentation and computer technology.

II. Representations of Material Behavior

Although the terms have been mentioned in general before, it is necessary to be more specific with respect to the concepts of *fluidity* and *solidity*. A body is said to be *fluid* if application of any anisotropic or inhomogeneous force system results in flow. Upon release of the force system, the body will not return to its initial configuration. A *perfect fluid* is defined as an incompressible substance in which the pressure in the interior of an infinitesimal element must always be directed perpendicularly to any section which passes through the element. A *perfectly viscous fluid*, on the other hand, is described as a compressible substance in which the tangential shearing stress is linearly related to the flow velocity gradient by a constant called the *coefficient of viscosity*. In this fluid, the work done on the material is immediately dissipated as heat. An inviscid fluid is one in which the coefficient is zero.

A body that behaves as a *perfectly elastic solid* has been defined in Chapter I-1. If the body is a *perfectly plastic solid*, it deforms as a perfectly elastic solid until a critical value of loading is reached, called a *yield point* in $n = 1$ (one dimension), a *yield curve* in $n = 2$, and a *yield surface* in $n = 3$; upon exceeding this value, the body then flows so that the applied force system remains constant. The assumption is made that the bulk modulus remains constant, *i.e.*, the material is incompressible, in the plastic domain.

These relationships are representative of what are termed *classical* bodies. They are just abstract mathematical idealizations or concepts which describe the mechanical behavior of real materials in limiting cases. In the literature, these idealizations have been assigned names in honor of the investigators who used them either originally or extensively. Hence, a *Pascalean* fluid is an inviscid fluid, a *Newtonian* fluid is a linear viscous liquid, and a *Hookean* solid is an elastic solid. An *Euclidean* solid is a special case of a Hookian solid in which the body is considered rigid and incompressible, *i.e.*, $\mu = k = \infty$.

Mathematical representations of material behavior can be conveniently grouped and collectively called *equations of state*. An equation of state can be defined as some cause-and-effect relationship among various parameters that describes a given medium's equilibrium response to prescribed loading stimuli. There is no basic restriction on the parameters that can be considered in formulating such an equation. They can include variables and functions that are kinematic, dynamic, thermodynamic, and electromagnetic in nature; for example, an expression for the state equation can incorporate factors of stress (or pressure), strain (or dilatation), temperature (or internal energy), entropy, and time.

Equations of state can be conveniently discussed in terms of their being essentially "thermodynamic" or "mechanical" in nature. This division is arbitrary (similar to the division we made between quasistatic and dynamic) since it is evident that the latter is dependent upon the former (and vice versa), and complete differentiation between the two is not possible. We mean by this statement that attention is directed to those mathematical models that idealize and emphasize primarily either thermodynamic phenomena or mechanical phenomena caused by mechanical loading, *e.g.*, stress and strain fields and strain rate behavior.

There are many examples of particularized thermodynamic equations of state. For fluid flow, in thermodynamic equilibrium, a relation among the several thermodynamic variables (only two of which are independent) can be written,

$$p = p\,(\rho,s) \tag{II-1.1}$$

where p is the pressure, ρ is the density, and s is the specific entropy. Equation (II-1.1) is called a *caloric* equation of state.

If viscosity and heat conduction are neglected, the specific entropy of the moving particle remains constant, and the resulting expression is termed the *adiabatic* equation of state. (Often such a relationship is used for a solid.) The word *isentropic* would perhaps be more appropriate here than adiabatic, but this term is usually reserved to describe an equation of state if entropy is held constant throughout the medium.

For *ideal* gases, the equation of state is

$$\frac{p}{\rho} = R\,\theta \tag{II-1.2}$$

where R is the universal gas constant R_0 divided by the effective molecular weight of the given gas. If a further assumption is made that the internal energy in this ideal gas is proportional to the temperature, the gas is called *polytropic* and the caloric relation becomes

$$p = A\,\rho^{\gamma} \tag{II-1.3}$$

where the coefficient A is a function of the entropy and γ is a constant called the *adiabatic exponent*.

Mechanical equations of state, as we have defined them, are often called *constitutive* relations, or more generally, *rheological* equations of state. Hooke's law is an example of a rate-independent constitutive relation. Rheology is the science of *deformation* and *flow* of matter. A body is said to flow if its degree of deformation changes continually with time (we have discussed deformation in connection with strain in Chapter I-1, Section II-B). Rheology is more closely associated with solid and fluid behavior rather than with the mechanics of gas flow.

In general, the response of a real material includes to some extent *all* the above-mentioned types of behavior as well as intermediate and composite types. In fact, the physical distinction of whether the aggregation of matter is in the solid, liquid, or gaseous state is sometimes confusing. For example, under certain extreme conditions of loading the body may become a fluid, although its rest state may be in the form of a solid. More definitively, the second axiom of rheology states that *every* material possesses *all* rheological properties, although in varying degrees (see Reiner [2], p. 11).

Constitutive equations can be constructed from the several classical idealizations discussed above to characterize in a reasonable fashion the mechanical behavior of some real rheological bodies. For example, a class of materials exists whose principal response is partly determined by viscous fluid characteristics and partly by elastic solid behavior. Hence, energy is neither governed by dissipative mechanisms as in the former nor totally conserved as in the latter. Such materials are called *viscoelastic* substances. The most important phenomenon that distinguishes these materials from perfectly viscous fluids is the occurrence of strain recovery upon release of the imposed stress system. The most important phenomena that distinguish these materials from perfectly elastic solids are the presence of time-dependent and dissipative effects; and from perfectly plastic solids, the most important phenomenon is the presence of rate considerations. One usually observes in the response of viscoelastic material the decrease of stress at a finite rate upon cessation of change of deformation; this is called *stress relaxation*. Viscoelastic media are also usually characterized by the gradual increase in strain with constant load; this is called *creep*.

One phenomenological mathematical model, representative of most of the types of behavior observed in viscoelastic systems, consists in the combination of various springs (for the elastic characteristics) and dashpots (for the viscous characteristics) into some functional relationship. To give a more quantitative representation of real viscoelastic bodies a more complex model is often used, one composed of a large number of spring and dashpot units with different response times, *i.e.*, relaxation times. If desired, the number of these units can be extended effectively to infinity using the *Boltzmann superposition principle*.

In a similar fashion, using these classical "building blocks," constitutive equations can be developed to represent other real rheological substances. The books by Reiner [2,3] and Fredrickson [4] discuss in considerable detail these various possibilities. Essentially then, the mechanical response of materials of physical interest is bounded by a Euclidean solid at one extreme and by a Pascalean fluid at the other. The subject of nonelastic deformation does not include Euclidean or Hookean solids.

A final comment on the philosophy of constitutive equation construction is appropriate. A constitutive relation can be either based upon or generated by an "experimental" approach or a "rational" approach. In the former technique experiments are conducted, with appropriate consideration given to the basic concepts of continuum mechanics, to observe the response of the material under specific, controlled test conditions. Then, ideally, conclusions are induced from the observations made, and a tractable mathematical model is developed (using, hopefully, the classical models above) to represent the real material behavior within specified limits. It is often difficult to achieve this goal completely because, in addition to material response, conditions of the test, such as the shape and size of the specimen, its method of loading and edge effects, are also reflected in the results to some extent.

The rational approach, on the other hand, postulates a constitutive relation on the basis of analytical simplicity, *e.g.*, these classical models, without experimental results. Then the limits of physical applicability are tested, and, ideally, verification of the relation is deduced over the desired limits. Again, difficulty is often experienced here resulting from the obvious problem in the selection of an appropriate constitutive relation *a priori*, and from the well-known nature of physical theories in that they are frequently couched most conveniently in terms of parameters difficult to measure.

Separately, both techniques contain obvious practical weaknesses. The most fruitful method to obtain a reasonable description of the material's behavior is with the use of both at the same time in which experimental and rational mechanics rely on each other in some combination. Generally, more emphasis is directed to experimental efforts, partly because there is a larger proportion of investigators in this field so inclined. The model (or models) that is developed is more or less continuously modified as further data and experience are obtained. It is interesting to note that Hooke's law as applied to small deformations is an exception with respect to simultaneous use of experimental and rational mechanics. The original form of the law was first founded on the basis of experimental observations, although the generalization as given by Eqs. (I-1.29) then utilized the rational approach.

III. Regimes of Stress Wave Propagation

The subject of stress wave propagation in materials can be conveniently divided into several subsections: (1) *elastic waves* in which the stresses are such that the medium obeys Hooke's law; (2) *viscoelastic waves* in which time-dependent viscous or frictional stresses act in addition to elastic Hookean stresses; (3) *plastic waves* in which the medium is stressed beyond the elastic limit; and (4) *shock waves*. Several general articles ([5]-[8]) on this subject are listed at the end of the section.

Waves propagating in viscoelastic materials are characterized by having the strain components lagging in time relative to the varying stress components. This lag depends on a time associated with the interval required for the viscous stresses to tend to equalize. This period is called the *relaxation time* and is defined as the time required for a process to proceed to within $1/e$ of its equilibrium value, where e (= 2.718···) is the base used in the natural system of logarithms. As a consequence, it is found that from the point of view of wave propagation, the material is dispersive and that sinusoidal waves of high frequency have greater propagation velocities than low-frequency waves.[2] It is also found that the internal friction results in the higher frequency waves being attenuated more rapidly. Hence as a result of this velocity dispersion and frequency-dependent attenuation, the shape of a mechanical wave as it is propagated through a viscoelastic medium changes. With respect to standing waves (although such waves are of lesser interest to us), the effect of friction is to damp out free vibrations and to change slightly the allowed frequencies (eigenfrequencies). Also because of the energy loss, the amplitude of higher frequency harmonics become negligible in comparison with those of the lower frequency components causing, in time, an initially nonsinusoidal composite vibration to approach often a sinusoidal form. As a generalization, temporal attenuation is of greater importance in standing wave phenomena, whereas spatial attenuation is of greater importance in progressive wave phenomena. These effects are in obvious contrast to waves in (unbounded) elastic solids in which the pulse shapes remain the same.

The type of viscoelastic response on which most experimental and theoretical effort has been expended is that termed linear viscoelastic behavior, *i.e.*, a behavior for which the stress components and the strain components are related by linear differential equations involving the stress, strain, and their time derivatives.

A linear viscoelastic solid may be defined in alternative ways which can be shown to be mathematically equivalent. One of the simplest ways to examine stress propagation in linear viscoelastic solids is with an infinite train of longitudinal sinusoidal waves with a single relaxation frequency. Many problems, however, do not lend themselves to such idealized treatment. Since the principle of superposition is applicable, a given nonsinusoidal disturbance may be considered as the sum of a spectrum of Fourier components (or relaxation frequencies) and treated according to the techniques mentioned in Chapter I-2, Section V.

Beyond a certain stress level (and usually at a higher strain rate loading), solids are subjected to permanent (or plastic) deformation. The theory of plastic wave propagation adopts many of the concepts of the static theory of plasticity (see Section II). Isothermal conditions are ordinarily assumed; that is, one uses only mechanical constitutive relations. One of the earliest significant efforts (in the 1940's and early 1950's) suggested that an increment of plastic strain would propagate along a semi-infinite rod at a characteristic velocity governed by the slope of the static stress-strain curve at the given plastic strain. Based upon this idea, a formal solution was given of the propagation of a disturbance along this semi-infinite rod. Effort directed toward the experimental verification of this theory met with moderate success. It was found that much of the discrepancy could be attributed to the use of static stress-strain relations, that is, a strain-rate effect was observable.

In plastic wave propagation, the stress-strain curve is concave toward the strain axis. When the stress-strain relation curves instead toward the stress axis, the velocity of propagation is greatest for those parts of the wave whose amplitudes are greatest, and the front of the disturbance becomes steeper and steeper as the pulse propagates through the medium. The final shape of the wave is then determined by the dissipative processes. Such waves are called shock waves, and can occur in solids, fluids, and gases.

Often under the conditions of very high-amplitude stress shock loading, the behavior can be taken to approximate that of a compressible fluid. That is, the deviatoric stress component (shear component of the stress tensor) can be treated as being of no consequence relative to the spherical (pressure) component.[3] A complete theory of hydrodynamic shock loading must take account of the thickness of the shock front and heat conduction and viscosity (which becomes important only in the front). However, useful results can be obtained simply by applying the laws of conservation of mass, momentum, and energy across the shock front without regard to its thickness. This approach is used in the development of the Rankine-Hugoniot relations. In actual fact, however, in the shock stress regime below roughly 50 times the static yield stress of the solid being investigated, the role of material cohesive strengths cannot generally be ignored. The compressive shock disturbance can often support a wave structure, so that the front does not have just a single steep-fronted shape. Chapter II-3 contains more explicit discussion of these consequences in shock-wave propagation.

At higher ranges of induced stress levels, generally instantaneous vaporization of the particles in or near the target area result because of the tremendous rates of energy transfer involved. A collision of a meteorite with a heavy atmosphere is an example of this type of loading.

IV. Conditions of Loading

A variety of possible conditions of loading exist in which to conduct experimental investigations. Two have been alluded to in Chapter I-5 and are important for our purposes, namely, the condition of one-dimensional stress and the condition of one-dimensional strain. In one-dimensional stress, there is no restriction to lateral motion of the material, and thus there is only one (macroscopic) stress component. This component corresponds to a principal stress in an isotropic medium. In one-dimensional strain the restriction is such that this lateral motion is zero, and thus there are three stress components. In the case of transverse isotropy, the lateral stresses are equal but different in general than the stress in the direction of applied loading (Chapter II-3, Section II) and must be inferred from the properties of the solid.

These two mechanical stress states are significant in that, while they are the simplest conditions to produce experimentally, they also represent the extreme bounds of material behavior which encompass most other states of stress. Moreover, these well defined loading conditions simplify analyses and investigation of new theories and/or effects. The clear distinction made between uniaxial stress and uniaxial strain loading is quite important in experiments involving wave effects. In quasistatic work distinction is also of importance, although tests are usually conducted under conditions of one-dimensional stress only.[4] Both states of loading are equally applicable to either elastic or nonelastic behavior.

Dynamic one-dimensional tensile, compressive, torsion, and shear stress loading can be investigated experimentally by the study of the propagation of disturbances in long bars or in thin-walled tubes of suitable dimensions so that radial stresses can be neglected. Thus, observations cannot be made over (relatively) very short times or at very high strain rates.

The dynamic one-dimensional strain condition is most often investigated using plate impact experiments generated, for example, using a gun or explosive plane-wave lens system [see expression (I-3.26) and the associated footnote]. In distinction to uniaxial stress examination, the observation periods here are limited to very short times (a few microseconds) and to regions unaffected by undesired stress pulse reflections, *i.e.*, to regions unaffected by edge effects. Thus, the material under study can be considered as being effectively unbounded laterally. Under this condition of loading, the highest strain rates and stress levels that are physically possible can be achieved.

Because of the obvious diversity of interest and large differences between testing methods for the two conditions of loading, attempts at correlation of the experiments have been rather infrequent and only relatively recently reported (*e.g.*, see Wasley and Walker [9]).

The ultimate goal of the mechanical measurement of such uniaxial stress and uniaxial strain stress waves from our point of view is the development of curves giving stress, strain, and particle displacement or velocity at any point in the medium, all as functions of time. Determination of these stress-strain-particle velocity curves is often difficult because of the rapid variation of the various parameters with time and shortness of duration of the event. In some experiments the magnitude of the stresses involved, even though they may remain elastic within the solid, create problems of measurement. Conventional mechanical gauges are usually inappropriate either because of their low-frequency response or because of the perturbations and distortions introduced from their use. Electronic and optical measuring techniques are most often employed today, and they are essentially always used in one-dimensional strain experimental work.

V. Summary

Often, the classical analyses (in particular, elastic analyses of solid media) are not adequate for obtaining a realistic description of the response of materials that behave nonelastically. Then a more general treatment is required, necessitating the introduction of a more complicated equation of state somewhere in the analysis. An equation of state is defined as some cause-and-effect relationship among various parameters that describes a given medium's equilibrium response to prescribed loading stimuli. It can be conveniently discussed in terms of it being essentially thermodynamic or mechanical in nature.

Mechanical equations of state are often called constitutive relations, or more generally, rheological equations of state. Included in mechanical equations of state are various idealizations. For example, a perfectly viscous fluid is described as a compressible substance in which the tangential shearing stress is linearly related to the flow velocity gradient. If a body is a perfectly plastic solid, it deforms as a perfectly elastic solid until a critical value of loading is reached, and upon exceeding this value, the body then flows so that the applied force system remains constant. No rate effects are present in either a perfectly elastic or perfectly plastic solid.

Constitutive equations for nonelastic behavior of real rheological bodies can be constructed from these and other idealizations. For example, a class of materials exists whose principal response is partly determined by viscous fluid characteristics and partly by elastic solid behavior; such materials are called viscoelastic substances, and can be represented in a phenomenological mathematical model by the combination of various springs and dashpots combined into some functional relationship.

The study of stress-wave propagation can be conveniently, although arbitrarily, divided into four classes: (1) elastic waves in which stresses are such that the medium obeys Hooke's law and internal friction is negligible; (2) viscoelastic waves in which time-dependent viscous or frictional stresses act in addition to elastic Hookean stresses; (3) plastic waves in which the medium is stressed beyond the elastic limit; and (4) shock waves in which the cohesive strength of the material often becomes of negligible significance in governing the nature of the motion, and hydrodynamic theory can be used to approximate its mechanical behavior. At intermediate shock stress levels, below approximately 50 times the static yield strength of the solid under study, material strength cannot generally be ignored and the treatment becomes more involved.

Of the variety of possible conditions of loading which exist for conducting experimental investigations two are important for our purposes, namely, the condition of one-dimensional stress and the condition of one-dimensional strain. In the former, there is no restriction to lateral motion of the specimen (and thus there is only one stress component); in the latter, the restriction is such that this lateral motion is zero (and thus there are three stress components).

Notes

[1]Exceptions do exist, for example, in the treatment of materials of very high porosity.
[2]This behavior is opposite that of electromagnetic waves in dispersive media.
[3]Mathematically, the separation of the general stress tensor into a deviatoric component and into a hydrostatic component is defined as

$$S_{ij} = S_{ij}^{(d)} + \frac{S_{kk}}{3} \delta_{ij} \tag{II-1.4}$$

where $S_{ij}^{(d)}$ is the deviatoric tensor, S_{kk} is the *trace* of the tensor (an invariant), and $S_{kk}/3 = -p$ is the hydrostatic pressure. Also see the comments associated with Eq. (9) in the Introduction, and with the discussion of the concept of principal stress in Chapter I-1, Section II-A. The general strain tensor can be separated in a similar fashion into dilatation and deviatoric strain tensors.
[4]Some special quasistatic tests are conducted under conditions of biaxial and triaxial stress.

References

[1] E. H. Lee, "Plastic-Wave Propagation Analysis and Elastic-Plastic Theory at Finite Deformation," in *Proc. 17th Sagamore Conf. on Shock Waves and the Mechanical Properties of Solids*, (J. J. Burke and V. Weiss, eds.), Syracuse University Press, New York, 1971.

[2] M. Reiner, *Deformation, Strain and Flow*, 2nd ed. (principally Chapters I, VII, XIV), Interscience, New York, 1960.

[3] M. Reiner, *Lectures on Theoretical Rheology*, 3rd ed., pp. 1, 21-23, 31-33, 37-43, 79-103, North-Holland, Amsterdam, 1960.

[4] A. G. Fredrickson, *Principles and Applications of Rheology*, pp. 1-3, 5, 6, 20-28, 30,

31, 64, 65, 71, 72, 98, 99, 118-122, 175-179, Prentice-Hall, Englewood Cliffs, N. J., 1964.

[5] W. Goldsmith, *Appl. Mech. Rev.*, **16**, 855 (1963).

[6] H. G. Hopkins, *Appl. Mech. Rev.*, **14**, 417 (1961).

[7] H. Kolsky, *Appl. Mech. Rev.*, **11**, 465 (1958).

[8] H. Kolsky, *J. Sound Vib.*, **1**, 88 (1964).

[9] R. J. Wasley and F. E. Walker, *J. Appl. Phys.*, **40**, 2639 (1969).

ONE-DIMENSIONAL STRESS WAVE INVESTIGATIONS

I. Pressure Bar Techniques

A. Hopkinson Pressure Bar

The first significant contribution toward measuring transient stresses was advanced by B. Hopkinson in 1914. [1] His apparatus, which has become known as the *Hopkinson pressure bar*, is an application of the elementary theory of uniaxial stress propagation of elastic disturbances in a cylinder where the wavelength of the disturbance (or some measure of the effective wavelength) is great compared with the lateral dimensions. An assumption implicit in this technique is that the pressure in the stress pulse is uniformly distributed over the (constant) cross section of the cylinder (see Chapter I-4, Section II-B).

The original apparatus that Hopkinson used consisted of a circular steel bar of appreciable length and small diameter which was suspended horizontally using threads of low inertia. At the end of the bar opposite the firing end, a finely lapped short cylindrical disk of the same diameter and type of steel as the main bar (and known as the "time piece") was positioned and held firmly in place using a thin layer of grease.

When a projectile impinges upon the firing end of such a bar, or when an explosive is detonated in contact with it (or perhaps in contact with a suitable attenuator), a compression disturbance is propagated down the cylinder. As has been shown in Chapter I-4, the pulse will travel without any appreciable distortion, *i.e.*, travel as a nondispersive wave at a constant velocity, so long as $R/\Lambda < 0.1$ and the material of the cylinder is not stressed beyond its proportional limit. Such an elastic compression pulse is transmitted through the joint between the cylinder and the time piece without change (the impedance change from the thin grease layer is negligible) and is reflected at the free end of the time piece as a tension pulse. The reflected tension disturbance propagates backward through the incident compressive pulse. Since the stress at any section of the bar is the sum of the stresses due to the incident and reflected pulses, as soon as the net loading at the joint between the bar and the time piece becomes tensile,[1] the latter separates (or spalls) with the momentum "trapped" in it. Using Hopkinson's method, this momentum in the time piece is measured with a balistic pendulum; the momentum remaining in the bar is determined from the amplitude of the swing of the bar. Ballistic pendulum devices were widely used in the pre-electronic era, and analysis involves simple momentum and energy considerations.

If A denotes the area of cross section of the time piece and L its length, and if S_{zz} is the stress in the pulse at time t, the momentum M trapped in the time piece is

$$M = A \int_0^{2L/c_0} S_{zz} dt. \qquad (II-2.1)$$

Thus, by measuring the momentum in time pieces of different lengths, the areas of the stress-time curves for the various time intervals can be obtained. In general, it is not possible to construct the precise form of this stress-time curve itself from such measurements. The reason for this lack of complete definition results from the fact that the method gives no indication of the exact time when the maximum stress (or any given stress for that matter) occurs. Thus, any number of shapes of the pulse would be consistent with one set of observations.

Although the exact shape of the stress-time disturbance cannot be deduced, the duration of the pulse and the maximum value of the stress $S_{zz_{max}}$ can always be found. These data are of great importance in practical situations. The duration is $2L_0/c_0$, where L_0 is the length of time piece required just to catch all of the momentum so that the bar remains at rest. A very short time piece is used to determine $S_{zz_{max}}$ such that the time interval $2L_0/c_0$ brackets the instant of maximum stress. Note that if $S_{zz_{max}}$ is known, the maximum particle velocity $\partial u_z/\partial t$ can be computed by Eq. (I-4.65). Furthermore, if U is the velocity of travel of the time piece, its momentum is $\rho A L U$, and from expression (II-2.1), $S_{zz_{max}} = \frac{1}{2}(\rho c_0 U)$.

In point of fact, the theory of the Hopkinson bar applies to any kind of medium, elastic or nonelastic, provided that stress waves are propagated with a constant velocity and the stress and particle velocity are related by an equation similar to (I-4.65). However, the original pressure bar was applied only to systems undergoing elastic response.

There are several difficulties associated with the Hopkinson pressure bar which detract from its usefulness, the principal ones having been mentioned. Nevertheless, the method enables the first scientific investigation on a laboratory scale into this phenomenon of transient pressures. This approach is still utilized in many places, *e.g.*, universities, as a teaching implement because it provides a clear exposition of the basic mechanical principles involved, it is simple to construct, and it is economical.

B. *Modifications of the Hopkinson Pressure Bar*

Many modifications of the original Hopkinson equipment have been introduced, two examples of which are discussed below. Essentially all of these modifications retain some part of the original pressure bar concept, but involve the added use of electronic instrumentation and/or high-speed photography. The main purpose of introducing such changes was to obtain continuous — or near continuous — records of the dynamic events to determine the shape of the pulse, something that the original Hopkinson pressure bar could not accomplish.

Apparently, Davies [2] in 1948 was responsible for the first significant improvement, one in which he devised electrical apparatus to give a continuous record of the longitudinal (or radial) displacement produced by a stress pulse at the free end of a long cylinder. For a uniaxial stress wave propagating in a cylinder in which the condition $R/\Lambda < 0.1$ prevails, Eq. (I-4.65) shows that the longitudinal stress S_{zz} is linearly proportional to the particle velocity $\partial u_z / \partial t$. The Davies bar makes use of the fact that because of reflection, the free end of the cylinder has twice this particle velocity (see Chapter I-3, Sections III-A and III-B and Rinehart [3]). Thus, if the free surface velocity is monitored as a function of time, the calculation of S_{zz} follows readily. With the Davies technique, the displacement-time curve, rather than the free surface velocity, is obtained; however, by differentiation of this curve, the stress-time relationship may be determined. Alternatively, if we measure the radial displacement at some position along the cylinder instead of the longitudinal displacement of the end of the bar, we find from Hooke's law that this radial surface motion is expressed by $\sigma S_{zz} R/E$, where the quantities have been defined previously (Chapter I-4, Section II-A). Thus, the longitudinal stress-time curve can be obtained from the radial displacement-time curve by multiplying the ordinates by $\sigma R/E$. Using his results, Davies found the dispersive effects predicted by the Pochhammer-Chree theory (Chapter II-5, Section III).

The method Davies used to measure the longitudinal displacement of the end surface of the cylinder essentially involved the use of a plane parallel plane capacitor. One plate consisted of the plane free end of the cylinder and the other was a charged conductor "fixed" in space. Because of the motion of the free surface, a potential difference was generated which could be related either theoretically or experimentally to the longitudinal displacement. The results could be displayed on an oscilloscope and recorded permanently with appropriate photography.

The use of this "condensor microphone" technique is not limited to the measurement of free surface motion caused by elastic waves. In fact, techniques to measure free surface velocity have developed to an extremely high degree of sophistication because of their pertinence in high-pressure shock wave study (see Chapter II-3, Section III).

A relatively recent improvement that is now in wide use was first suggested by Kolsky [4] in 1949. This technique forms the basis for nearly all mechanical dynamic stress-strain-strain rate studies under conditions of uniaxial stress loading conducted today. A short specimen is sandwiched between two elastic pressure bars and is loaded by a single pulse (usually, but not necessarily, compression) traveling through the system – and hence the name, *split Hopkinson pressure bar*. The pressure bars act to apply the load to the specimen and are instrumented with transducers (*e.g.*, strain gauges or condensor microphones) so as to obtain continuous values of stresses or displacements at the faces of the specimen. Such an arrangement allows the determination, averaged in the sample, of stress versus strain, with strain rate as a parameter, of a constant amplitude loading pulse of the order of 100-μsec duration.

A significant and very important feature of the split pressure bar is that the specimen itself need not remain elastic, as is required for the incident and transmitter bars. Indeed, the usual condition for investigations of material behavior using this technique is that the sample is stressed beyond its proportional limit so that its nonelastic stress-strain-strain rate behavior is also obtained.

One form of the apparatus is shown schematically in Fig. II.1. Strain rates as high as 10^4 sec^{-1} can be achieved using this method.

In his original work, Kolsky did not use a movable striker bar to generate his pulse loadings; rather, he used an explosive detonator in direct contact with an anvil which in turn was placed in contact with the incident pressure bar. He also used condensor microphones in his equipment instead of strain gauges as shown in the figure (although his attempt to use a cylindrical capacitor to obtain the longitudinal displacements of the incident bar did not prove reliable). Krafft *et al.* [5] were apparently the first investigators to introduce the use of strain gauges and a laterally guided striker bar propelled from a gun or ram – the technique in common use at present and illustrated in Fig. II.1. Nevertheless, Kolsky is generally credited with having originated the design for this basic modification.

For ease in visualization, let us consider the initial loading pulse to be compressive. Tension and shear loading studies are also feasible (see Hauser [6] and Hoge [7]), and torsion loading is discussed specifically later in the section. When this compressive pulse reaches the specimen, a disturbance is reflected from the interface back into the incident bar while another is propagated through the specimen to the transmitter bar. The relative magnitudes of these incident,

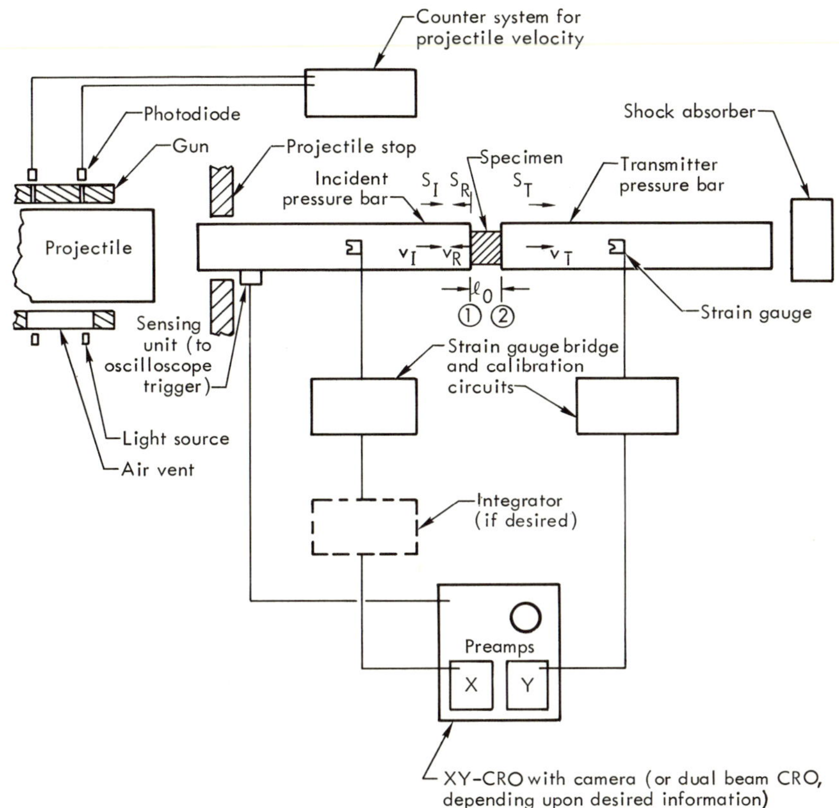

Fig. II.1. Schematic of split Hopkinson pressure bar.

reflected, and transmitted pulses depend upon the impedance properties of the specimen (neglecting internal friction and attenuation effects). The specimen length is chosen "short" so that the wave transit time within the sample is small compared with the duration of the loading pulse. Thus, numerous internal reflections occur in the specimen; the stress distribution therein rapidly equilibrates so that, at least for short samples, wave propagation effects can be safely neglected. (Additional comments on this assumption are given at the end of the section.)

The assumption is made that the measurement of the surface strain by a strain gauge represents the strain in the interior of the bar on which the gauge is positioned. Such an assumption is consistent with the condition that the stress is taken to be uniformly distributed over the cross section of the cylinder for the test situation that $R/\Lambda < 0.1$. That the strain gauges record accurately the conditions of stress within the rods is also assured with the proper calibration of the system, e.g., by use of a load cell standard in conjunction with a screw testing machine. These statements are confirmed in the work reported by Graham and Ripperger [8] and by Habberstad *et al.* [9].

We have excluded from discussion such practical considerations of strain gauging as time of response, appropriate sizes (when compared with either the lateral dimension of the bar and/or the effective wavelength of the disturbance), and similar problems. Adequate elaboration, however, is given in the references cited.

Now, having continuous records of the incident, reflected, and transmitted pulses (denoted S_I, S_R, and S_T, respectively), the stress-strain-strain rate behavior averaged in the specimen can be obtained from the force and displacement boundary conditions at the sample faces and from the uniaxial stress theory of elastic wave propagation.

Several essentially equivalent methods are available to derive the governing equations. We follow the development suggested by Hauser [6] and Hoge [10]. One can refer to Lindholm [11,12] for another development that uses strains rather than stresses.

At interface 1 in Fig. II.1, there is continuity of motion and force between the incident bar and the specimen. Assuming the specimen to have an acoustic impedance less than that of the two pressure bars, the reflected wave is a tension pulse, and we can write

$$v_{s_1} = v_I + v_R$$

and   (II-2.2)

$$S_{s_1} A_s = (S_I - S_R) A$$

where v_{s_1} is the particle velocity in the specimen at interface 1, v_I and v_R are the particle velocities associated with the incident and reflected waves, respectively, and A_s and A are the cross-sectional areas of the specimen and bars, respectively. From Eq. (I-4.65), $S_{zz} = \rho\, c_0(\partial u_z/\partial t) = \rho c_0\, v_z$, and we have

$$v_{s_1} = (S_I + S_R)/\rho c_0$$

and   (II-2.3)

$$S_{s_1} = \frac{A}{A_s}(S_I - S_R)$$

where ρc_0 refers to the impedance of the pressure bars. At interface 2, similar reasoning leads to

$$v_{s_2} = S_T/\rho c_0$$

and (II-2.4)

$$S_{s_2} = \frac{A}{A_s} S_T.$$

By averaging conditions at the interfaces, average specimen stress and strain rates can be obtained. Thus,

$$S_s = \frac{A}{2A_s}(S_I - S_R + S_T) \tag{II-2.5}$$

and

$$\dot{\epsilon}_s = \left(v_{s_1} - v_{s_2}\right)/\ell_0$$

$$= (S_I + S_R - S_T)/\rho c_0 \ell_0. \tag{II-2.6}$$

Strain at any time t can now be determined by integrating the strain rate from 0 to t, namely

$$\epsilon_s = \int_0^t [(S_I + S_R - S_T)/\rho c_0 \ell_0]\, dt. \tag{II-2.7}$$

Details of an analysis of a split Hopkinson pressure bar experiment are given in the next section.

This experimental method has been generally accepted as providing valid data, although assumptions that are required with respect to the consideration of inertial effects and uniformity along the specimen have received criticism (Conn [13]; Bell [14]). A few comments are thus in order with regard to certain experimental precautions that must be observed to assure the results obtained are, in fact, valid (see also Rand and Jackson [15]; Wasley *et al.* [16]; and Jahsman [17]). In particular, uncertainty arises in calculations at times immediately after the specimen is loaded because of the nonuniform stress and strain conditions existing throughout the sample. The time to reach equilibrium conditions is primarily a function of the modulus-to-density ratio, *i.e.*, its wave velocity. One must therefore be careful in obtaining elastic modulus data. Thus uncertainty in early time information may be reduced or removed by averaging the calculated stresses on each face of the specimen at every instant of time and by choosing small length-to-diameter ratios. Use of quartz crystal pressure transducer instrumentation in addition to strain gauges is particularly helpful in this regard (see below).

It is necessary to interject a comment here regarding terminology. Strictly speaking, an elastic modulus is a physical constant of the material. In fact, however, the modulus for many materials, particularly nonmetals, is a function of strain rate in the "elastic" regime, and thus the term *stiffness parameter* is sometimes used. Moreover, in practice, the stress-strain curve may not be precisely linear in this regime, and distinction must often be made between the *secant* elastic modulus and the *tangent* elastic modulus. The former is measured by the slope of the chord to the stress-strain curve at any given stress; the latter is measured by the slope of the tangent to the stress-strain curve at any given stress. The two coincide if the material is perfectly linear-elastic.

Another factor that contributes to the uncertainty of calculations is the effect of the lateral inertia of the impacted specimen. Radial stress is generated, creating a biaxial stress condition (see the papers by Malkov [18] and Baganoff [19] for discussion), and some unknown increase in longitudinal stress results. Inertial effects can be shown to be relatively unimportant, provided reasonable length-to-diameter ratios are chosen for the specimen and strain rates are held relatively constant. These conditions are not difficult to achieve in most experiments. Hollow bars and/or samples are often used for such purposes.

A third factor results from friction between the loaded specimen ends and bar faces. These restraints, if present, would create an undesired biaxial or triaxial stress state which changes along the specimen length and would be expected both to raise the magnitude of the measured stress over the stress which would occur if the material were unrestrained and to cause the sample to deform in a "barreling" mode. For samples of practical proportions, friction effects are not serious insofar as the stress-strain results are concerned — say less than 5% — but are relatively important in causing differences in shape during deformation. In any event, a thin layer of lubricant, *e.g.*, molybdenum disulfide, is generally placed on the loading interfaces to minimize this problem. Note that in this case, and in distinction to above, we cannot choose a length-to-diameter ratio that is too small or else end frictional effects become proportionally to large, and in the limit we approach a condition of one-dimensional strain. Thus, in practice we are required to achieve a balance; it is usually arrived at experimentally. As a first approximation, we can try $\ell/d \approx 0.5$ for solid specimens.

The split Hopkinson pressure bar suffers from the inherent theoretical complication that for $R/\Lambda > 0.1$, (fundamental) longitudinal elastic waves propagated in the cylinders show dispersion, causing elastic stress pulses to distort as they travel. Thus, one must be careful in applying elementary elastic bar theory in certain cases (although, in practice, the conditions of satisfactory application are not difficult to meet — see Lindholm [12]). We have already mentioned (Chapter I-4, Section II-A) the possibility of using the fact that no dispersion occurs for the fundamental in torsional waves in a circular cylinder in developing a technique to measure transient pressures. Some early experimental work in this area was reported by Davies and Owen [20]. They observed the angular displacement caused by torsional waves of a small, optically polished flat on the end surface of the bar by using a high-speed streaking camera. Recent effort has been reported by Phillips [21] in which torsional waves with very short risetimes were introduced in the main loading rod of a torsional split Hopkinson bar by means of the simultaneous detonation of two explosive charges at the ends of long "pre-load" bars in contact with the main loading rod. The torsional pulse is monitored by use of (semiconductor) strain gauge rosettes suitably positioned. Such a technique also avoids many of the problems associated with radial inertia and barreling effects. A somewhat similar experimental arrangement reported by Duffy *et al.* [22] gives results on 1100-0 aluminum.

Several other refinements have been added to the split Hopkinson pressure bar technique that extend its usefulness in special situations and make its use more relevant to present material study. For example, the split-bar instrumented with both a set of piezoelectric x-cut quartz crystal pressure transducers (operating in the integrated charge mode) and a set of strain gauge transducers, was developed to enable investigation of the stress-strain rate behavior of brittle materials (*e.g.*, certain rocks) and/or low-strength materials (*e.g.*, plastic foams). Details are given in Wasley *et al.* [16] and in Hoge and Wasley [23].

A fast-rise pressure bar has been developed (Jones [24]) in which the bar is constructed of beryllium and the stress is sensed using a polarized ferroelectric disc. It is found that submicrosecond risetimes can be obtained, as opposed to risetimes in the standard split-bar with strain gauge instrumentation of several microseconds.

A method using a special furnace has been invented to determine stress-strain curves at various strain rates up to $550°C$ (Chiddister and Malvern [25]). Special lubricants are required at the higher temperatures.

C. Application of the Split Hopkinson Pressure Bar Technique to a Real Material

In experimental work involving mechanical loading of high explosives, it is often desirable for safety reasons to substitute for the reactive solid a material that simulates its quasistatic and dynamic mechanical response. One such material has a composition 70.5 wt% cyanuric acid, 14.5 wt% barium nitrate, and 15 wt% polytetrafluoroethylene, a thermoplastic resin. The initial nominal weight density[2] is 1.86 g/cm^3 = 0.0675 lb/in.^{3}. This solid can be broadly characterized as a strain-rate-independent, highly filled polymer; it exhibits relatively ductile behavior.

A normal example of a dual-beam oscilloscope trace of a split Hopkinson pressure bar of the above highly filled polymer is shown in Fig. II.2. The values of S_I(= 11,100 psi), S_R, and S_T are indicated on the figure. The specimen length is 0.2 in.; its diameter is 0.5 in. (nominal). The projectile velocity was measured at 232 in./sec.

The first step in the data reduction process is to check approximately the value of the input stress S_I measured by the strain gauge. This check is accomplished using Eq. (I-4.65), with the knowledge that cross-sectional area, and hence the impedance, of the projectile is considerably greater than that of the incident pressure bar. (Recall that in the situation of uniaxial stress, impedance strictly also includes the area.) Thus, we write

$$S_{zz} = S_I = \rho c_0 v_z$$

$$\approx 0.9\, \rho c_0 v_{proj} \tag{II-2.8}$$

where ρ and c_0 of the aluminum incident pressure bar are 0.000259 lb sec^2/in.^{4} and 2×10^5 in./sec, respectively, and the factor 0.9 (in this test) accounts for the fact that although the impedance of the projectile is much greater than that of the input bar, it is not infinitely greater. Substituting the appropriate values

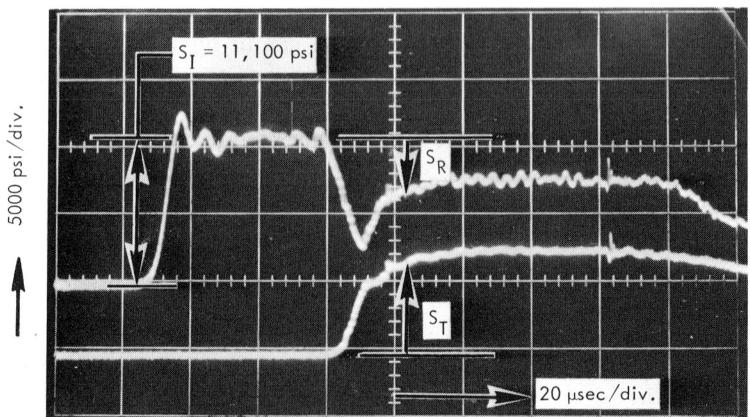

Fig. II.2. Typical oscilloscope record of a split Hopkinson bar test of a highly filled polymer.

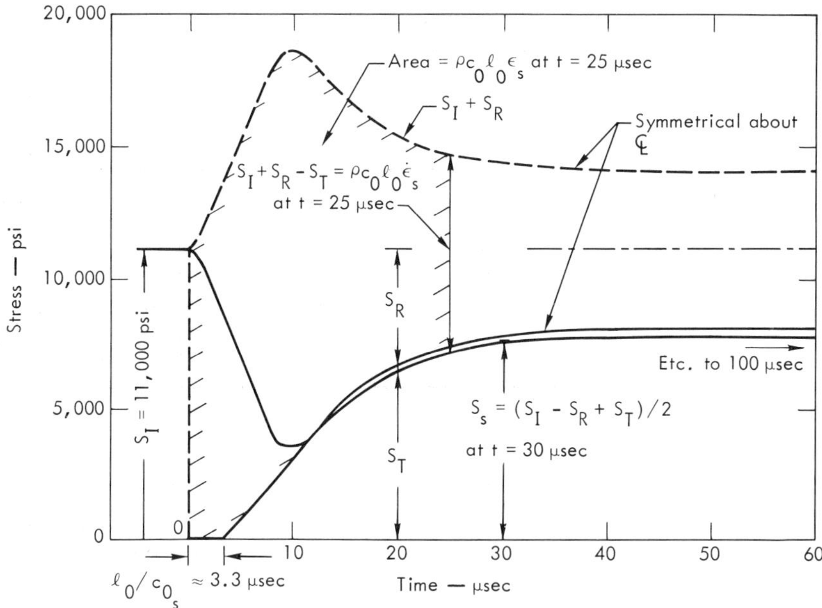

Fig. II.3. Data reduction scheme for split Hopkinson bar test.

into Eq. (II-2.8), we get $S_I = 11{,}200$ psi, which confirms that the value of 11,100 psi from the strain gauge measurement is a valid reading.

The oscilloscope traces are then replotted on a suitable piece of graph paper. This process is illustrated essentially to scale in Fig. II.3. Note that the transmitted wave S_T is positioned in time so that its start is later than the start of the reflected wave S_R by an amount equal to the specimen transit time, ℓ_0/c_{0_s}, where c_{0_s} is the rod velocity in the sample. The precise placement of these two curves is not critical; this is fortunate since c_{0_s} is not usually well known. The rod velocity for the material under study is about 60,000 in./sec.

Graphical techniques are now employed in applying Eqs. (II-2.6), (II-2.7), and (II-2.8) to obtain the stress-strain-strain rate curve. A curve of $(S_I + S_R)$ is drawn (shown dashed in the figure). The ordinate $(S_I + S_R - S_T)$, which is proportional to the average specimen strain rate $\dot{\epsilon}_s$, can then be scaled at any time t. The area that this ordinate develops from time 0 to time t can be determined using a planimeter or by counting squares (the latter technique was used in this example). The value obtained is proportional to the average specimen strain ϵ_s at time t. The average stress in the specimen S_s at time t is scaled by graphically averaging the two curves $(S_I - S_R)$ and S_T.

The results are given in Table II.1. The column headings are self explanatory.

Table II.1
Results of split Hopkinson bar test.

μsec	$S_I + S_R - S_T$ (psi)	$\dot{\epsilon}$ (sec^{-1})	No. squares	ϵ_s (%)	S_s (psi)
10	15,600	1,510	54	1.31	3,300
15	11,900	1,150	81	1.97	5,200
20	9,100	880	102	2.47	6,500
30	7,400	715	133	3.22	7,500
40	6,700	650	161	3.90	7,900
50	6,400	620	186	4.50	7,900
60	6,300	610	211	5.11	7,900

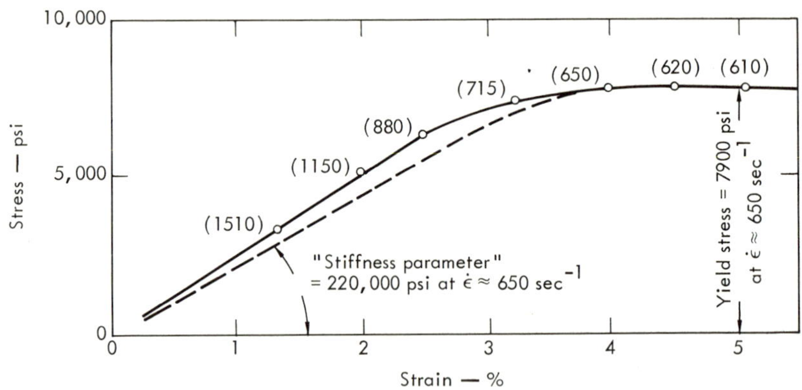

Fig. II.4. Stress-strain-strain rate curve. Numbers in parentheses indicate strain rates. The dashed line gives estimate of stress-strain curve at constant strain rate. The ultrasonically determined "stiffness parameter" = 2.0 × 10^6 psi.

The quantities $\rho_0 c_0 \ell_0$ and the area of one square in Fig. II.3 are 10.33 lb sec/in.^{2} and 2500 lb/in.^{2} (μsec), respectively. Thus the percent specimen strain at time t can be calculated as 0.0242 times the number of squares at that time t.

The stress-strain-strain rate curve is plotted in Fig. II.4. Since the total strain rate is not constant in this dynamic test, a correction must be made to the curve to obtain the usual (constant strain rate) stress-strain relation. A cross-plotting routine (Karnes and Ripperger [26]) can be used to generate such curves from the experimental information. The technique is first to draw stress-strain rate curves at a constant strain from the original information and then, using these results, construct stress-strain curves at constant strain rates. One must use care in applying this technique to a strongly strain-rate-dependent material, however. Since it is not possible to use this method effectively with only one experimentally determined curve, for purposes of this example we simply estimate the desired stress-strain-constant strain rate relation with a dashed line in Fig. II.4. Complete results for this material at a variety of strain rates can be found in Wasley *et al.* [16]. It is shown therein that the modulus of elasticity E (or stiffness parameter as suggested in Section I-B) obtained ultrasonically forms the upper end of a "fan" of moduli determined with the same material at lower strain rates and higher strain levels with the split Hopkinson pressure bar.

II. Summary

Methods are examined for the measurement of stress waves under conditions of one-dimensional stress.

The first significant contribution toward measuring transient stresses was advanced by B. Hopkinson in 1914. His apparatus, which has become known as the Hopkinson pressure bar, is an application of the elementary theory of uniaxial stress propagation of elastic disturbances in a cylinder where the wavelength of the disturbance is great compared with the lateral dimensions.

Hopkinson's apparatus is primarily of historical interest now. However, it did serve as the motivation and foundation for many subsequent designs that enabled a more complete characterization of the behavior to be obtained. A relatively recent improvement now in wide use was first basically suggested by H. Kolsky in 1949. This technique, or slight modification thereof, is the approach in use for nearly all mechanical dynamic stress-strain-strain rate studies under conditions of uniaxial stress loading conducted today. A short specimen is sandwiched between two pressure bars and is loaded by a single pulse (usually compression) traveling through the system – and hence the name, split Hopkinson pressure bar. The pressure bars act to apply the load to the specimen and are instrumented with transducers (*e.g.*, strain gauges) so as to obtain continuous values of stresses or displacements at the faces of the specimen. Such an arrangement allows the determination, averaged in the sample, of stress versus strain, with strain rate as a parameter, of a constant-amplitude loading pulse of the order of 100 μsec duration. Strain rates as high as 10^4 sec^{-1} can be achieved.

An important feature of the split pressure bar is that the specimen itself need not remain elastic as is required for the incident and transmitter bars. Thus, nonelastic stress-strain-strain rate behavior is also obtained.

An application of the split-Hopkinson pressure bar technique to a real material is given. The material simulates the quasistatic and dynamic mechanical behavior of certain filled polymeric high explosives.

Notes

[1]For most observations, except perhaps those at very low input stress levels, the small adhesion introduced because of the grease need not be considered.

[2]To preserve consistency with other investigators, we use the foot-pound-second system of units that is normally employed in Hopkinson pressure bar work.

References

[1] B. Hopkinson, *Roy. Soc. Phil. Trans.*, **A213**, 437 (1914).

[2] R. M. Davies, *Roy. Soc. Phil. Trans.*, **A240**, 375 (1948).

[3] J. S. Rinehart, *On Fractures Caused by Explosions and Impacts*, Chap. 3, Quarterly of the Colorado School of Mines, Golden, Colo., 1960.

[4] H. Kolsky, *Proc. Phys. Soc. London*, **62B**, 676 (1949).

[5] J. M. Krafft, A. M. Sullivan and C. F. Tipper, *Roy. Soc. Phil. Trans.*, **A221**, 114 (1954).

[6] F. E. Hauser, *Exptl. Mech.*, **6**, 395 (1966).

[7] K. G. Hoge, *Explosivstoffe*, **2**, 39 (1970).

[8] R. A. Graham and E. A. Ripperger, "A Comparison of Surface Strains to Average Strains in Longitudinal Elastic Wave Propagation," in *Proc. Fourth Midwestern Conf. on Solid Mechanics*, University of Texas, Austin, 1959.

[9] J. L. Habberstad, K. G. Hoge and J. E. Foster, *An Experimental and Numerical Study of Elastic Strain Waves on the Center Line of a 6061-T6 Aluminum Bar*, University of California Lawrence Livermore Laboratory, Report UCRL-72717, 1970.

[10] K. G. Hoge, "The Behavior of Plastic-Bonded Explosives Under Dynamic Compressive Loads," in *High-Speed Testing*, Vol. VI, Interscience, New York, 1967.

[11] U. S. Lindholm, *J. Mech. Phys. Solids*, **12**, 317 (1964).

[12] U. S. Lindholm, "High Strain Rate Tests," in *Measurement of Mechanical Properties*, Vol. 5, Part I in series, *Techniques of Metals Research*, (R. F. Bunshah, ed.), Interscience, New York, 1971.

[13] A. F. Conn, *J. Mech. Phys. Solids*, **13**, 311 (1965).

[14] J. F. Bell, *J. Mech. Phys. Solids*, **14**, 309 (1966).

[15] J. L. Rand and J. W. Jackson, "The Split Hopkinson Pressure Bar," in *Behavior of Dense Media under High Dynamic Pressures*, IUTAM, Paris, 1967; Gordon and Breach, New York, 1968.

[16] R. J. Wasley, K. G. Hoge and J. C. Cast, *Rev. Sci. Instr.*, **40**, 889 (1969).

[17] W. E. Jahsman, *J. Appl. Mech.*, **38**, 75 (1971).

[18] M. A. Malkov, *Soviet Phys. – Doklady*, **8**, 209 (1963).

[19] D. Baganoff, "Pressure Gauge for Shock Reflection Studies," in *Proc. 5th Intern. Shock Tube Symp.*, U. S. Naval Ordnance Laboratory, White Oak, Md., 1965.

[20] R. M. Davies and J. D. Owen, *Proc. Roy. Soc.*, **A204**, 17 (1950).

[21] J. W. Phillips, *A Method for Determining Material Properties at High Rates of Shearing Strain*, Brown University, Division of Applied Mathematics, Technical Report No. 9, 1969.

[22] J. Duffy, J. D. Campbell and R. H. Hawley, *J. Appl. Mech.*, 38, 83 (1971).

[23] K. G. Hoge and R. J. Wasley, "Dynamic Compressive Behavior of Various Materials," in High Speed Testing, Vol. VI, Interscience, New York, 1969.

[24] I. R. Jones, *Rev. Sci. Instr.*, 37, 1059 (1966).

[25] J. L. Chiddister and L. E. Malvern, *Exptl. Mech.*, 3, 81 (1963).

[26] C. H. Karnes and E. A. Ripperger, *J. Mech. Phys. Solids*, **14**, 75 (1966).

NONELASTIC (SHOCK) ONE-DIMENSIONAL STRAIN WAVE INVESTIGATIONS

I. General

Discussion of the concepts, definitions, and theory of shock-wave phenomena is given in Section II, presented according to the natural order of events involved first in shock-wave loading and then in unloading. The dynamic behavior is "mapped" in the stress-strain plane, and the various curves are interrelated. The topics considered are only a portion, although a basic portion, of the work in shock-wave physics. The response of an organic, polycrystalline solid is examined that serves as an example of the application of many of the points treated. Discussion of other specialized interests — for example, magnetic and electrical effects, phase transitions, and polymerization processes — can be found in the references.

In Section III we examine some of the more recent experimental techniques that are used in measuring shock-wave parameters, and apparent trends in instrumentation and methods are indicated. At present, this information is only partially collected in various symposia proceedings, most of it being rather widely scattered in the literature. The possibility that such a presentation will quickly become dated in this rapidly developing field is reduced by focusing comment upon the purposes and intentions of the techniques presented, rather than only upon their engineering and physical aspects.

II. Analytical Discussion; Concepts Applied to Behavior of a Real Material

A definition of a shock disturbance in a solid appropriate to our considerations takes such a disturbance as a rapid, almost discontinuous, finite increase in stress, density, particle velocity, temperature, and entropy that propagates at supersonic velocity in the material. This statement does not exclude investigation of a multiple wave structure, although it must be modified when we consider an elastic precursor in which the disturbance velocity is usually sonic and in which density, temperature, and entropy changes are generally negligible. The definition does not require the disturbance profile to be steady (in a moving frame of reference), although it is for most considerations.

We do not examine very strong shock waves (say >10 Mbar[1]) in which the solid may melt or vaporize, the molecules may disintegrate, and even free electrons may exist (see Alder [1]). Under these conditions, more of the total work done by the shock then goes into temperature rise of the material rather than into its volume decrease. This temperature rise has associated with it a pressure increase, appropriately called a "thermal" pressure. This pressure is in contrast to the "elastic" or "cold" pressure caused by the repulsive forces among the atoms arising from the compression. Under these higher pressures, the Thomas-Fermi-Dirac theory (or its various modifications) is usually adequate. For the conditions of several megabars, the "cold" and "thermal" pressures are of comparable magnitude. For the conditions of several hundred kilobars and below, the "cold" or "elastic" pressure is dominant, and more of the internal energy acquired by the medium is concentrated in the form of potential energy. Although the higher pressure regime is of great importance in today's technology, we are more concerned with lower stress levels in which material strengths and rigidity behavior must be considered, and a hydrodynamic model (see Chapter II-1, Section III) is usually not adequate for satisfactory description. Analysis can then often be made with the help of an elastic-plastic model (discussed below) to account for the anisotropic stress configuration.

Figure II.5 shows an idealized compressive wave structure in the stress-time plane that can be observed (with varying approximations to this idealization) in a variety of solids. This figure shows an elastic precursor moving ahead of the main compression wave. The amplitude of the precursor, if one does indeed form in the solid, is designated by Point 1 and is called the *Hugoniot elastic limit (HEL)*. It is generally a constant, and depends upon the dynamic shear and yield strength properties of the material in which the shock disturbance is traveling. The velocity of the precursor is given by the longitudinal elastic wave velocity, Eq. (I-2.10). These are the usual conditions, although it is shown in the example below that neither the shape of the precursor is as well defined as illustrated in the figure nor is its velocity exactly predicted by Eq. (I-2.10). Moreover, the amplitude of the elastic precursor may occasionally be a function of driving stress, *i.e.*, a strain-rate effect can also be observed in the "elastic" regime (also shown below). The risetimes of such elastic precursor compressive waves (after compensation for the shock tilt and the recording system is made) depend upon the material being studied, but for many metals, times on the order of 10^{-8} sec can be expected.

Following this elastic precursor there is the main compression wave, appropriately called a *plastic follower*, in which the stress rises to its final or *driving* value (Point 2). Similar to the precursor wave, the plastic follower is not necessarily as well defined as shown in the idealized figure. Its risetime depends upon the material being studied and also upon the driving stress; times of the order of 10^{-7} sec can be expected for many metals. Since the propagation velocity of the follower increases for increasing driving stress, an elastic precursor arising from material shear failure will not form if the final stress level exceeds a critical value in a given medium.

The shock-loading structure in which elastic-plastic effects exist has been examined in a number of materials. An example of a material in which such a study has been made is the high explosive, pressed TNT (Wasley and Walker [2]). The initial nominal density is 1.648 g/cm^3. This material is a brittle solid and can be broadly classified as being strain-rate-dependent, organic and polycrystalline.

Experimental procedures are discussed more thoroughly in Section III, but basically a gun was used to provide the planar, symmetrical (usually) impacts. Tests were conducted at an atmospheric pressure of about 10^{-2} Torr. The instrumentation included polarized ferroelectric pin probes and quartz crystal pressure transducers. The crystal pressure transducers were attached to the rear surface of the TNT samples and were used to record the voltage- (and hence stress-) time profiles and the structure of the wave emerging from the impacted

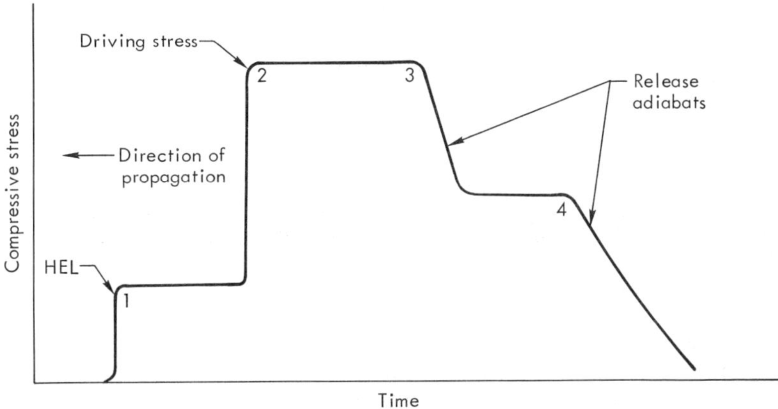

Fig. II.5. Idealized loading and unloading wave structure.

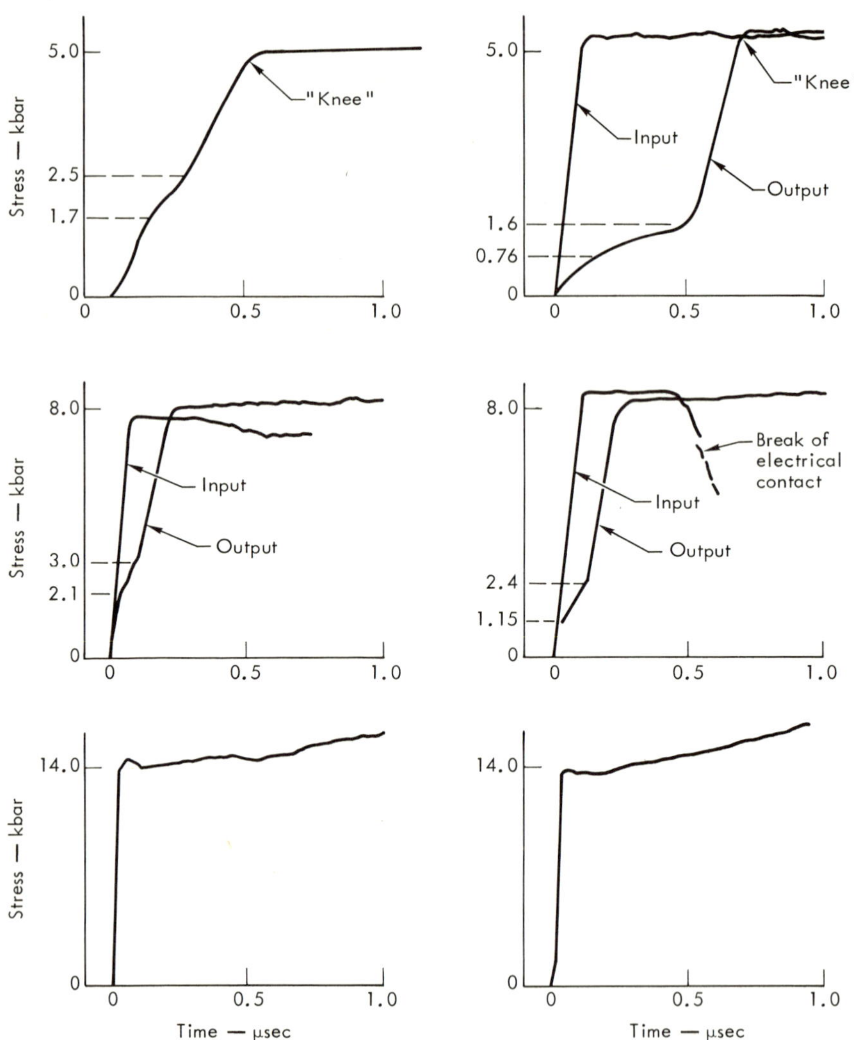

Fig. II.6. Selected transducer traces of input and output uniaxial strain disturbances for pressed TNT. (a) sample thickness, 6.3 mm (nominal); (b) sample thickness, 19.1 mm (nominal).

sample. In several experiments, transducers mounted on the projectile also monitored the profile of the incident disturbance.

Preliminary reduction of the transducer data was necessary to determine the stress-time history in the crystal from the voltage-time information recorded

(Graham *et al.* [3]). The stress-time history in the TNT sample was determined from the impedance mismatch between it and the gauge (Jones *et al.* [4]). Small time corrections were applied to the records to compensate for the projectile tilt (~0.0004 rad) and for the warping of the lapped TNT surfaces caused by changes in temperature during a test.

Under all test conditions reported here, there was no observed release of chemical energy during impact time periods, and the material was treated as macroscopically inert. The subject of the shock initiation of the explosive is considered separately (Walker and Wasley [5]).

Selected traces of the stress-time profiles are illustrated in Fig. II.6. Several interesting inferences can be drawn from the information shown in the figure. First with regard to the various input disturbances observed, there is apparently a rate-limiting mechanism governing their risetimes. Using data from the several experiments instrumented to record such disturbances, no correlation of risetime with any obvious test parameter, *e.g.*, driving stress, was observed; rather, uniform risetimes of the order of 0.1 μsec were found in these experiments after corrections were made for impact tilt. This risetime is considerably larger than that inherent in the crystal and the associated electronic measuring apparatus.

Second, an "elastic precursor" wave forms within the TNT specimens, as noted in several of the monitored emerging stress waves. It is followed by a "plastic follower" wave. The elastic precursor will not form in this organic material above an impact stress level of about 10 kbar. At these higher stresses, the follower wave propagates at a velocity greater than that of the precursor.

The general structure of the elastic-plastic disturbance is not one that can be associated with the more classical observations of elastic-plastic wave formation, as has been noted above in connection with Fig. II.5. The length, shape, magnitude, and risetime of the precursor and the risetime of the follower appear to be functions of both sample thickness and driving stress. Similar traces are found in recent work with metals, *e.g.*, Taylor [6] and Graham *et al.* [7]. Details of the results of the investigation of the shock-loading structure for pressed TNT are contained in Wasley and Walker [2].

A multiple compressive shock structure in the medium can also form if electronic of first- or second-order phase transitions occur. In general, complications are introduced if the material exhibits significant strain-rate effects, work-hardening behavior, and stress-relaxation characteristics. The latter situation is often evident in the early formation stages of an elastic precursor and can be understood using a model based on (microscopic) dislocation mechanics; recent contributions are given by Jones and Mote [8] and by Gillis *et al.* [9].

The shock loading information given in Fig. II.5 can also be displayed and examined in the stress- (or pressure-) strain (or relative volume) plane. Figure II.7 is a schematic illustration of the general shapes and relative positions

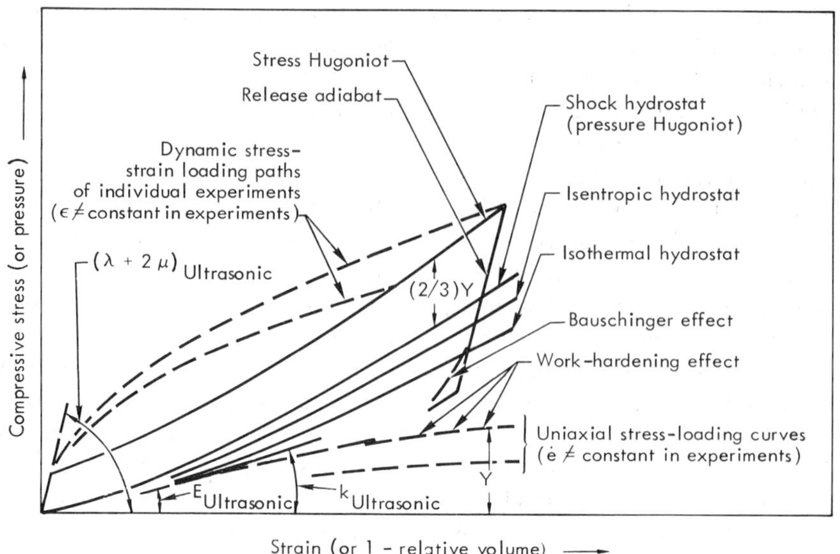

Fig. II.7. Idealized loading and unloading curves.

of various curves for a typical solid in this plane. The relative differences among
the curves have been exaggerated for clarity.

The average stress-strain paths followed at a given location in the solid under
study during various uniaxial strain experiments can be deduced from stress-time
profiles of the type illustrated in Fig. II.5. One analysis procedure (see Wasley
and Walker [10]) for generating these compression curves involves approxi-
mating the continuous stress-time record by a series of small arbitrary stress
jumps. Each jump in the approximation can be attributed to the arrival of a
small shock and treated according to the Rankine-Hugoniot conservation
relations (discussed below), enabling the stepwise calculation of the normal
stress σ versus specific volume V behavior. The strain ϵ can then be obtained
from the specific volume calculations, using a form of the *natural* definition of
strain,

$$\epsilon_n = \sum_{j=1}^{n} (V_{j-1} - V_j)/V_j, \tag{II-3.1}$$

in which the index n refers to the nth stress jump in the above approximation.
Such a definition, as opposed to the *engineering* definition (see Chapter I-2,

Section II) makes it possible to consider large strains in a satisfactory and practical fashion. The total strain rates taken to be associated with these strains are defined and calculated in the conventional manner, *i.e.*, $\dot{\epsilon} \approx \Delta\epsilon/\Delta t$. The elastic and plastic strain rates in a test are not constant as the material is loaded. As the disturbance passes a given position in the specimen, the strain rate increases to a maximum and then decreases as the peak stress is reached.

For some materials, this approach in itself is sufficient to adequately establish the curves. However, because of strain rate effects and complex wave interactions which may exist (the latter depending partly upon the measuring system used), iteration processes using certain numerical methods are sometimes also used (*e.g.*, see Barker *et al.* [11]; Munson and Barker [12]; or Guess [13].

The *stress Hugoniot* in the shock stress-strain plane is defined as the locus of (macroscopic) final, *i.e.*, equilibrium, stress-strain points from a series of experiments that can be reached by the shock loading of the material. It is called a *principal Hugoniot* if the material is loaded from some standard or normal (although arbitrary) ambient conditions of temperature, pressure, and relative density. Note that the Hugoniot is distinctly different from the dynamic rate-dependent, stress-strain curves discussed above. Neither is it a complete stress-volume-energy thermodynamic equation of state, *per se*, although it does represent a locus of states on an equation-of-state surface for a material. This latter situation can be seen in Fig. II.8 in which a qualitative pressure-relative, volume-specific, internal-energy surface is drawn for an arbitrary material with no phase changes. Note that the shock hydrostat (defined below) crosses a series of constant energy pressure-volume curves. Two other curves are also illustrated on the figure. The "atmospheric pressure" curve is the locus of points on the equation-of-state surface obtained by heating the material without external mechanical confinement, *i.e.*, simple volume thermal expansion occurs. The "constant initial volume" curve on the surface shows the behavior of the material when it is maintained at a particular volume and heated.

The stress Hugoniot can be generated by two different (although not completely independent) methods. First, the driving (final) stress values on the various measured wave profiles (see Fig. II.5) and their associated strain values can be defined as points on the stress Hugoniot. The strain rates connected with these driving stresses are comparatively low, thus making these stress-strain points appropriate for selection as values.

Second, two of the three Rankine-Hugoniot equations can be invoked to calculate these same Hugoniot points. The two equations that we can use to determine the stress and strain values are straightforward applications of the conservation of momentum and mass across a shock front, assuming that steady-state conditions exist and that the initial and final states are in mechanical and thermodynamic equilibrium. Within the framework of these assumptions, the relations are completely general and are valid irrespective of the

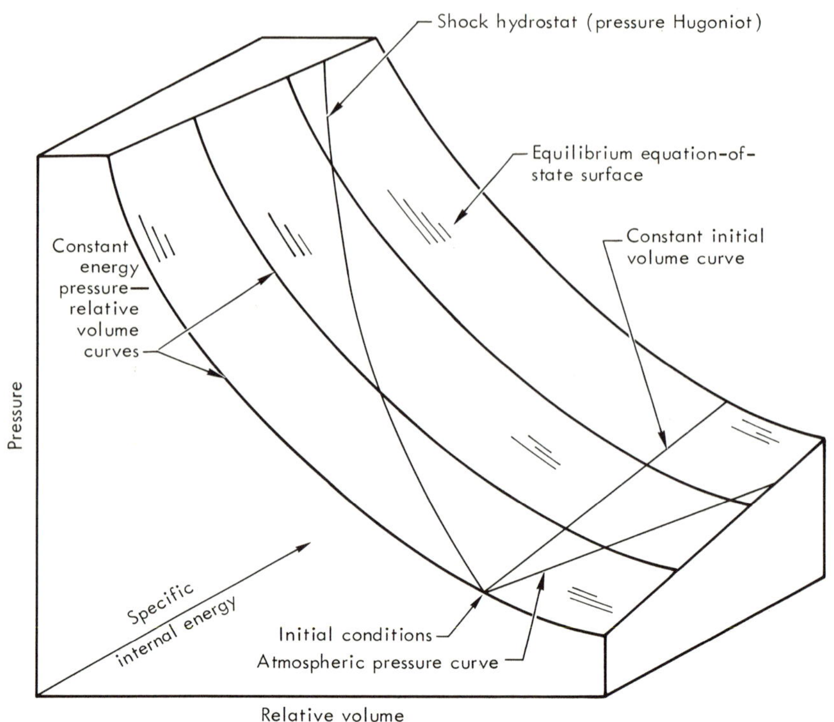

Fig. II.8. Equation-of-state surface for an arbitrary material (schematic).

medium, the thickness of the shock front, and the strength of material effects (the last being true as long as the kinetic and kinematic variables are all taken parallel to the direction of the shock motion).

Let a shock disturbance with velocity u_s move into a solid at rest. There results a sudden (although as we have indicated, not immediate) increase in stress σ, in particle velocity u_p, in density ρ $(= 1/V)$, and in specific internal energy E. Then we can write for the two equations

$$\sigma_1 - \sigma_0 = \rho_0 u_s u_p$$

and   (II-3.2)

$$\rho_0/\rho_1 = 1 - u_p/u_s$$

where the subscript 0 refers to the state immediately ahead of the shock front (in this case, an undisturbed condition) and the subscript 1 the state behind it.

If we want to consider a two (or more) shock-wave structure, such as shown in Fig. II.5, Eqs. (II-3.2) can be extended easily to the form

$$\sigma_2 - \sigma_1 = \rho_1 u_{s_{21}} u_{p_{21}}$$

and   (II-3.3)

$$\rho_2/\rho_1 = 1 - u_{p_{21}}/u_{s_{21}}$$

where the single numerical subscripts relate to absolute states ahead of and behind the shock front (as before), and the double numerical subscripts relate to the conditions considered at the first subscript location, *i.e.*, 2, when referenced to the (disturbed) conditions at the second subscript location, *i.e.*, 1.

It is apparent from the examination of expressions (II-3.2) that the dynamic measurement of only two variables enables one to generate stress-strain information (recalling that strain can be related to relative density or to relative volume values). This fact is important later when we discuss experimental techniques. It is also important for the reader's perspective to recall the following: Stress propagation in a continuum is described by the equations of motion, continuity, and conservation of energy, and by an equation of state among internal energy and stress, strain, and their derivatives (see also Chapter II-1, Section III). Expressions (II-3.2) and (II-3.3) are simply the first two of these four equations integrated across a (near) discontinuity. It is interesting to note that the hydrodynamic equations of an inviscid and nonconducting fluid do, in fact, admit the existence of mathematically discontinuous, or shock, solutions.

In general, the stress Hugoniot differs from the *shock hydrostat* or *pressure Hugoniot* (Fig. II.7). A point on the stress Hugoniot represents a state of stress composed of a deviatoric (shear) component of the stress tensor in addition to the spherical (pressure) component (see Eq. (II-1.4)), the latter being a point on the shock hydrostat. Thus, the shock hydrostat is just the Hugoniot of an ideal fluid.

Having the stress Hugoniot curve, one method to obtain the stress deviator, and hence the shock hydrostat, requires we first relate the stress Hugoniot with a particular uniaxial-strain, stress-strain curve at a constant total strain rate[2] (so that application of simple elastic-plastic theory becomes possible), and then estimate the shock hydrostat using data from an appropriate uniaxial-stress, stress-strain curve (again at a constant strain rate) and an elastic-plastic theory. The pertinent equations in the relatively simple but frequently used elastic, perfectly plastic model formulated several years ago by Wood [14] are

$$p_H = \sigma_H - (2/3)\, Y$$

and   (II-3.4)

$$\epsilon = (3/2)\, e - Y/6k.$$

In the equations, p_H is the pressure on the shock hydrostat, σ_H is the stress on the stress Hugoniot, Y is the flow stress in uniaxial stress (see Fig. II.7), ϵ and e are strains in uniaxial strain and uniaxial stress, respectively, and k is the coefficient of bulk modulus [Eq. (I-1.44)].[3] Several assumptions are required in the development of expressions (II-3.4): (1) The total strain is composed of both elastic and plastic components; (2) the Hooke's law applies to the behavior of the elastic strain; (3) no volume change occurs with the plastic strain; (4) the Tresca or Von Mises yield criteria apply; (5) equal plastic work defines equivalent conditions in uniaxial strain and uniaxial stress; and (6) k is independent of average pressure.

It is apparent that a prerequisite in the computational process is to obtain experimentally (or assume) the dynamic rheological compressive stress-strain-strain rate behavior under conditions of uniaxial stress. Techniques to generate such information have been described in Chapter II-2, Section I-B. Typical, although qualitative, curves are illustrated in Fig. II.7. The flow stress (or simple yield stress) Y is shown in the figure for an arbitrary curve and at an arbitrary strain.

Much of the work to date using this elastic-plastic model has been confined to the study of metals using certain additional assumptions: Strains are "small"; uniaxial stress behavior is elastic, perfectly-plastic; and the material is strain-rate-independent. However, it is noted that a satisfactory degree of approximation can be achieved up to about 4% or 5% strain and with rate-dependent materials if in both stress states comparable constant rates of straining are used (see the following example with pressed TNT). Work-hardening effects that affect the material uniaxial stress yield strength Y (so that the mechanical response is not perfectly plastic) and that can arise because of the dependence of Y upon the hydrostatic pressure, the temperature, and/or the plastic strain history, can also be introduced into this theory (see Butcher and Karnes [15]).

There are other analyses that can be used in addition to the simple approach discussed above. For example, an elastic-hydrodynamic model has been used (Doran and Linde [16]) with some success. Such a model requires a Hugoniot transition region directly from elastic to hydrodynamic behavior. A general continuum analytical theory of elastic-plastic deformation has been developed (Lee and Liu [17]) that includes both consideration of finite elastic and plastic strain components and of change in temperature due to thermomechanical coupling effects. Similar analytical effort (Clifton [18]) has included time-dependent behavior in the plastic (finite) strain regime. Numerical computer techniques are available (Wilkins [19] and Herrmann *et al.* [20]) to calculate the elastic-viscous-plastic shock response of materials; computational reproduction of experimentally determined wave profiles in aluminum similar to Fig. II.5 is satisfactory.

The shock hydrostat differs from the *isentropic hydrostat* or *isentrope* (see Fig. II.7) in that there is a thermodynamic effect of an increase in entropy in the shock front which causes the material to be at a higher temperature and pressure at a given relative density on the shock hydrostat than on the isentrope.

One method to correct the shock hydrostat to isentropic conditions involves the determination of the increase in specific internal energy E_H to various points on the shock hydrostat by integrating $p_H dV$ along this curve to the points in question. The subscript H refers to conditions on the shock hydrostat. The Mie-Grüneisen equation of state (see Rice *et al.* [21]), which is applicable to states of hydrostatic pressure, can then be used to construct the isentropic hydrostats. The form of the Mie-Grüneisen equation is

$$p_S - p_H = (\gamma/V)(E_S - E_H) \tag{II-3.5}$$

where γ is the Mie-Grüneisen parameter and the subscript S refers to conditions on the isentropic hydrostat. All the quantities in Eq. (II-3.5) are related to the same specific volume. The Mie-Grüneisen parameter as a function of volume has been determined semi-empirically or estimated for a number of materials (*e.g.*, see McQueen and Marsh [22] and Stephens [23] for some values and discussion). Since the entropy effect is not usually pronounced at pressures below 100 kbar,[4] one may often approximate the Mie-Grüneisen parameter by the zero-pressure value and assume it is to remain constant. Then, starting at zero reference pressure and identical internal energies and setting $E_S = \int p_S dV$, curves of p_S versus V or ρ can be generated from Eq. (II-3.5) using numerical/computer techniques.

The data can be treated further to calculate the *isothermal hydrostat* or *isotherm*, which is the stress strain path followed by loading the material under isothermal, hydrostatic conditions. The temperature along the isentrope can be found using standard thermodynamic techniques (*e.g.*, see Walsh and Christian [24] or Skidmore [25]). Then using the temperature difference between the isentrope and the isotherm, together with the heat capacity at constant volume (usually assumed constant), the energy difference between the two curves can be determined. The energy difference is used in the Mie-Grüneisen equation of state [Eq. (II-3.5) with isothermal pressure and energy p_T and E_T substituted for p_H and E_H] to obtain the pressure-volume or pressure-density curve for the isothermal hydrostat. Such calculations allow direct comparison between dynamic and quasistatic measurements of pressure-volume relationships.

There are several additional parameters shown on Fig. II.7 that require comment. The Lamé constants, λ and μ, the adiabatic coefficient of bulk modulus k, and the modulus of elasticity E (or preferably, stiffness parameter — see Chapter II-2, Section I-B) can be determined ultrasonically [see Chapter I-5, Section II-A and Eqs. (I-1.42), (I-2.10) and (I-2.11)] from measurements of the longitudinal and shear velocities of the material, assuming that conditions of isotropy and homogeneity exist for such measurements. These values can in turn be related to the slopes of three lines in the stress-strain plane, one with slope $(\lambda + 2\mu)$ which should coincide with the elastic portion of the uniaxial strain curve, one with slope k which should coincide with the initial portion of the isothermal hydrostat,[5] and one with slope E which can be considered as essentially the upper limit of a "fan" (if the material is rate-dependent) of slopes of stress-strain curves under uniaxial stress conditions, i.e., a "fan" of stiffness parameters (see the end of Chapter II-2, Section I-C).

Now, as before in the discussion of the shock-loading structure, we illustrate many of the above points with an example material. Again, we used pressed TNT (Wasley and Walker [2]). The experimental arrangement and the general test procedures used are as previously described.

It was desired first to determine stress-strain-strain rate curves for the solid under study, i.e., the loading paths taken at some location in the sample during various uniaxial strain compressions. The position chosen for investigation in these experiments was the specimen-quartz interface. These curves were developed from the TNT stress-time transducer records that were obtained in each of the experiments using the technique described earlier.

Several selected resulting stress-strain rate curves are shown in Figs. II.9 and II.10. Enlargements of the curves at lower stresses are also included for clarification. Note that natural strain is used as the abscissa. As discussed above, the total strain rate in a test was not constant as the material was loaded. This situation applies at any point in the sample, but for these experiments we are concerned with the specimen-quartz interface. The logarithms of the calculated total strain rates are indicated on the curves.

The "knees" of the various transducer records are shown in Figs. II.6, II.8 and II.9. They are represented by open circles in the latter two figures. This knee was taken to be the stress-strain value associated with a point on the Hugoniot curve. At higher driving stresses there was a sharply defined break; however, below about 5 kbar, the traces were generally curved near the driving stress and some judgment was required to determine a unique knee.

Acoustic tests were conducted to enable calculation of $(\lambda + 2\mu)$ and the coefficient of bulk modulus k. Results are shown in the figures. Different methods and several specimen geometries were used to ascertain that assump-

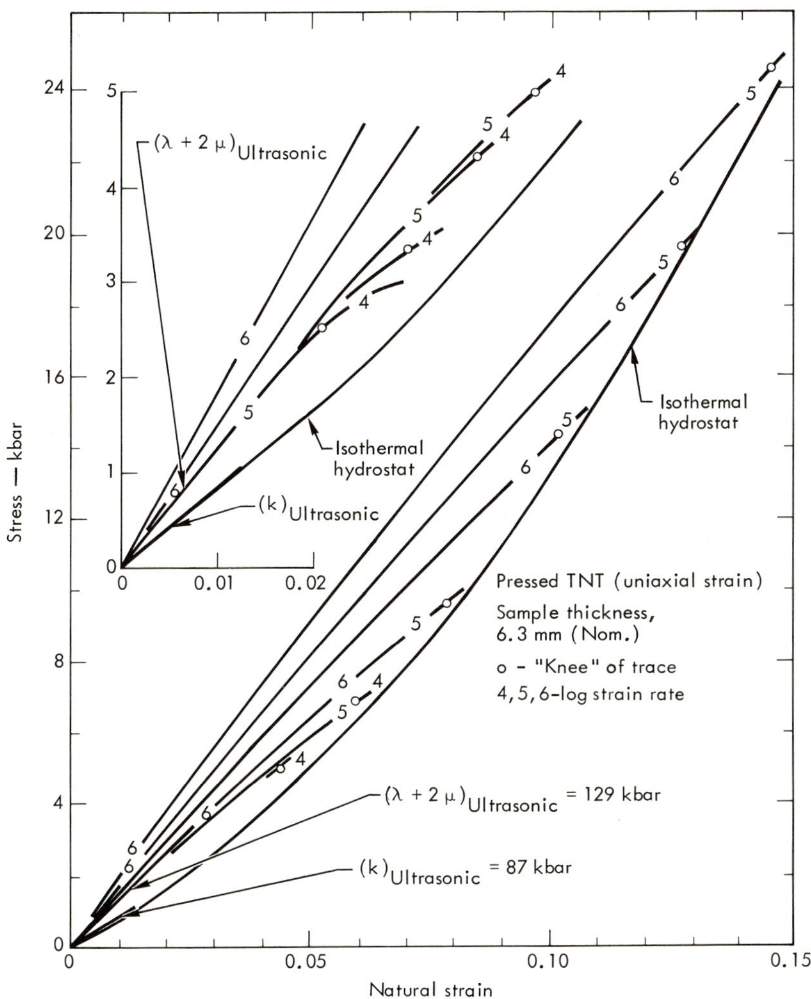

Fig. II.9. Selected stress-strain curves at variable strain rates (6.3 mm nominal).

tions of isotropy and homogeneity were valid and that conditions of one-dimensional strain and linear response existed. No effect of material rate dependence was detected under the ultrasonic test conditions used.

The experimentally determined isothermal (18°C) hydrostat for pressed TNT (ρ_0 = 1.63 g/cm^3) (Vasil'ev *et al.* [26]) is given in the figures for purposes of comparison. Note that the ultrasonically deduced bulk modulus k agrees well with the beginning slope of this curve.

It is seen that rate effects do not noticeably affect the paths of loading for the various tests in the "elastic" regime (the HEL can be taken at about 2.5 kbar) until the driving stress exceeds about 10 kbar. Although there may indeed exist some slight negative curvature of the initial portions of the stress-strain lines (refer below to the variation of elastic wave velocity in Fig. II.13), they correspond closely to the ultrasonically determined modulus $(\lambda + 2\mu)$ within the elastic regime under this driving stress limitation.

The stress Hugoniots for the two thicknesses of TNT specimens were deter-

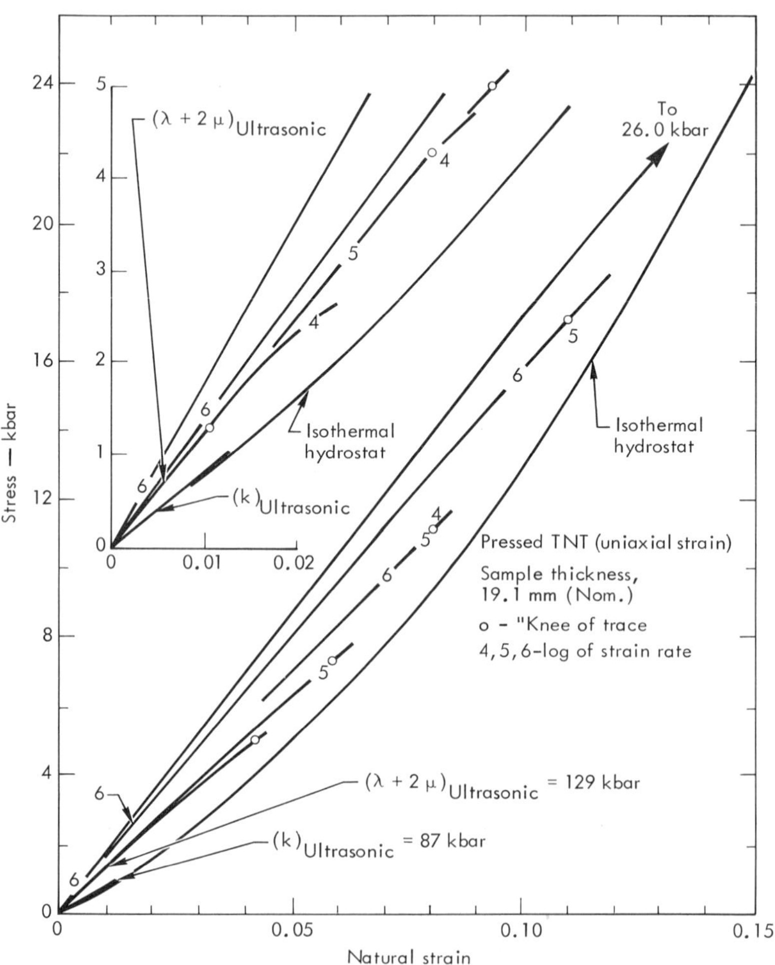

Fig. II.10. Selected stress-strain curves at variable strain rates (19.1 mm nominal).

mined using the two techniques described previously. A two-wave model [Eqs. (II-3.3)] was used in the Rankine-Hugoniot calculations. In the particular experimental work under discussion in which a gun was employed, additional expressions were required before these equations could be applied. First, the specimen particle velocity in an absolute frame of reference u_{p20} was obtained from its relationship with the projectile velocity v_{proj}, as expressed by the equation

$$u_{p20} = \frac{\left(\rho_0 u_{s20}\right)_P}{\left(\rho_0 u_{s20}\right)_P + \left(\rho_0 u_{s20}\right)_S} v_{proj} \qquad \text{(II-3.6)}$$

where $\left(\rho_0 u_{s20}\right)$ is the shock impedance and the subscripts P and S refer to the projectile and specimen, respectively. Relation (II-3.6) can be readily derived on the basis on conservation of momentum, using the assumptions of equality of normal stress and particle velocity across the interface (Wasley and O'Brien [27]). The assumption is made in the application of this equation that the particle velocity remains constant across the specimen thickness. If the impact is symmetrical, as was usually the situation in these experiments, Eq. (II-3.6) becomes

$$u_{p20} = \frac{1}{2} v_{proj}. \qquad \text{(II-3.7)}$$

The quantities u_{s21} and u_{p21} in Eqs. (II-3.3) are determined by the expressions

$$u_{s21} = u_{s20} - u_{p10}$$

and $\qquad\qquad\qquad\qquad\qquad\qquad\qquad\qquad\qquad\qquad\qquad\qquad\qquad$ (II-3.8)

$$u_{p21} = u_{p20} - u_{p10}$$

where u_{s20} is a calculable number from the test data and u_{p10} is obtained from the quartz transducer results and the use of the first of Eqs. (II-3.2).

The use of the information contained in the gauge records in these experiments presented a problem in deciding exactly which points in time on a given trace represented the arrival of the elastic precursor (if one is present) and of the driving stress at the specimen-crystal interface. As one method of solution, the two transit time calculations (corrected for tilt) were based on the points on the trace corresponding to one-half the Hugoniot elastic limit stress (σ_1) and one-half the difference between the driving stress (σ_2) and this elastic limit.

The reduced Hugoniot data in the stress-strain plane are plotted in Figs. II.11 and II.12. Figure II.11 illustrates the Hugoniots of both 6.3- and 19.1-mm-thick samples, determined on the basis on the knees of the various quartz crystal traces. Figure II.12 shows the Hugoniots for the samples determined on the basis

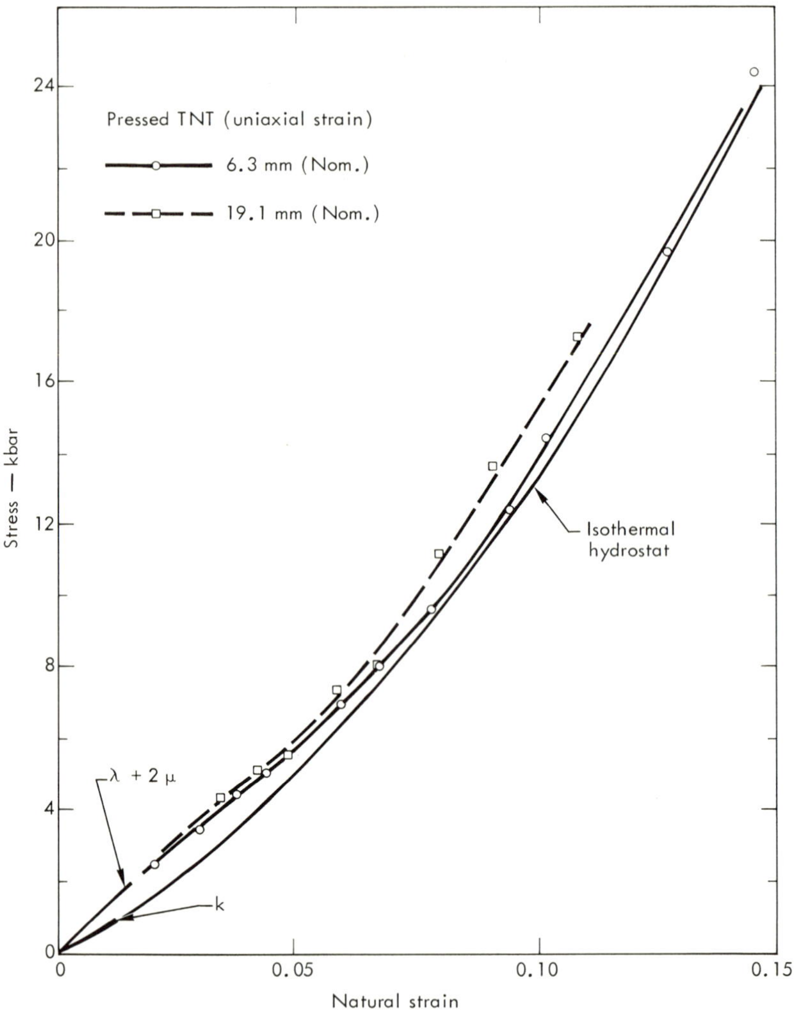

Fig. II.11. Hugoniot curves: shock stress-strain plane (based on "knee" of transducer traces).

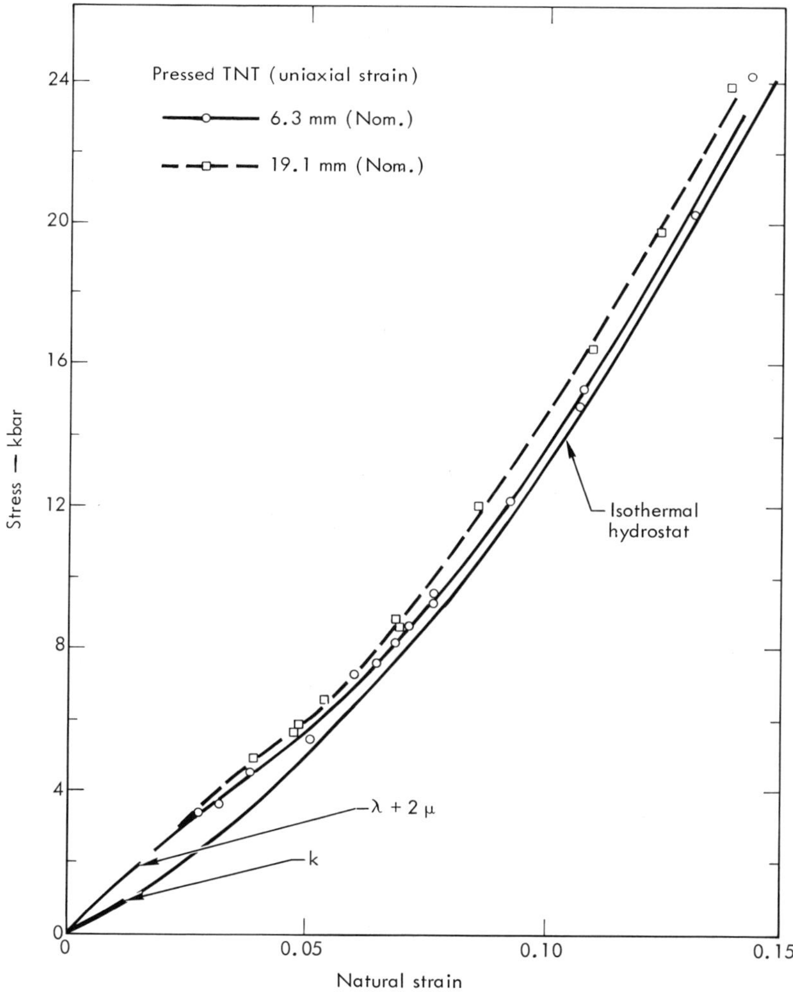

Fig. II.12. Hugoniot curves: shock stress-strain plane (based on Rankine-Hugoniot calculations).

of calculations with the Rankine-Hugoniot conservation equations. Again, the experimental isothermal (18°C) hydrostat is given in the figures for comparison.

It appears from Figs. II.11 and II.12 that the Hugoniot for TNT is some function of sample thickness and that a single curve does not describe the response of this material at stresses above the elastic regime. The thicker sample generally has a greater slope above about 3 kbar. The four Hugoniots illustrated in the two figures exhibit a point of inflection at about 6 kbar, below which there exists a range of stresses in which a negative second stress derivative exists $(\partial^2\sigma/\partial^2\epsilon < 0)$. Thus, in this range an unsteady shock wave results, so the fact that the Hugoniot varies in some fashion with specimen thickness is not surprising (Ivanov *et al.* [28]). It can be shown theoretically (Cowperthwaite [29]) that this instability arises because certain states do satisfy the conservation laws of mass, momentum, and energy for an irreversible shock process, but do not satisfy the second law of thermodynamics.

The Hugoniot curves are also plotted in the shock-velocity — particle-velocity plane in Fig. II.13. It appears that the elastic wave velocity decreases with increasing sample thickness and approaches the longitudinal ultrasonic velocity at small values of particle velocity, indicating the possible existence of some energy dissipation (Taylor and Rice [30]) or relaxation mechanism in this regime.

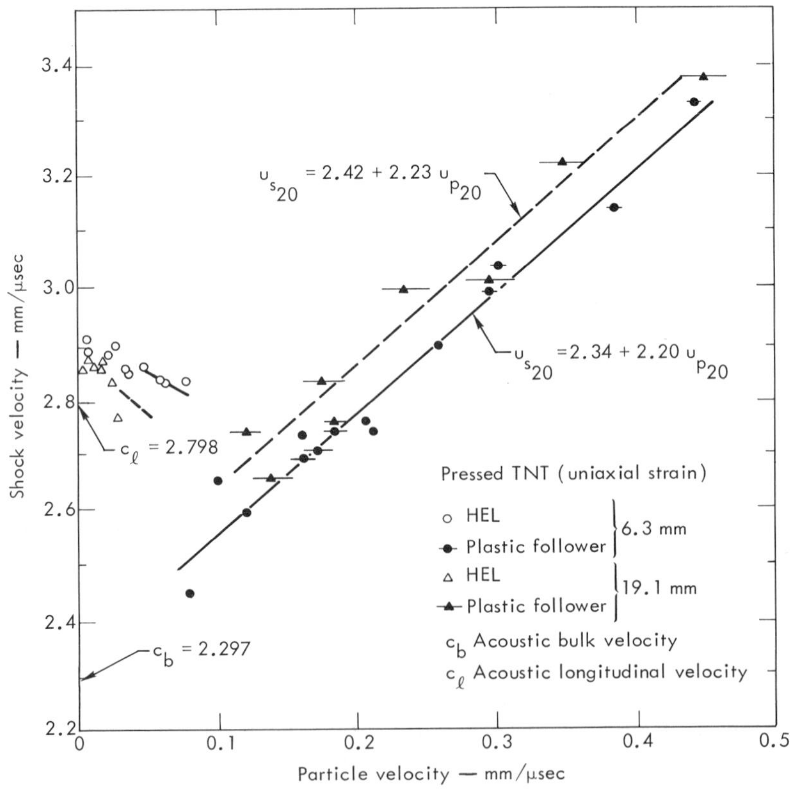

Fig. II.13. Hugoniot curves: shock-velocity – particle-velocity plane.

The two plastic shock-velocity lines in the figure are drawn on the basis of a linear least-squares regression analysis.[6] The standard deviations of these lines do not overlap. It is seen that the plastic shock velocity is apparently greater for the thicker specimen at a given particle velocity.

The *bulk* (or *acoustic*) sonic velocity c_b is also shown on the figure. It represents the sound velocity in an ideal fluid, and can be determined by the equations

$$c_b^2 = \frac{k}{\rho} = c_\ell^2 - \frac{4}{3} c_s^2. \qquad \text{(II-3.9)}$$

This bulk sonic velocity was not used in the above regression analyses, although often it is so used.

Although the observations with respect to the plastic shock velocity are consistent with interpretations based upon Figs. II.11 and II.12, one might have expected the positions of the two curves to be reversed. It is conceivable that this situation could be related in some fashion to the release of some small quantity of chemical energy from the TNT at or near the shock front, but for reasons advanced in the work of Wasley and Walker [2], this possibility is not strong.

The shock, isentropic, and isothermal hydrostats are determined using the techniques described above. Figures II.14 and II.15 give the results of the calculated (estimated) shock hydrostats. The shock hydrostats are also labeled with an estimate of the temperature rise (Walsh *et al.* [31] and Walker [32]). Short extrapolations to these hydrostats are made for illustration with the assumption that the maximum deviatoric stress component $2Y/3$ does not change appreciably, *i.e.*, perfectly plastic behavior is assumed. The dynamic uniaxial stress data shown in the two figures were obtained by interpolation and extrapolation of the work of Wasley *et al.* [33].

The calculated isothermal hydrostats[7] and a comparison with experimental results are also illustrated in Figs. II.14 and II.15. Agreement is statistically reasonable between these estimated isothermal hydrostats and the experimental curve (recalling that two different laboratories were involved in the efforts and that materials of two different nominal densities were used).

In any event, it appears that the above method represents a reasonably satisfactory approach, useful in engineering analyses for shock loading of at least two widely different classes of solids, namely, metals and polycrystalline organic materials.

To this point, only loading phenomena have been examined. An equally important part of the problem is the subsequent decompression of the material. A wave that reduces the level of shock stress is called a *rarefaction* or *release*

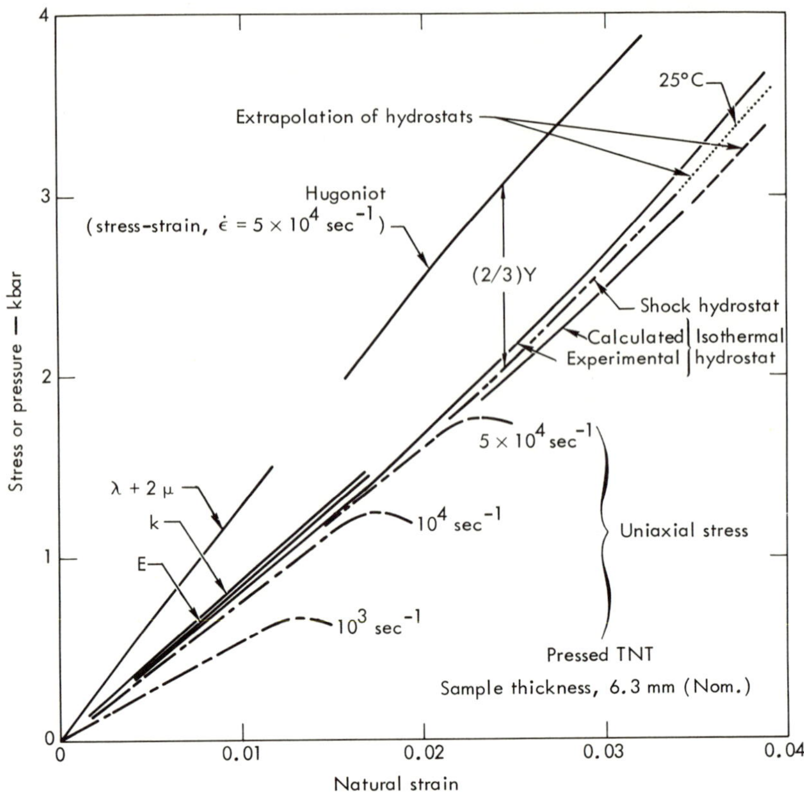

Fig. II.14. Estimated shock hydrostat **and** isothermal hydrostat (6.3 mm nominal).

wave. The unloading process or path is usually referred to as a *release adiabat* and is partially shown schematically in Fig. II.7. Rarefactions can result from geometry limitations (see comments in Chapter I-2, Section I), but in controlled laboratory experiments, they are introduced deliberately (if they are so desired in the particular test) from the rear in the direction of shock propagation, *e.g.*, with the use of a thin driver plate. The velocity of propagation of a rarefaction wave (and also a shock disturbance) is a classical characteristics problem (Chapter I-1, Section IV-C). As the stress level drops the various parts of the release wave move at different velocities, and the wave, in general, lengthens as it propagates. Note that this effect is in contrast to a shock-loading disturbance in which the wave steepens as it propagates.

One reason for the importance of shock unloading arises because we can obtain (in principle) experimental high-pressure mechanical behavior data that

can be generated in no other way. For example we can deduce, all as functions
of stress and rate, the uniaxial stress yield strength Y, the shear modulus μ, the
adiabatic bulk modulus k, Poisson's ratio, sound wave velocities (longitudinal,
shear, and bulk or acoustic), and work-hardening and Bauschinger effects (see
Fig. II.7 for illustrations of the last two effects).

Figure II.5 illustrates an idealized structure of the unloading process of the
kind seen in many solids and shows first an elastic release (Point 3) and then a
plastic release (Point 4). The ideal fluid (hydrodynamic) model does not predict
such a two-part release structure. The simple elastic-plastic theory discussed
above with respect to loading response can also be applied to the determination
of release adiabats. The model indicates that initially the material unloads along
an elastic release curve until the difference in the stresses are large enough to
cause yielding. Thereafter, the release curve is essentially that of a hydrostati-

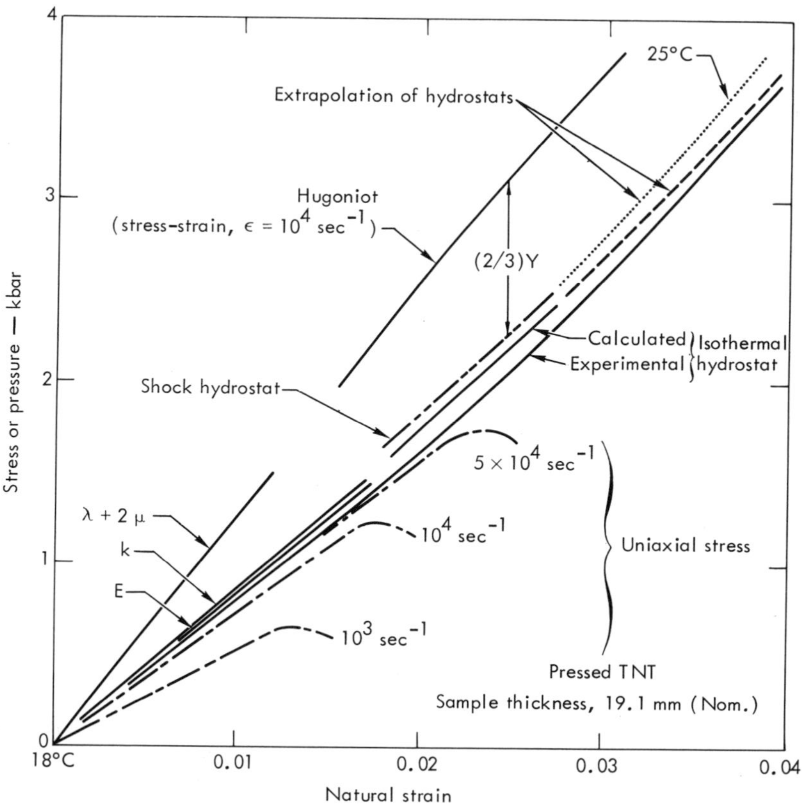

Fig. II.15. Estimated shock hydrostat and isothermal hydrostat (19.1 mm nominal).

cally compressed material, and the slope is given by the bulk modulus. Figure II.7 illustrates this behavior.

Upon release of the driving stress to zero, the material is still left with a net strain. This is caused from lateral elastic stresses which are not relieved in this process and which maintain the material at the yield point. These lateral stresses can only be relieved by lateral expansion which is not allowed with this idealized model.

Experimental work in this field is very recent; one of the first efforts was reported by Al'tshuler *et al.* [34] in 1960. Techniques are still in the early stages of development.

An experimental release curve can be constructed once the sound velocity is obtained as a function of pressure. More specifically, the unloading characteristics $c + u_p$ are first obtained with some detection system (say a pressure transducer). The quantity u_p is the local particle velocity, and c is the desired sound velocity in the particle velocity coordinate system. We can then calculate u_p from the *Riemann equation*

$$\frac{du_p}{dp} = \left(-\frac{dV}{dp}\right)^{1/2} = \frac{1}{\rho c} \tag{II-3.10}$$

and solve for c by iteration. A numerical code can be written to make such calculations as well as to construct the release curves from the calculated c values.

Although the experimental investigation of shock unloading phenomena is still in its infancy, the observations made to date (*e.g.*, see Kusubov and van Thiel [35]) indicate, at least for some materials, the need for a more elaborate model than the previously proposed elastic-plastic model which has been fairly successful in describing loading phenomena. For example, the Baushinger effect illustrated in Fig. II.7 is not accounted for in this simple release model. Viscous effects also appear to be prominent (Wilkins [19]).

A logical extension of unloading of the material is to allow it to pass into tension so that dynamic fracture or spall can result (see also Chapter I-3, Section III-A and Chapter II-2, Section I-A). There are very important practical ramifications that follow from this phenomenon, and the reader is referred to the monograph by Rinehart [36] and a recent paper by Charest *et al.* [37].

When consideration is given to shock-pulse propagation, the amplitude degrades and the shape disperses with distance of travel, primarily because the shock front moves more slowly than the following (lengthening) rarefaction. In addition, there is continuous internal energy dissipation into the form of heat (sometimes called "waste heat"), thus contributing to the pulse attenuation. These effects are more pronounced with pulses transmitted in porous media (see Wasley *et al.* [38]).

III. Experimental Methods

As pointed out in connection with Eqs. (II-3.2) and (II-3.3), the dynamic measurement of only two of the four variables, u_s, u_p, ρ and σ (or p) permit us to generate stress-strain information and Hugoniot curves. The experiments necessary to obtain data on these variables are limited only by the imagination of the experimenter, governed, of course, by certain physical limitations and rules. We divide the subject into several parts, treating measurement of each of the above four variables in turn (which are in general order of difficulty).

The parameter that is usually measured in all shock experiments is the (steady-state) shock velocity. The parameter can be measured electronically or optically by recording the time required for a shock to pass through a known thickness of material. One of the simplest methods, in principle, is with the use of *electrical discharge switches*, or more colloquially, "pins." A disturbance that arrives at the location of such a pin causes two electrodes to short-circuit either by metallic contact or by conduction through the surrounding gas ionized by the associated shock. This in turn causes an RC network to short-circuit and to give a voltage pulse on a time-calibrated (usually rasterized) recording oscilloscope. Two principal types of electrical discharge pins are available: single conductor (bare wire) and coaxial. The former has been used extensively,[8] but it is unsuitable for use with nonconducting materials. Use of a coaxial pin overcomes this limitation, but it creates difficulty in consistently giving a satisfactory signal at lower stresses, say below 10 kbar. Recent designs of coaxial pins using vapor-plating techniques (Wasley *et al.* [39]) avoid this problem.

Another pin design incorporates the use of a polarized ferroelectric material [*e.g.*, lead-zirconate-titanate (PZT) or barium titanate ($BaTiO_3$)] in its tip (Steffes and Bowen [40]). The passage of a shock causes the material to lose its polarization and release a charge; when discharged through a resistor, a high-amplitude voltage pulse results that can be recorded on an oscilloscope. Recent consideration is also being given to the use of certain phenomena encountered in solid-state physics for very fast switching, *e.g.*, field breakdown of a Schottky barrier induced by a pressure pulse.

The times measured with such switches are customarily determined to less than 0.01 μsec. Since the associated pin positions are usually measured to 0.001 mm, the corresponding shock velocities can be deduced with considerable accuracy.

Essentially all optical methods to determine shock velocity use a *streak camera* as the recording instrument. In such cameras, the field of view is restricted to a narrow slit so as to admit only a thin line of the image. This line is oriented perpendicularly to the centerline of the film, and the image is swept along the film parallel to this centerline at a predetermined constant sweep rate by means of a rotating mirror. The resulting film record consists of a time-resolved picture of what is seen within the field of view of the slit, generally a series of light and dark events.

For this type of camera, writing speeds on the film are possible up to 20 mm/μsec so that the resolution in time is comparable to times measured with the above electronic pin switches. The consequent short exposure times require that the observed event either be self-luminous, such as detonation waves and certain gas shocks, or that it be illuminated by an auxiliary light source, such as the intense short-duration light emission caused by the explosively induced shock in argon gas.

The method of using *flasher gaps* is one of the earliest optical techniques developed to measure shock velocity, originating at about the same time as bare wire pins. The gaps are typically 0.01 mm thick and are filled with a gas such as argon. The incident shock wave transmitted into the gap causes the compression and heating of the gas to brilliant luminosity which can be recorded on the streaking camera film. This method is of somewhat limited value in observing multiple shock systems, such as discussed above.

Many polished surfaces appear to change their reflectivity when shock waves are incident upon them. This change in reflectivity provides the basis for several other optical methods since the consequent change in light intensity can be observed by a streaking camera. Another similar scheme involves the use of transparent mirrors silvered on their inside faces which are placed in flat contact with a surface of the specimen. When a shock strikes such an interface, the reflectivity is suddenly altered.

Other types of cameras are available for recording high-speed transient phenomena, but are of less applicability for our purposes here. One camera is called a *framing camera* in which the image is swept around the film track by a rotating mirror, and individual images are recorded through suitable framing lenses. This type of camera can record several million frames per second. Occasionally, the need arises for an ultrafast shutter. The *Kerr cell shutter* can then be used, which is opened by applying a voltage pulse to rotate a plane-polarized light beam through 90 deg. Using appropriate polaroid filters, the light can be switched on and off.

A disadvantage of all these cameras, as compared with the use of pins, is their high initial cost.

The next parameter often measured is that of free-surface velocity, from which particle velocity can be deduced. If entropy effects are negligible and porous materials are not considered, the free-surface velocity is just twice the particle velocity. More specifically, this relationship is accurate to within about 1% for compressions in solids less than about 15% (see Rice *et al.* [21]) and thus for our considerations quite satisfactory.

A desirable recording system is capable of measuring the velocity of a free surface continuously in time. As in the measurement of shock velocity, we have available both electronic and optical methods. We can again use pin switch techniques, although this method does not satisfy the condition of continuous time measurement. The *slanted resistor* technique (which in a sense can be considered as an infinite number of pins infinitely closely spaced) does allow a continuous time record (Barker and Hollenbach [41]). In this method, a small diameter resistance wire (or ribbon) is stretched between two supports at a predetermined angle to the free surface. As the conducting surface strikes the wire, the portion of the wire up to the point of impact is electrically short-circuited, and the position of the point of impact, and hence the free-surface displacement, is monitored by measurement of the voltage drop as a function of time.

If the sample has an electrically conducting plane free surface, a portion of this surface can be constructed as one plate of a parallel plate *condensor microphone* (Taylor and Rice [30] and Ivanov *et al.* [28]; see also Chapter II-2, Section I-B). Any surface motion results in a change in capacitance which can be measured in time by standard electrical methods. The capacitor technique has been improved in time resolution to the point where it is especially suited for measuring multiple shock behavior.

The *inclined mirror* method is the optical counterpart of the slanted resistor technique described above. The mirror is positioned on the specimen free surface and is inclined at a small angle. The free surface of the specimen is shock-accelerated very rapidly, essentially discontinuously, and begins to collide with the inclined mirror. The arrival of the plane shock at the mirror is indicated by an abrupt change in the intensity of the reflected light. The apparent (phase) velocity of the point of collision, which is related to the particle velocity, is recorded continuously on the film of a streak camera.

If the surface of the sample to be observed can be polished, and if reasonable reflectivity is retained upon shock loading, the familiar principle of the *optical lever* can be used. The polished free surface of the specimen is positioned at some angle to the incident shock wave, causing small rotational displacements of the free surface to occur upon arrival of the shock. Then by appropriate reflection geometry, reflections of light from point sources that are incident on the same free surface can be effectively displaced a much greater amount. These reflections can be recorded on film by a streak camera as a function of time as a series of light streaks.

Another optical technique, somewhat more sophisticated than the above, is the *reflected wire* method (Davis and Craig [42]). A thin wire is positioned a small distance from the polished free surface of the specimen, and the streak camera views both the wire and its image reflected in the surface. When the shock strikes the free surface, the image appears to move toward the actual wire with a velocity twice that of the free-surface velocity. The camera records, with appropriate illumination, only a component of that velocity because of the viewing angle.

This technique has been modified for simultaneous measurement of shock and free-surface velocities using multiple wires (Wasley and Cast [43]). Figure II.16 illustrates a typical experimental set-up for a two-wire system. The reflective driver plate and sample, wires, and objective lens can be positioned on

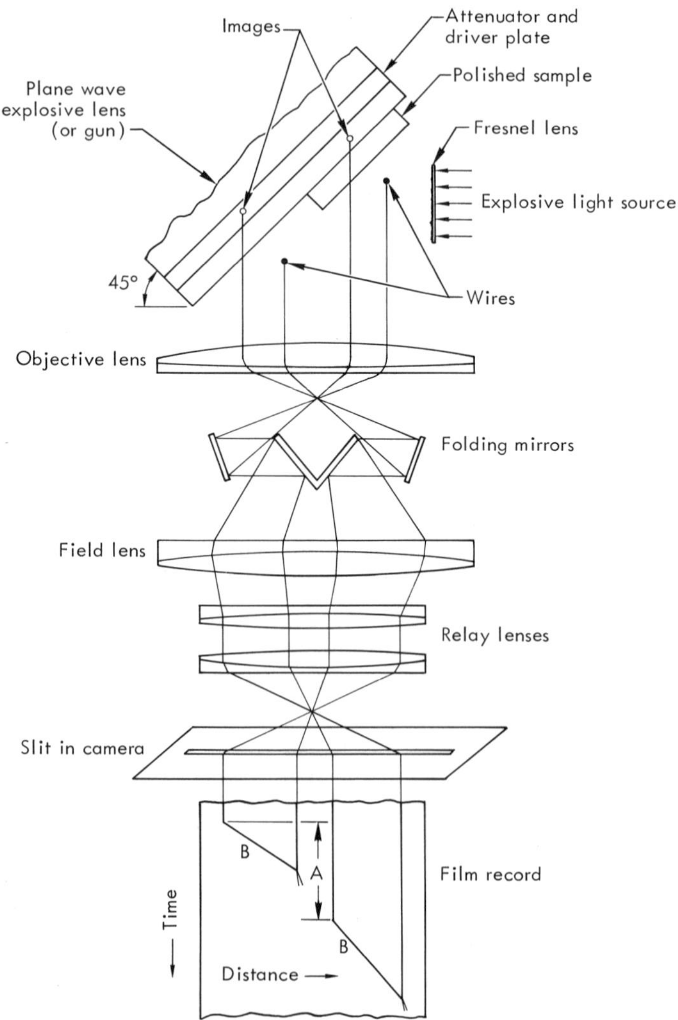

Fig. II.16. Schematic drawing of experimental technique for a two-wire system. A represents the sample shock transit time. B illustrates the slopes related to free-surface velocity.

an optical bench in the laboratory. At the firing site, this assembly, Fresnel lens, field lens, and relay lenses are aligned optically using a point light source projected through the camera. The folding mirrors are then inserted into the system and adjusted. The Fresnel lens is located so that the light during the experiment completely fills the objective lens. As shown in the figure, the two sets of wires and their images are placed adjacent to each other in the slit plane so that there will be no overlap in the film plane during motion. The wires are situated so that the lengths of their optical paths are the same.

Note that since the particle velocity of the driver plate can be obtained if its Hugoniot is known, the driving stress follows readily and the transmitted stress into the sample can be found from impedance mismatch calculations. Because of its inherent accuracy and relative simplicity, the method has particular value in experiments using thin samples.

A recently described technique (Barker [44]) uses a *laser interferometer* to measure the velocity history of a reflecting surface of a shock-loaded specimen. Essentially, the Doppler shift induced in the laser beam is monitored by suitable instrumentation. A schematic of the experimental configuration is shown in Fig. II.17. The laser beam is focused at the target free surface. The first of the two beam splitters causes part of the light to pass into a delay leg. The second beam splitter recombines the two rays so that the photomultiplier records a brightness which depends upon the relative phases of the recombined rays. In this way, fringes are generated at the photomultiplier tube in proportion to the velocity attained by the free surface, and the frequency of fringes generated (or beat frequency) is proportional to its acceleration. The photomultiplier output can be recorded on an oscilloscope. Excellent resolution is possible using this technique, both in the loading and the subsequent unloading regimes. It is possible to obtain a response time of about $0.002\ \mu\mathrm{sec} = 2$ nsec.

It should be noted that the optical methods we have described usually have a limitation in that there exists a general requirement for a mirror finish on the specimen surface, the quality of which will be reasonably maintained during shock loading. Since vapor-plated surfaces do not in general produce reliable results with these techniques, we are limited for the most part to metal specimens.

Sophisticated methods have recently been developed (see Dremin and Adadurov [45] and Petersen *et al.* [46]) to measure particle velocity directly rather than by inference from free-surface velocity measurements. Essentially, the motion of thin conducting foils embedded in a nonconducting specimen is monitored by placing the specimen in a magnetic field and measuring the electromotive force induced in the foil as the shock wave passes through the specimen. Such an approach assumes the foils do not perturb significantly the specimen's response. At present, this technique is limited to the study of dielectric specimens, such as ceramics and rocks.

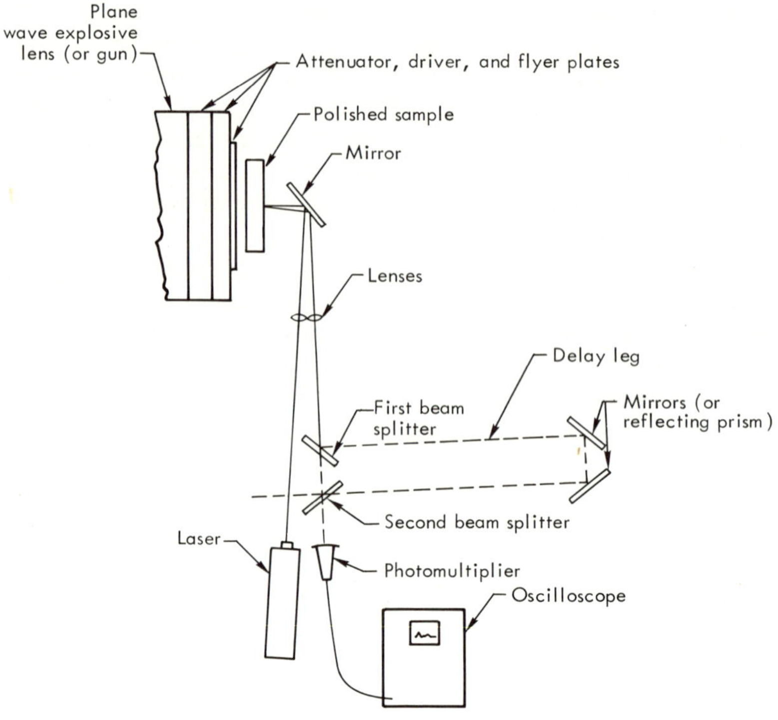

Fig. II.17. Schematic drawing of velocity interferometer.

Application of elastic theory in association with shock-wave studies can be seen in the work reported by Dennin [47] investigating the shock compression of granite. A shock was introduced into a wedge-shaped sample using an explosive plane wave generator. The free-surface velocity was determined from the motion of the profile of the surface by means of a framing camera. Because of the configuration of the experiment and since the specimen supported a two-wave structure, the reflection of the elastic precursor from the free surface generated two waves back into the specimen, a longitudinal and a shear wave. These waves interacted with the oncoming driving (plastic) wave. Vector diagrams were constructed showing the material velocities associated with each of these waves and their relation to the free-surface angles. The surface angles arising from the various elastic waves were determined using a direct application of the theory developed in Chapter I-3, Section III-B on reflection of oblique elastic waves at a free boundary.

High-speed (flash) radiography can be used to enable direct measurement of the density under shock loading. Machines are available that emit very intense pulses of x rays lasting for less than 0.1 μsec, each with radiation sufficient to penetrate several centimeters of many metals. The equipment takes pictures similar to those in medical work. At present, the direct measurement of density from such a method is not very accurate since density gradients occur at the edges of the experiment and make interpretation difficult. This technique can also be used to determine the location of shocks, and thus their velocity, within the body of an experimental system. In addition, the direct observation of particle velocity can be made very accurately if foils of greater opacity to x rays than the sample material are imbedded within the specimen.

For the past several years, transducers have been developed to measure the stress or pressure in the shock wave directly; the results show considerable promise. An ideal arrangement for making these direct stress measurements would embed in the sample a stress-sensing device that had the same impedance characteristics as the specimen, and of such dimensions so as not to perturb the shock to be measured (*e.g.*, from gaps) or be influenced by rarefactions from edge effects.

Several materials that exhibit electrical transducing properties have been investigated and developed for use as stress transducers. One of the most successful, and one that has been used extensively within the past several years uses the piezoelectric properties of quartz (see Jones *et al.* [4] or Halpin *et al.* [48]) so that it operates as a current generator. An x-cut quartz disk is placed in contact with the specimen that is to be shock loaded. As the shock wave passes from the specimen into the quartz, electric polarization is produced which increases monotonically with pressure.

A number of assumptions are required to generate the relevant equations governing the response of quartz as a stress transducer. For example, (1) the stressed region of the quartz must be in a state of one-dimensional strain; (2) the electric fields produced by the piezoelectric effect are one-dimensional; (3) a short circuit must exist between the two faces of the gauge; (4) the conductivity in the quartz is zero; (5) the dielectric permittivity of the quartz does not change with finite shock stress and finite electric field; and (6) the piezoelectric polarization is directly proportional to the stress component normal to the plane shock front. Using these (and several other) assumptions, it can be shown (Graham *et al.* [3]) that the target-quartz interface stress normal to the shock front is given by

$$\sigma = \frac{i\,\ell}{f\,A\,u_s} = \frac{E\,\ell}{f\,A\,R\,u_s}, \quad 0 < t < \ell/u_s \tag{II-3.11}$$

where i is the current produced by the piezoelectric effect, ℓ is the thickness of the quartz gauge, f is the piezoelectric polarization coefficient (which is weakly dependent upon the shock stress), A is the active area of the gauge, u_s is the shock velocity in quartz (a constant equal to 5.72 mm/μsec), E is the voltage, R is the load resistance of the circuit, and t is the wave transit time through the gauge.

One method of "packaging" the transducer to use in experiments is illustrated schematically in Fig. II.18 (O'Brien and Wasley [49]). In the design shown, a 50-ohm resistive load is generally used as the shunt resistance for the inner electrode of the crystal, *i.e.*, the active area inside the guard ring on the back face of the transducer. The load resistance is given by the shunt resistor and the cable termination resistor in parallel. The load resistance used for the outer electrode of the crystal, *i.e.*, the area outside the guard ring, is selected so that the potential on the two electrode areas is kept approximately the same. The purpose of the guard ring is to ensure one-dimensional strain and electric fields in the central (active) portion of the transducer.

An alternate method of transducer construction (Jones and Halpin [50]) that is somewhat simpler than the above mentioned and yet for practical purposes apparently gives the same results involves shorting the outer electrode directly to ground by coating the entire outer periphery of the gauge with the conducting material. However, this technique eliminates the generation of a signal from the outer electrode, which signal can often be useful as a timing mark on other recording instruments.

Note that the useful time interval of measurement is given by the limits in Eq. (II-3.11). This is in contrast to more conventional method (*e.g.*, see Wasley *et al.* [2]; refer also to Chapter II-2, Section I-B) of observing the piezoelectric charge on a time scale that is long compared with the wave transit time through the gauge.

Quartz transducers of the sort described above give valid stress measurements up to about 30 kbar in the quartz. At sufficiently high polarization, there is an internal electrical breakdown which limits the magnitude of current generated by the crystal. The time resolution of these gauges is about 0.01 μsec. Recall that since the stress configuration is usually anisotropic in solid materials, the component of the stress in the direction of the shock propagation is measured by the transducer.

A disadvantage of the quartz transducer is that it generally introduces an impedance mismatch between the specimen and transducer, and one must make appropriate calculations to determine the stress-time history in the sample itself. This is a very important practical consideration when studying waves with multiple shocks — for example, when an elastic precursor forms; reflections at the sample-quartz interface arising from the precursor interaction can introduce complications in the analysis of the plastic follower wave in the sample.

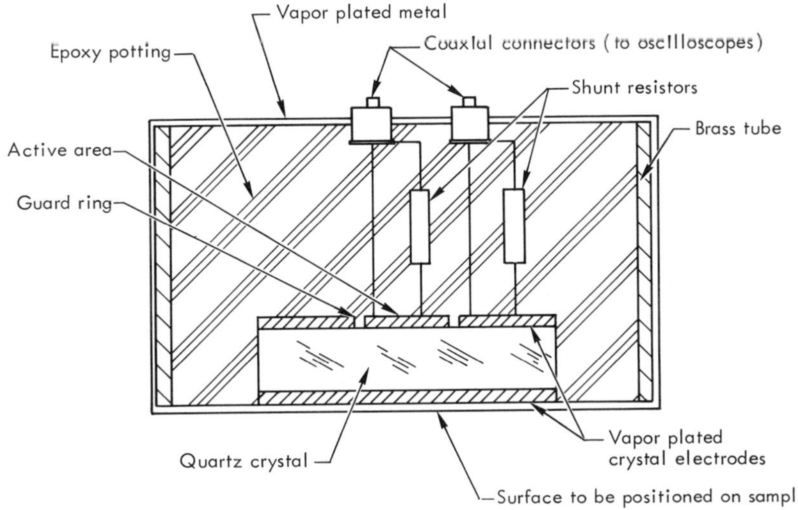

Fig. II.18. Schematic of transducer package.

Another stress gauge recently developed which appears to hold the most versatility is one made of manganin as the sensing element,[9] an alloy consisting of about 84% copper, 12% mangenese, and 4% nickel (Kusubov and van Thiel [35] ; Bernstein and Keough [51]). This is a piezoresistive gauge such that its change in electrical resistance is directly proportional to the applied shock stress. It approaches the ideal transducer in that it can be embedded directly in the specimen being shocked.

An advantage results from the significant stress range over which valid readings can be obtained — much greater than that possible with the quartz transducer. The manganin gauge can be used from several kilobars to several hundred kilobars. The gauge is insensitive to temperature changes, and it can also be employed to examine unloading phenomena, *i.e.*, to trace the entire loading-unloading stress-time history of a shock event (see Fig. II.5). Using this technique, it is now possible to resolve very fine features of shock behavior, and it appears to provide one of the most fruitful directions for future transducer research.

Finally, mention should be given of several sources in which complications of Hugoniot data and related information are collected. One can refer to the works by Kohn [52], van Thiel *et al.* [53] and Rinehart [54].

IV. Summary

Shock disturbance propagation in condensed media (mostly solids) is examined under conditions of macroscopic uniaxial strain. Both analytical and experimental discussion is presented, with emphasis directed toward the latter. Examination of the response of an organic, polycrystalline solid illustrates many of the points treated.

Loading and unloading (release) behavior is studied by examining various dynamic curves in the shock stress-strain plane. First, the average stress-strain-strain rate paths followed at a given location in the solid can be deduced from the experimental stress-time profiles with the help of the Rankine-Hugoniot conservation relations. These conservation relations are discussed in particular because of their importance in experimental work.

Next, the stress Hugoniot curve is defined as the locus of essentially final equilibrium stress-strain points from a series of experiments that can be reached by the shock loading of the material. In general, the stress Hugoniot differs from the shock hydrostat (or pressure Hugoniot) curve because a point on the former represents a state of stress composed of a deviatoric stress component and a pressure component, while a point on the latter represents only a state of pressure. The stress deviator can be obtained using elastic-plastic theory together with appropriate uniaxial stress, stress-strain-strain rate data.

The shock hydrostat differs from the isentropic hydrostat because there is a thermodynamic effect of an increase in entropy in shock loading. One method of obtaining the isentrope from the shock hydrostat is by using the Mie-Grüneisen formalism. The data can be further treated by thermodynamic techniques to calculate the isothermal hydrostat, which is the stress-strain path followed by the material loaded under isothermal, hydrostatic conditions. These results can be directly compared to data generated from quasi-static experiments.

According to simple elastic-plastic theory, the material unloads initially along an elastic release curve until yielding occurs, and thereafter, the release adiabat unloads plastically. Experimental observations indicate, however, that at least for some materials, a more elaborate release model is required. Baushinger and viscous effects, which cannot be accounted for by elastic-plastic theory, appear to be important.

A number of experimental methods are discussed, more from a point of view of their purposes and intentions rather than from their engineering and physical aspects. Techniques are examined that enable the generation of Hugoniot and loading-unloading stress-strain curves. The subject is divided into several parts, treating measurement of each of the variables, u_s, u_p, ρ and σ (or p) in turn. The available methods are varied, but are usually based upon ultrafast electronics and/or optics.

Notes

[1] 1 Mbar = 10^3 kbar = 10^6 bars.

[2] Experimental dynamic stress-strain test data are not usually obtained at constant strain rates. The cross-plotting routine discussed in Chapter II-2, Section I-C with respect to one-dimensional stress wave investigations can also be used to generate such curves in one-dimensional strain.

[3] The adiabatic bulk modulus is often used here.

[4] In fact, often the effect is negligible. If so, then the assumption can be made that the isentrope corresponds to the Hugoniot. This assumption results in what may be called a purely mechanical theory or the *weak shock* approximation.

[5] The adiabatic and isothermal bulk coefficients essentially coincide since strain-rate effects under spherical (hydrodynamic) loading are generally negligible.

[6] The spread in particle velocities in the plastic regime in a given test arises from the difference in the two methods of calculation used. The average of this spread is the point that is plotted.

[7] The isentropic hydrostats are not shown on Figs. II.14 and II.15. Although these curves were determined and temperature rises at selected stresses were calculated, the lines fall very near the calculated isothermal hydrostats and are difficult to graph distinctly.

[8] Bare wire pins were the type used in the first experimental shock-wave studies in the late 1940's at Los Alamos Scientific Laboratory.

[9] Other very recently investigated piezoresistive materials, *e.g.*, ytterbium and carbon, are also proving satisfactory.

References

[1] B. J. Alder, "Physics Experiments with Strong Pressure Pulses," in *Solids Under Pressure*, (W. Paul and D. M. Warschauer, eds.), McGraw-Hill, New York, 1963.

[2] R. J. Wasley and F. E. Walker, *J. Appl. Phys.*, **40**, 2639 (1969).

[3] R. A. Graham, F. W. Neilson and W. B. Benedick, *J. Appl. Phys.*, **36** 1775 (1965).

[4] O. E. Jones, F. W. Neilson and W. B. Benedick, *J. Appl. Phys.*, **35**, 3224 (1962).

[5] F. E. Walker and R. J. Wasley, *Explosivstoffe*, **17**, 9 (1969).

[6] J. W. Taylor, *J. Appl. Phys.*, **36**, 3146 (1965).

[7] R. A. Graham, D. H. Anderson and J. R. Holland, *J. Appl. Phys.*, **38**, 223 (1967).

[8] O. E. Jones, and J. D. Mote, *J. Appl. Phys.*, **40**, 4920 (1969).

[9] P. P. Gillis, K. G. Hoge and R. J. Wasley, *J. Appl. Phys.*, **42**, 2145 (1971).

[10] R. J. Wasley and F. E. Walker, *A Method for the Numerical Analysis of Pressure Transducer Records*, University of California, Lawrence Livermore Laboratory, Report UCRL-50233 (1967).

[11] L. M. Barker, C. D. Lundergan and W. Herrmann, *J. Appl. Phys.*, **35**, 1203 (1964).

[12] D. E. Munson and L. M. Barker, *J. Appl. Phys.*, **37**, 1652 (1966).

[13] T. R. Guess, *Some Dynamic Mechanical Properties of Armco 21-6-9 Stainless Steel*, Sandia Laboratories, Albuquerque, New Mexico, Report SC-RR-69-761 (1969).

[14] D. S. Wood, *J. Appl. Mech.*, **19**, 521 (1952).

[15] B. M. Butcher and C. H. Karnes, *J. Appl. Phys.*, **37**, 402 (1966).

[16] D. G. Doran and R. K. Linde, "Shock Effects in Solids," in *Solid State Physics*, Vol. 19, (F. Seitz and D. Turnbull, eds.), Academic, New York, 1966.

[17] E. H. Lee and D. T. Liu, *J. Appl. Phys.*, 38, 19 (1967).

[18] R. J. Clifton, "On the Analysis of Elastic/Visco-Plastic Waves and the Mechanical Properties of Solids," in *Proc. 17th Sagamore Conf. on Shock Waves and the Mechanical Properties of Solids*, (J. J. Burke and V. Weiss, eds.), Syracuse University Press, New York, 1971.

[19] M. L. Wilkins, "The Calculation of Elastic-Viscous-Plastic Effects in Materials," in *Proc. 17th Sagamore Conf. on Shock Waves and the Mechanical Properties of Solids*, (J. J. Burke and V. Weiss, eds.), Syracuse University Press, New York, 1971.

[20] W. Herrmann, R. J. Lawrence and D. S. Mason, *Strain Hardening and Strain Rate in One-Dimensional Wave Propagation Calculations*, Sandia Laboratories, Albuquerque, New Mexico, Report SC-RR-70-471 (1970).

[21] M. H. Rice, R. G. McQueen and J. M. Walsh, "Compression of Solids by Strong Shock Waves," in *Solid State Physics*, Vol. 6, Academic, New York, 1958.

[22] R. G. McQueen and S. P. Marsh, *J. Appl. Phys.*, 31, 1253 (1960).

[23] D. R. Stephens, *Calculations of Adiabats and Hugoniots from Isotherms*, University of California, Lawrence Livermore Laboratory, Report UCID-15459 (1969).

[24] J. M. Walsh, and R. H. Christian, *Phys. Rev.*, 97, 1544 (1955).

[25] I. C. Skidmore, *Appl. Matl. Res.*, p. 131, July 1965.

[26] M. Ya. Vasil'ev, D. B. Balashov and L. N. Mokrousov, *Russian J. Phys. Chem.*, 34, 1159 (1960).

[27] R. J. Wasley and J. F. O'Brien, "Low-Pressure Hugoniots of Solid Explosives," in *Preprints, Vol. I, Fourth Symp. on Detonation*, U.S. Naval Ordnance Laboratory, White Oak, Maryland (1965).

[28] A. G. Ivanov, S. A. Novikov, and V. A. Sinitsyn, *Soviet Phys.-Solid State*, 5, 196 (1963).

[29] M. Cowperthwaite, *J. Franklin Inst.*, 285, 275 (1968).

[30] J. W. Taylor and M. H. Rice, *J. Appl. Phys.*, 34, 374 (1963).

[31] J. M. Walsh, M. H. Rice, R. G. McQueen and F. L. Yarger, *Phys. Rev.*, 108, 196 (1957).

[32] F. E. Walker, *A Fortran Program (TEMCAL) for Calculation of Hugoniot Temperatures, Isotherms, and Release Adiabats from Hydrodynamic Data*, University of California, Lawrence Livermore Laboratory, Report UCRL-50210 (1967).

[33] R. J. Wasley, K. G. Hoge and J. C. Cast, *Rev. Sci. Instr.*, 40, 889 (1960).

[34] L. V. Al'tshuler, S. B. Kormer, M. E. Brazhnik, L. A. Vladimirov, M. P. Speranskaya and A. I. Funtikov, *Zh. Eksperim Teor. Fiz.*, 38, 1061 (1960).

[35] A. S. Kusubov and M. van Thiel, *J. Appl. Phys.*, 40, 3776 (1969).

[36] J. S. Rinehart, "On Fractures Caused by Explosions and Impacts," *Quarterly of the Colorado School of Mines*, Golden, Colorado (1960).

[37] J. A. Charest, D. E. Horne and B. D. Jenrette, *Phenomenological Considerations for Spall Measurements*, EG&G, Santa Barbara, California, Technical Paper S-56-TP (1970). Paper presented at the Amer. Phys. Soc. 1969 Winter Meeting, Los Angeles, California,

[38] R. J. Wasley, E. J. Nidick, Jr., R. H. Valentine and K. G. Hoge, *J. Appl. Polymer Sci.*, 15, 2303 (1971).

[39] R. J. Wasley, J. F. O'Brien and D. R. Henley, *Rev. Sci. Inst.*, 35, 466 (1964).

[40] F. J. Steffes and J. R. Bowen, *Rev. Sci. Instr.*, 39, 1092 (1968).

[41] L. M. Barker and R. E. Hollenbach, *Rev. Sci. Instr.*, 35, 742 (1964).

[42] W. C. Davis and B. G. Craig, *Rev. Sci. Instr.*, 32, 579 (1961).

[43] R. J. Wasley and J. C. Cast, *Rev. Sci. Instr.*, 35, 1727 (1964).

[44] L. M. Barker, "Fine Structure of Compressive and Release Wave Shapes in Aluminum

Measured by the Velocity Interferometer Technique," in *Behavior of Dense Media Under High Dynamic Pressure*, IUTAM, Paris, 1967, Gordon and Breach, New York, 1968.

[45] A. N. Dremin and G. A. Adadurov, *Soviet Phys.-Solid State*, **6**, 1379 (1964).

[46] C. F. Petersen, W. J. Murri, and R. W. Gates, *Dynamic Properties of Rocks*, Stanford Research Institute, Menlo Park, California, Project PGU-6273, DASA 2298, 1969.

[47] R. S. Dennin, *J. Appl. Phys.*, **41**, 5309 (1970).

[48] W. J. Halpin, O. E. Jones and R. A. Graham, "Submicrosecond Technique for Simultaneous Observation of Input and Propagated Impact Stresses," in *Symposium on Dynamic Behavior of Materials*, ASTM Special Technical Publication No. 336, Am. Soc. Testing Matls. (1963).

[49] J. F. O'Brien and R. J. Wasley, *Rev. Sci. Instr.*, **37**, 331 (1966).

[50] G. A. Jones and W. J. Halpin, *Rev. Sci. Instr.*, **39**, 258 (1968).

[51] T. Bernstein and D. D. Keough, *J. Appl. Phys.*, **35**, 1471 (1964).

[52] B. J. Kohn, *Compilation of Hugoniot Equations of State*, Air Force Weapons Laboratory, Albuquerque, New Mexico, Technical Report AFWL-TR-69-38 (1969).

[53] M. van Thiel (gen. ed.), A. S. Kusubov and A. C. Mitchell (asst. eds.), *Compendium of Shock Wave Data*, Volumes I and II, University of California, Lawrence Livermore Laboratory, Report UCRL-50108 (1967).

[54] J. S. Rinehart, *Compilation of Dynamic Equation of State Data for Solids and Liquids*, U.S. Naval Ordnance Test Station, California, Report NOTS-TP-3798 (1965).

OTHER SELECTED REFERENCES

I. Notation; Tensors

R. Aris, *Vectors, Tensors, and the Basic Equations of Fluid Mechanics*, Prentice-Hall, Englewood Cliffs, New Jersey, 1962.

D. E. Christie, *Vector Mechanics*, 2nd ed., McGraw-Hill, New York, 1964.

II. Numerical Solution of Wave Problems; Elastic and Nonelastic

L. M. Barker, *SWAP-9: An Improved Stress Wave Analyzing Program*, Sandia Laboratories, Albuquerque, New Mexico, Report SC-RR-69-233 (1969).

S. E. Benzley, L. D. Bertholf and G. E. Clark, *TOODY II-A: A Computer Program for Two-Dimensional Wave Propagation*, Sandia Laboratories, Albuquerque, New Mexico, Report SC-DR-69-516 (1969).

L. D. Bertholf, *Longitudinal Elastic Wave Propagation in Finite Cylindrical Bars*, Washington State University, Shock Dynamics Laboratory, Pullman, Washington, Report 66-03 (1966).

L. D. Bertholf, *J. Appl. Mech.* **34**, 725 (1967).

J. D. Campbell and D. B. Taylor, "On the Numerical Solution of a Wave Propagation Problem in the Theory of Dislocation Motion," in *Stress Waves in Anelastic Solids*, IUTAM, Brown University, 1963, (H. Kolsky and W. Prager, eds.,) Springer-Verlag, Berlin, 1964.

P. Holzhauser and R. J. Lawrence, *WONDY-III-An Improved Program for Calculating Problems of Motion in One-Dimension*, Sandia Laboratories, Albuquerque, New Mexico, Report SC-DR-68-217 (1968).

III. Elasticity

A. Quasistatic Behavior

R. G. Dong, *Procedures for Experimental Identification of Appropriate Constitutive Theory for Materials*, University of California, Lawrence Livermore Laboratory, Report UCRL-71095 (1968).

W. Flügge, *Viscoelasticity*, Blaisdell Publishing Company, Waltham, Mass., 1967.

R. F. S. Hearmon, *Applied Anisotropic Elasticity*, Oxford University Press, London, 1961.

N. I. Muskhelishvili, *Some Basic Problems of the Mathematical Theory of Elasticity*, 4th ed., P. Noordhoff, Groningen, The Netherlands, 1963.

L. E. Nielsen, *Mechanical Properties of Polymers*, 3rd printing, Reinhold, New York, 1965.

J. J. Stoker, *Nonlinear Elasticity*, Gordon and Breach, New York, 1968.

B. Dynamic Behavior

1. Vibration Phenomena.

R. E. D. Bishop, G. M. L. Gladwell and S. Michaelson, *The Matrix Analyses of Vibration*, Cambridge University Press, London, 1965.

J. P. Den Hartog, *Mechanical Vibrations*, 4th ed., McGraw-Hill, New York, 1956.

W. T. Thompson, *Vibration Theory and Applications*, 3rd printing, Prentice-Hall, Englewood Cliffs, New Jersey, 1965.

S. Timoshenko and D. H. Young, *Vibration Problems in Engineering*, D. van Nostrand, New York, 1955.

2. Wave Phenomena.

C. Glickstein, *Basic Ultrasonics*, John F. Rider, New York (a division of Hayden), 1960.

O. E. Jones and F. R. Norwood, *Axially Symmetric Cross-Sectional Strain and Stress Distributions in Suddenly Loaded Cylindrical Elastic Bars*, Applied Mechanics Conf., Pasadena, California, 1967, Paper No. 67-APM-33.

J. Miklowitz, "Elastic Wave Propagation" in *Applied Mechanics Surveys*, Spartan Books, Washington, D.C., 1966.

IV. General (Mechanical) Wave Propagation

C. A. Coulson, *Waves*, 7th ed., 3rd printing, Oliver and Boyd, London, 1965.

R. M. Davies, *Brit. J. Appl. Phys.*, **7**, 203 (1956).

N. Feather, *The Physics of Vibrations and Waves*, 2nd printing, Aldine, Chicago, 1963.

F. G. Friedlander, *Sound Pulses*, Cambridge University Press, London, 1958.

L. E. Kinsler and A. R. Frey, *Fundamentals of Acoustics*, 2nd ed., 4th printing, Wiley, New York, 1966.

W. P. Mason (ed.), *Physical Acoustics*, Academic, New York, Vol. 1, Parts A and B, 1964; Vol. 2, Parts A and B, 1965.

J. Miklowitz, *Appl. Mech. Rev.*, **13**, 865 (1960).

J. Miklowitz (ed.), *Wave Propagation in Solids*, ASME Winter Annual Meeting, ASME, Los Angeles, 1969.

M. Redwood, *Mechanical Waveguides*, Pergamon, New York, 1960.

R. V. Sharman, *Vibrations and Waves*, Butterworth, London, 1963.

R. W. B. Stephens and A. E. Bate, *Acoustics and Vibrational Physics*, 2nd ed., Edward Arnold, London, 1966.

R. A. Waldron, *Waves and Oscillations*, D. van Nostrand, Princeton, New Jersey, 1964.

V. Nonelastic Wave Propagation; Shock Waves

A. One Dimensional Stress Loading

F. E. Hauser, J. A. Simmons and J. E. Dorn, "Strain Rate Effects in Plastic Wave Propagation," in *Response of Metals to High Velocity Deformation*, (P. G. Shewmon and V. F. Zackay, eds.), Interscience, New York, 1961.

H. G. Hopkins, "Mechanical Waves and Strain-Rate Effects in Metals," in *Stress Waves in Anelastic Solids*, IUTAM, Brown University, 1963 (H. Kolsky and W. Prager, eds.), Springer-Verlag, Berlin, 1964.

U. S. Lindholm, *Some Experiments with the Split Hopkinson Pressure Bar*, Southwest Research Inst., San Antonio, Texas, Report No. 1, Project No. 02-1102 (1964).

E. A. Ripperger and H. Watson, Jr., "The Relationship Between the Constitutive Equation and One-Dimensional Wave Propagation," in *Symposium on Mechanical Behavior of Materials Under Dynamic Loads*, San Antonio, Texas, 1967 (U. S. Lindholm, ed.), Springer-Verlag, New York, 1968.

R. J. Wasley, *Propagation of Stress Disturbances in Elastic-Plastic Cylinders: One-Dimensional Theories*, University of California, Lawrence Livermore Laboratory, Report UCRL-14617 (1965).

B. One Dimensional Strain Loading

L. V. Al'tshuler, *Soviet Phys. – Uspekhi*, **8**, 52 (1965)

W. Band and G. E. Duvall, *Am. J. Phys.*, **20**, 780 (1961).

J. N. Bradley, *Shock Waves in Chemistry and Physics*, Wiley, New York, 1962.

B. M. Butcher and D. E. Munson, "The Application of Dislocation Dynamics to Impact-Induced Deformation Under Uniaxial Strain," in *Dislocation Dynamics*, (A. R. Rosenfield *et al.*, eds.), McGraw-Hill, New York, 1968.

M. Cowperthwaite, *Am. J. Phys.*, **34**, 1025 (1965).

A. N. Dremin and O. N. Breusov, *Russian Chem. Rev.*, **37** 392 (1968).

G. E. Duvall, *Appl. Mech. Rev.*, **15**, 849 (1962).

G. E. Duvall, *International Science and Technology*, April 1963, p. 45.

G. E. Duvall and G. R. Fowles, "Shock Waves," in *High Pressure Physics and Chemistry*, (R. S. Bradley, ed.), Academic, New York, 1963.

G. E. Duvall, "Propagation of Plane Shock Waves in a Stress-Relaxing Medium," in *Stress Waves in Anelastic Solids*, IUTAM, Brown University, 1963 (H. Kolsky and W. Prager, eds.), Springer-Verlag, Berlin, 1964.

J. O. Erkman and A. B. Christensen, *J. Appl. Phys.*, **38**, 5395 (1967).

G. R. Fowles, *J. Appl. Phys.*, **32**, 1475 (1961).

SUBJECT INDEX

Other books
of interest
to you...

Because of your interest in our books, we have included the following catalog of books for your convenience.

Any of these books are available on an approval basis. This section has been reprinted in full from our **material science** catalog.

If you wish to receive a complete catalog of MDI books, journals and encyclopedias, please write to us and we will be happy to send you one.

MARCEL DEKKER, INC.
95 Madison Avenue, New York, N.Y. 10016

material science

including
Polymers, Plastics, Fibers, and Coatings
Metals and Metallurgy
Ceramics and Glass
Vacuum Science

ALTGELT and SEGAL
Gel Permeation Chromatography

edited by KLAUS H. ALTGELT, *Chevron Research Company, Richmond, California*, and LEON SEGAL, *South Regional Research Laboratory, U.S.D.A., New Orleans, Louisiana*

672 pages, illustrated. 1971

Demonstrates the manifold applications of gel permeation chromatography in the field of polymer chemistry. Directed to all research, quality-control, and analytical chemists working with conventional and unconventional polymers and other large molecules in the fields of polymer, cellulose, and petroleum chemistry.

CONTENTS: The sizes of polymer molecules and the GPC separation, *F. W. Billmeyer, Jr. and K. H. Altgelt.* Gel permeation chromatography column packings — types and uses, *D. J. Harmon.* Chromatographic instrumentation and detection of gel permeation effluents, *E. M. Barrall, II and J. F. Johnson.* Peak resolution and separation power in gel permeation chromatography, *D. J. Harmon.* A review of peak broadening in gel chromatography, *R. N. Kelley and F. W. Billmeyer, Jr.* Mathematical methods of correcting instrumental spreading in GPC, *L. H. Tung.* Comparison of different techniques of correcting for band broadening in GPC, *J. H. Duerksen.* Separation mechanisms in gel permeation chromatography, *W. W. Yau, C. P. Malone, and H. L. Suchan.* Gel permeation chromatography and thermodynamic equilibrium. *E. F. Casassa.* Calibration of GPC columns, *H. Coll.* Data treatment in GPC, *L. H. Tung.* The overload effect in gel permeation chromatography, *J. C. Moore.* Gel permeation chromatography using a bio-glas substrate having a broad pore size distribution, *A. R. Cooper, J. H. Cain, E. M. Barrall, II, and J. F. Johnson.* High resolution gel permeation chromatography — using recycle, *K. J. Bombaugh and R. F. Levangie.* Gel permeation chromatography with high loads, *K. H. Altgelt.* Fast gel permeation chromatography, *J. N. Little, J. L. Waters, K. J. Bombaugh, and W. J.*

Pauplis. Extension of GPC techniques, *G. Meyerhoff.* Phase distribution chromatography (PDC) of polystyrene, *R. H. Casper and G. V. Schulz.* Apparent and real distribution in GPC (experiments with PMMA samples), *K. C. Berger and G. V. Schulz.* The instrument spreading correction in GPC. I: The general shape function using a linear calibration curve, *T. Provder and E. M. Rosen.* The instrument spreading correction in GPC. II: The general shape function using the Fourier transform method with a nonlinear calibration curve, *E. M. Rosen and T. Provder.* Behavior of micellar solutions in gel permeation chromatography: A theory based on a simple model, *H. Coll.* Gel permeation analysis of macromolecular association by an equilibrium method, *B. F. Cameron, L. Sklar, V. Greenfield, and A. D. Adler.* Gel filtration chromatography, *B. F. Cameron.* Determination of polymer branching with gel permeation chromatography. Abstract of a review, *E. E. Drott and R. A. Mendelson.* Fractionation of linear polyethylene with gel permeation chromatography. Part III, *N. Nakajima.* Application of GPC in the study of stereospecific block copolymers, *R. D. Mate and M. R. Ambler.* Composition of butadiene-styrene copolymers by gel permeation chromatography, *H. E. Adams.* A direct GPC calibration for low molecular weight polybutadiene, employing dual detectors, *J. R. Runyon.* Quantitative determination of plasticizers in polymeric mixtures by GPC, *D. F. Alliet and J. M. Pacco.* Evaluation of pulps, rayon fibers, and cellulose acetate by GPC and other fractionation methods, *W. J. Alexander and T. E. Muller.* Characterization of the internal pore structures of cotton and chemically modified cottons by gel permeation, *L. F. Martin, F. A. Blouin, and S. P. Rowland.* Application of GPC to studies of the viscose process. I: Evaluation of the method, *L. H. Phifer and J. Dyer.* Application of GPC to studies of the viscose process. II: The effects to steeping and alkali-crumb aging, *J. Dyer and L. H. Phifer.* Gel permeation chromatography calibration. I: Use of calibration curves based on polystyrene in THF and integral distribution curves of elution volume to generate calibration curves for polymers in 2,2,2-trifluoroethanol, *T. Provder, J. C. Woodbrey, and J. H. Clark.* Modification of a gel permeation chromatograph for automatic sam-

(continued)

ALTGELT and SEGAL *(continued)*

ple injection and on-line computer data recording, *A. R. Gregges, B. F. Dowden, E. M. Barral, II, and T. T. Horikawa.* Characterization of crude oils by gel permeation chromatography, *H. H. Oelert, D. R. Latham, and W. E. Haines.* Separation and characterization of high-molecular-weight saturate fractions by gel permeation chromatography, *J. H. Weber and H. H. Oelert.* Fractionation of residuals by gel permeation chromatography, *E. W. Albaugh, P. C. Talarico, B. E. Davis, and R. A. Wirkkala.* Combined gel permeation chromatography–NMR techniques in the characterization of petroleum residuals, *F. E. Dickson, R. A. Wirkkala, and B. E. Davis.* A rapid method of identification and assessment of total crude oils and crude oil fractions by gel permeation chromatography, *J. N. Done and W. K. Reid.* Gel permeation analysis of asphaltenes from steam stimulated oil wells, *C. A. Stout and S. W. Nicksic.* GPC separation and integrated structural analysis of petroleum heavy ends, *K. H. Altgelt and E. Hirsch.*

AMERICAN VACUUM SOCIETY
Experimental Vacuum Science and Technology

edited by THE AMERICAN VACUUM SOCIETY EDUCATION COMMITTEE
288 pages, illustrated. 1973

A collection of experiments, which are graded from simple procedures to sophisticated vacuum processes, and designed to aid instructors and introduce students to the basic concepts and techniques of the field of vacuum science. Includes an extensive bibliography to stimulate further investigation. Especially useful for all students and teachers in the many fields of the basic sciences and engineering where vacuum methods and techniques are important.

CONTENTS: **Section 1: Procedures in Vacuum Production and Measurement,** *W. Brunner and H. Patton.* **Section 2: Experiments which Illustrate the Characteristics of the Vacuum Environment:** Demonstration of the outgassing of different vacuum materials, *F. Rosebury.* Comparison of gas evolution phenomenon from glass and metal system envelopes during baking, *R. Lawson.* Determination of the net quantity of gas flowing through a cylindrical tube, *K. Busen.* **Section 3: Experiments which Illustrate the Dependence of the Physical Properties of Gases on Gas Density:** Measurement of the pumping action of an ionization gauge, *H. Farber.* Study of the linearity of an ionization gauge, *J. Miller, III.* Calibration of gauges, *C. Morrison.* **Section 4: Experiments which Examine Physical and Chemical Interactions at Surfaces:** Study of the sorption of gases for different gas–sorbent combinations, *K. Wear.* The use of sorbents as traps and pumps, *H. Farber.* Sorption of gases

by titanium, *H. Farber.* Investigation of the passage of oxygen across a silver barrier, *K. Busen.* **Section 5: Processes Requiring a Vacuum Environment:** Thin film evaporation, *M. Thomas.* Fabrication of a nichrome resistor, *R. Riegert and G. Breitweiser.* Sputtering, *P. Grosewald.* Ejection patterns in single crystal sputtering, *G. Wehner.* **Section 6: Special Projects:** Study of the sublimation of ice at various pressures, *W. Parker.* Study of friction, *P. McElligott.* Measurement of the mean free path of conduction electrons in silver, *R. Olson and J. Wilson.* Construction and use of a cathode ray tube, *B. Kendall and H. Luther.* Construction of a vacuum triode using solder glass techniques, *J. King and J. Orsula.* Experiments using solder glass techniques, *D. Whitcomb.* **Section 7: Speculations:** Original thought experiments, *M. Carbone.* Provocative ideas and questions, *N. Milleron.*

BEER   *Liquid Metals:* Chemistry and Physics

(Monographs and Textbooks in Material Science Series, Volume 4)

edited by SYLVAN Z. BEER, *Converta Enterprises, Inc., Syracuse, New York*

742 pages, illustrated. 1972

Presents a comprehensive review of the research done on the liquid state of metals, bringing together the latest advances, as well as data previously scattered among a wide variety of publications. Of prime importance to chemists, physicists, research metallurgists, metallurgical engineers, and materials scientists working in the areas of liquid-state theory, the theory of metals, process metallurgy involving liquid metals, and high-temperature chemistry.

CONTENTS: On the thermodynamic formalism of metallic solutions, *C. H. P. Lupis.* Kinetics of evaporation of various elements from liquid iron alloys under vacuum, *R. Ohno.* Relation between thermodynamic and electrical properties of liquid alloys, *D. N. Lee and B. D. Lichter.* The surface tension of liquid metals, *B. C. Allen.* Significant structure theory applied to liquid metals, *S. M. Breitling and H. Eyring.* Diffraction analysis of liquid metals and alloys, *C. N. J. Wagner.* The optical properties of liquid metals, *J. N. Hodgson.* Effect of pressure on the properties of liquid metals, *A. Rapoport.* Sound propagation in liquid metals, *R. T. Beyer and E. M. Ring.* The viscosity of liquid metals, *R. T. Beyer and E. M. Ring.* Magnetic properties of liquid metals, *R. Dupree and E. F. W. Seymour.* Diffusion in liquid metals, *N. H. Nachtrieb.* Electromigration in liquid alloys, *S. G. Epstein.* Electronic nature of liquid metals and liquid metal theory, *J. E. Enderby.* Structure and properties of noncrystalline metallic alloys produced by rapid quenching of liquid alloys, *B. C. Giessen and C. N. J. Wagner.*

BLACK and PRESTON *High-Modulus Wholly Aromatic Fibers*

(Fiber Science Series, Volume 5)

edited by W. BRUCE BLACK, *Monsanto Textiles Company, Pensacola, Florida* and JACK PRESTON, *Monsanto Textiles Company, Chemstrand Research Center, Durham, North Carolina*

304 pages, illustrated. 1973

Based on a symposium on high–modulus aromatic fibers held by the American Chemical Society in Boston on April 13, 1972. The first formal publication of research which shows the relationship of fiber properties to polymer structure. Extremely significant reading for all fiber scientists and material scientists; plastics scientists and engineers interested in fiber-reinforced plastics; spacecraft and aircraft oriented engineers; and scientists in the industrial fiber, sports equipment, and airframe fields.

CONTENTS: High-modulus wholly aromatic fibers: Introduction to the Symposium and historical perspective, *W. Black.* High-modulus wholly aromatic fibers. I. Wholly ordered poly-amide-hydrazines and poly-1,3,4,-oxadizole-amides, *J. Preston, W. Black and W. Hofferbert, Jr.* High-modulus wholly aromatic fibers. II. Partially ordered polyamide-hydrazides, *J. Preston, W. Black, and W. Hofferbert, Jr.* Self-regulating polycondensations. II. A study of the order present in polyamide-hydrazides derived from terephthaloyl chloride and p-aminobenz-hydrazide, *R. Morrison, J. Preston, J. Randall, and W. Black.* Self-regulating polycondensations. III. NMR analysis of oligomers derived from terephthaloyl chloride and p-aminobenz-hydrazide, *J. Randall, R. Morrison, and J. Preston.* Some physical and mechanical properties of some high-modulus fibers prepared from all-para aromatic polyamide-hydrazides, *W. Black, J. Preston, H. Morgan, G. Raumann, and M. Lilyquist.* Morphology and crystal structure of wholly aromatic all-para polyamide-hydrazide polymers, *V. Holland.* X-ray study of an all-para wholly aromatic polyamide-hydra-zide[a,b], *R. Miller.* Molecular weight characterization of wholly para-oriented, aromatic polyamide-hydrazides and wholly aromatic polyamides, *J. Burke.* Construction and properties of fabrics of high-modulus organic fibers useful for composite reinforcing, *M. Lilyquist, R. DeBrunner, and J. Fincke.* Mechanical properties of a high-modulus polyamide-hydrazide fiber in composites and of the polyamide-hydrazide fiber and fabric composites, *D. Zaukelies and B. Daniels.* Tire cord application of high-modulus fibers derived from polyamide-hydrazides, *G. Raumann and J. Brownlee.* The application of high-modulus fibers to ballistic protection, *R. Laible, F. Figucia, and W. Ferguson.* High-modulus wholly aromatic fibers. III. Random copolymers containing hydrazide and/or carbonamide linkages, *J. Preston, H. Morgan, and W. Black.*

BOLKER *Natural and Synthetic Polymers: An Introduction*

by HENRY I. BOLKER, *Department of Chemistry, McGill University, Montreal, Quebec*

in preparation. 1973

Presents a unified approach to polymer chemistry, with equal emphasis on natural and synthetic polymers, and is arranged in a logical sequence of topics based on increasing complexity of molecular architecture. Useful as a textbook for a first course in polymer chemistry and as a reference book for workers in the field.

CONTENTS: Introduction • Natural condensation polymers: The linear polysaccharides • Synthetic condensation (step-growth) polymers • Addition (chain-growth) polymers • Stereoregularity in addition polymers • Branched homopolymers: Synthetic and natural • Natural heteropolymers: I. Heteropolysaccharides • Natural heteropolymers: II. Nucleic acids • Copolymers and copolymerization • Cross-linking in synthetic polymers • Natural heteropolymers: III. Polypeptides and proteins • Lignins.

BROWNING *Analysis of Paper*

by B. L. BROWNING, *The Institute of Paper Chemistry, Appleton, Wisconsin*

352 pages, illustrated. 1969

Provides comprehensive coverage of methods for chemical analysis of paper. Is of value to manufacturers of paper and paper board, suppliers of components or of additives introduced into paper, converters and printers, purchasers and users, librarians, and others concerned with the properties, behavior, and applications of paper that are related to composition.

CONTENTS: Paper as a commodity • Sampling and preparation of sample • Determination of moisture • Fiber analysis • Fiber quality methods • Lignin • Rosin size • Starch • Proteins • Coatings • Waxes and oils • Fillers and white coating pigments • Dyes and colored pigments • Acidity and alkalinity • Residues and impurities • Biological control agents • General identification of additives in paper • Synthetic resins • Wet-strength agents • Polysaccharides and gums • Miscellaneous additives • Noncellulose fibers • Specks and spots • Permanence of paper • Paper in forensic science.

BUTLER, O'DRISCOLL, and SHEN *Reviews in Macromolecular Chemistry*

(Book Edition)

edited by GEORGE B. BUTLER, *Department of Chemistry, University of Florida,*

(continued)

BUTLER, O'DRISCOLL, and SHEN *(continued)*

Gainesville, and KENNETH F. O'DRISCOLL, *Department of Chemical Engineering, University of Waterloo, Ontario, Canada* and MITCHEL SHEN, *Department of Chemical Engineering, University of California, Berkeley*

Vol. 1 *out of print*

Vol. 2 388 pages, illustrated. 1968

Vol. 3 430 pages, illustrated. 1969

Vol. 4 428 pages, illustrated. 1970

Vol. 5, Part I see NEUSE and ROSENBERG

Vol. 5, Part II
250 pages, illustrated. 1970

Vol. 6 498 pages, illustrated. 1971

Vol. 7 314 pages, illustrated. 1972

Vol. 8 346 pages, illustrated. 1972

Vol. 9 380 pages, illustrated. 1973

Reviews of the currently published literature for those who wish to keep abreast of the new and rapidly advancing developments in macromolecular chemistry. Of interest to organic and physical chemists, biochemists, engineers, and all students and research workers in polymer chemistry and related fields.

CONTENTS:

Volume 1: Application of molecular orbital theory to vinyl polymerization, *K. F. O'Driscoll and T. Yonezawa.* Poly(alkylene oxides), *A. E. Gurgiolo.* Polyurethanes, *D. J. Lyman.* Uncatalyzed, uninhibited thermal oxidation of saturated polyolefins, *L. Reich and S. S. Stivala.* Double-strand polymers, *W. De Winter.* Biomedical polymers, *D. J. Lyman.* Gel permeation chromatography with organic solvents, *J. F. Johnson, R. S. Porter, and M. J. R. Cantow.*

Volume 2: Phosphorus-containing polymers: Introduction, *M. Sander and E. Steininger.* Linear polymers with phosphorus in side chains, *M. Sander and E. Steininger.* Linear polymers with phosphorus and carbon in the main chain, *M. Sander and E. Steininger.* Reassessment of the theory of polyesterification with particular reference to alkyd resins, *D. H. Solomon.* Symmetry considerations for stereoregular polymers, *A. M. Liquori.* Copolymerization of vinyl monomers with ring compounds, *R. A. Patsiga.* Application of high-resolution nuclear magnetic resonance to polymer structure determination, I., *K. C. Ramey and W. S. Brey, Jr.* Ten years of polymer single crystals, *D. A. Blackadder.* Thermal degradation of polystyrene, *G. G. Cameron and J. R. MacCallum.*

Volume 3: Phosphorous-containing resins, *M. Sander and E. Steininger.* Inorganic phosphorous polymers, *M. Sander and E. Steininger.* Phosphorylation of polymers, *M. Sander and E. Steininger.* Sulphur-containing polymers, *E. J. Goethals.* Polymer molecular weight distributions, *N. Amundson and D. Luss.* Heteroatom

ring-containing polymers, *A. D. Delman.* Molecular theories of rubber-like elasticity and polymer viscoelasticity, *M. Shen, W. F. Hall, and R. E. DeWames.* End-group studies using dye techniques, *S. R. Palit and B. M. Mandal.* Free-radical spin labels for macromolecules, *J. D. Ingham.* The synthesis of thermally stable polymers: A progress report, *J. I. Jones.*

Volume 4: Polymer enzymes and enzyme analogs, *A. S. Lindsey.* Stability of polycarbonate, *A. Davis and J. Golden.* Cross-linking — effect on physical properties of polymers, *L. E. Nielsen.* The synthesis of thermally stable polymeric azomethines by polycondensation reactions, *G. F. D'Alelio and R. K. Schoenig.* On the dehydrochlorination and the stabilization of polyvinyl chloride, *M. Onozuka and M. Asahina.* Recent advances in the development of flame-retardant polymers, *A. D. Delman.* Thermodynamics of polymerization. I, *H. Sawada.*

Vol. 5, Part II: Ring-chain equilibria, *H. Allcock.* Occupied volume of liquids and polymers, *R. Haward.* The application of ESR techniques to high polymer fracture, *H. Kausch-Blecken von Schmeling.* The science of determining copolymerization reactivity ratios, *P. Tidwell and G. Mortimer.* Block polymers and related heterophase elastomers, *G. Estes, S. Cooper, and A. Tobolsky.*

Volume 6: Proton magnetic resonance of molecular interactions in polymer solutions, *K.-J. Liu and J. E. Anderson.* Preparation and polymerization of vinyl heterocyclic compounds, *K. Takemoto.* Catalysis in isocyanate reactions, *K. C. Frisch and L. P. Rumao.* Thermodynamics of polymerization. II. Thermodynamics of ring—opening polymerization, *H. Sawada.* Copolymers of naturally occurring macromolecules, *I. C. Watt.* Molecular configuration and pyrolysis of phenolic-novolaks, *E. L. Winkler and J. A. Parker.* Physical properties of ionic polymers, *E. P. Otocka.* Synthesis and properties of polyphenyls and polyphenylenes, *J. G. Speight, P. Kovacic, and F. W. Koch.* Dependence of flow properties on molecular weight, temperature, and shear, *A. Casale, R. S. Porter, and J. F. Johnson.* Synthesis methods and properties of polyazoles, *V. V. Korshak and M. M. Teplyakov.*

Volume 7: Linear polyquinoxalines, *P. M. Hergenrother.* Nylons—known and unknown. A comprehensive index of linear aliphatic polyamides of regular structure, *H. K. Livingston, M. S. Sioshansi, and M. D. Glick.* Recent advances in polymer fractionation, *L. H. Tung.* Rheology of adhesion, *D. H. Kaelble.* Solvation of synthetic and natural polyelectrolytes, *B. E. Conway.* Hydrogen transfer polymerization with anionic catalysts and the problem of anionic isomerization polymerization, *J. P. Kennedy and T. Otsu.*

Volume 8: Polymerization by carbenoids, carbenes, and nitrenes, *M. Imoto and T. Nakaya.* Collagen and gelatin in the solid state, *I. V. Yannas.* Ring-opening polymerization of cycloolefins, *N. Calderon.* Thermodynamics of polymerization. III, *H. Sawada.* Polymerization of N-vinylcarbazole initiated by metal salts, *M. Biswas and D. Chakravarty.* Vibrational spectroscopy of polymers, *F. J. Boerio and J. L. Koenig.* Polymer compatibility, *S. Krause.*

Volume 9: Mechanical properties of polymers: The influence of molecular weight and molecular weight distribution, *J. Martin, J. Johnson, and A. Cooper.* On the mathematical modeling of polymerization reactors, *W. Ray.* Anionic cyclopolymerization, *C. McCormick and G. Butler.* Carbon-13 NMR of polymers, *V. Mochel.* Thermodynamics of polymerization. IV. Thermodynamics of equilibrium polymerization, *H. Sawada.*

CARROLL *Physical Methods in Macromolecular Chemistry*

a series edited by BENJAMIN CARROLL, *Rutgers—The State University, Newark, New Jersey*

Vol. 1 400 pages, illustrated. 1969

Vol. 2 384 pages, illustrated. 1972

A series which reviews why and how analytical methods are used in the study of macromolecules. Each method is critically discussed by experts in the field. Directed to researchers in polymer chemistry, biopolymers, and organic and inorganic chemistry.

CONTENTS:

Volume 1: Surface chemistry and polymers, *M. Rosoff.* Internal reflection spectroscopy, *J. K. Barr and P. A. Flournoy.* Electric properties of synthetic polymers, *E. O. Forster.* Assessing radiation effects in polymers, *P. Y. Feng and E. S. Freeman.* Fluorescence techniques for polymer solutions, *D. J. R. Laurence.* Insoluble polymers: Molecular weights and their distributions, *H. C. Cheung.*

Volume 2: Gel permeation chromatography in polymer chemistry, *D. D. Bly.* Interactions of polymers with small ions and molecules, *D. J. R. Laurence.* Electric properties of biopolymers: Proteins, *E. O. Forster and A. P. Minton.* Thermal methods, *E. P. Manche and B. Carroll.*

CARTER *Essential Fiber Chemistry*

(Fiber Science Series, Volume 2)

by MARY E. CARTER, *FMC Corporation, American Viscose Division, Marcus Hook, Pennsylvania*

232 pages, illustrated. 1971

Discusses the chemical and physical structure and properties of ten commercially important fibers. Useful to all chemists interested in the research and development of natural and man-made fibers.

CONTENTS: Cotton • Rayon • Cellulose acetate • Wool • Polyamide • Acrylic fibers • Polyethylene terephthalate • Polyolefins • Spandex • Glass.

CONLEY *Thermal Stability of Polymers*

In 2 Volumes

(Monographs in Macromolecular Chemistry Series)

edited by R. T. CONLEY, *Wright State University, Dayton, Ohio*

Vol. 1 656 pages, illustrated. 1970

Vol. 2 in preparation. 1974

CONTENTS: Introduction, *R. T. Conley.* Molecular structure and stability criteria, *R. T. Conley.* The relationship between the kinetics and mechanism of thermal depolymerization, *R. H. Boyd.* Random scission processes, *A. V. Tobolsky, A. M. Kotliar, and T. C. P. Lee.* Fundamental reactions in oxidation chemistry, *P. M. Norling and A. V. Tobolsky.* Thermal and oxidative degradation of polyethylene, polypropylene, and related olefin polymers, *R. H. Hansen.* Thermal and oxidative degradation of natural rubber and allied substances, *E. M. Bevilacqua.* Vinyl and vinylidene polymers, *R. T. Conley and R. Malloy.* Fluorocarbon polymers, *W. W. Wright.* Thermal and thermo-oxidative degradation of polyamides, polyesters, polyethers, and related polymers, *R. T. Conley and R. A. Gaudiana.* Thermosetting resins, *R. T. Conley.* Thermal and thermo-oxidative degradation of cellulosic polymers, *R. T. Conley.* Heterocyclic polymers, *G. P. Shulman.* Degradation of inorganic polymers, *J. Economy and J. H. Mason.*

D'ALELIO and PARKER
Ablative Plastics

edited by GAETANO F. D'ALELIO, *Department of Chemistry, University of Notre Dame, Indiana,* and JOHN A. PARKER, *NASA, Ames Research Center, Moffet Field, California*

504 pages, illustrated. 1971

Provides the comprehensive and rational approach required for the design and production of reliable head shields for future space missions. Includes discussions on the various aspects of heat-rejection mode as a function of heating rate; the nature of the heat transfer, both radiative and conductive; and the nature of degrading polymers. A valuable reference for all aerospace scientists, polymer chemists, physicists, and aerodynamic engineers.

CONTENTS: Ablative polymers in aerospace technology, *D. L. Schmidt.* Hypervelocity heat protection—a review of laboratory experiments, *N. S. Vojvodich.* A review of ablative studies of interest to naval applications, *F. J. Koubek.* Structural design and thermal properties of polymers, *G. F. D'Alelio.* Characterization of an epoxy-anhydride ablative system using com-

(continued)

D'ALELIO and PARKER *(continued)*

puter treatment of analytical results, *C. G. Taylor and E. L. Pendleton.* The synthesis and characterization of some potential ablative polymers, *R. Y. Wen, L. F. Sonnabend, and R. Eddy.* Thermal degradation and curing of polyphenylene, *D. N. Vincent.* Thermosetting polyphenylene resin—its synthesis and use in ablative composites, *N. Bilow and L. J. Miller.* Structural ablative plastics, *R. M. Lurie, S. F. D'Urso, and C. K. Mullen.* Prediction of heat shield performance in terms of epoxy resin structure, *G. J. Fleming.* Ablative resins for hyperthermal environments, *B. S. Marks and L. Rubin.* The development of polybenzimidazole composites as ablative heat shields, *R. R. Dickey, J. H. Lundell, and J. A. Parker.* Ablative degradation of a silicon foam, *T. McKeon.* Thermophysical characteristics of high-performance ablative composites, *M. L. Minges.* The design and development of a high-heating rate thermogravimetric analyzer suitable for use with ablative plastics, *A. M. Melnick and E. J. Nolan.* Pyrolysis kinetics of nylon 6-6, phenolic resin and their composites, *H. E. Goldstein.* Pyrolysis-gas chromatography as a tool for studying the degradation of ablative plastics, *R. M. Ross.* Nonequilibrium flow and the kinetics of chemical reactions in the char zone, *G. C. April, R. W. Pike, and E. G. del Valle.* Arc-image testing of ablation materials, *E. M. Liston.* Development and characterization of a radio frequency-transparent ablator, *E. L. Strauss.* Tailoring polymers for entry into the atmosphere of Mars and Venus, *R. G. Nagler.*

DIGGLE *Oxides and Oxide Films*
in multi-volumes

(The Anodic Behavior of Metals and Semiconductors Series)

edited by JOHN W. DIGGLE, *Research School of Chemistry, The Australian National University, Canberra*

Vol. 1 552 pages, illustrated. 1972
Vol. 2 424 pages, illustrated. 1973

Treats the anodic behavior of metals and semiconductors and peripheral areas in an authoritative and interdisciplinary manner. The initial volumes deal with the physics and chemistry of oxides and oxide films. Of great value for all those involved in electrochemistry, materials science, solid state physics, electrical engineering, metallurgy, corrosion science, semiconductors, and electrochemical technology.

CONTENTS:

Volume 1: Passivation and passivity, *V. Brusić.* Mechanisms of ionic transport through oxide films, *M. J. Dignam.* Electronic current flow through ideal dielectric films, *C. A. Mead.* Electrical double layer at metal oxide-solution interfaces, *S. Ahmed.*

Volume 2: Anodic oxide films. Influence of solid-state properties on electrochemical behavior, *A. Vijh.* Dielectric loss mechanism in amorphous oxide films, *D. M. Smyth.* Porous anodic films in aluminum, *G. C. Wood.* Dissolution of oxide phases, *J. W. Diggle.*

FOURT and HOLLIES
Clothing: Comfort and Function

(Fiber Science Series, Volume 1)

by LYMAN FOURT and NORMAN HOLLIES, *Gillette Research Institute, Rockville, Maryland*

272 pages, illustrated. 1970

A unified review of the present state of knowledge in the science of clothing. Of interest to textile scientists, fiber producers and marketers, textile converters, and garment makers.

CONTENTS: The factors involved in the study of clothing • Clothing considered as a system interacting with the body • Clothing considered as a structural assemblage of materials • Heat and moisture relations in clothing • Physiological and field testing of clothing by wearing it • Physical properties of clothing and clothing materials in relation to comfort • Differences between fibers with respect to comfort • Current trends and new developments in the study of clothing.

FRISCH and REEGEN *Ring-Opening Polymerization*

(Kinetics and Mechanisms of Polymerization Series, Volume 2)

edited by KURT C. FRISCH, and SIDNEY L. REEGEN, *Polymer Institute, University of Detroit, Michigan*

544 pages, illustrated. 1969

Covers the polymerization of important classes of cyclic monomers such as ethylene and propylene oxide, alkylenimines, and sulfides, lactones, lactams, cyclic silicone compounds, and cyclic nitrogen containing heterocycles. Of great interest to the industrial, commercial, and academic worlds as it has application to elastomers, coatings, fibers, films, and foams.

CONTENTS: 1,2 Epoxides, *Y. Ishii and S. Sakai.* 1,3 Epoxides and higher epoxides, *P. Dreyfuss and M. P. Dreyfuss.* Cyclic formals, *J. Furukawa and K. Tada.* Cyclic sulfides, *P. Sigwalt.* Alkylenimines, *M. Hauser.* Lactones, *R. D. Lundberg and E. F. Cox.* Lactams, *H. K. Reimschuessel.* Cyclic siloxanes and silazanes, *E. E. Bostick.* Nitrogen-containing heterocyclic compounds, *V. Kargin and V. Kabanov.* N-carboxy-α-amino acid anhydrides, *Y. Shalitin.*

FRISCH and SAUNDERS
Plastic Foams

(Monographs on Plastic Series, Volume 1)

edited by KURT C. FRISCH, *University of Detroit, Michigan,* and JAMES H. SAUNDERS, *Monsanto Company, Pensacola, Florida*

Part I 464 pages, illustrated. 1972

Part II 704 pages, illustrated. 1973

Gives an integrated picture of the fundamental principles, technology, and applications of foams, and offers a thorough treatment of specific types of plastic foams. Emphasis is placed on the newer trends in this science.

Of particular value to chemists and engineers engaged in research and development, and marketing and production personnel in the polymer and plastics industry.

CONTENTS:

Part I: Introduction, *K. C. Frisch.* The mechanism of foam formation, *J. H. Saunders and R. H. Hansen.* Flexible polyurethane foams, *G. T. Gmitter, H. J. Fabris, and E. M. Maxey.* Sponge rubber and latex foam, *R. L. Zimmerman and H. R. Bailey.* Polyolefin foams, *D. J. Sundquist.* Polyvinyl chloride foams, *A. C. Werner.* Silicone foams, *H. L. Vincent.* Testing of cellular materials, *R. A. Stengard.*

Part II: Rigid urethane foams, *J. K. Backus and P. G. Gemeinhardt.* Polystyrene and related thermoplastic foams, *A. R. Ingram and J. Fogel.* Phenolic foams, *A. J. Papa and W. R. Proops.* Urea-formaldehyde foams, *K. C. Frisch.* Epoxy-resin foams, *H. Lee and K. Neville.* New high-temperature-resistant plastic foams, *E. E. Hardy and J. H. Saunders.* Miscellaneous foams, *K. C. Frisch.* Inorganic foams, *M. Wismer.* Effects of cell geometry on foam performance, *R. H. Harding.* Thermal decomposition and flammability of foams, *P. E. Burgess, Jr. and C. J. Hilado.* Foams in transportation, *M. Kaplan and L. M. Zwolinski.* Architectural uses of foam plastics, *S. C. A. Paraskevopoulos.* Military and space applications of cellular materials, *R. J. F. Palchak.*

GARG, SVALBONAS, and GURTMAN
Analysis of Structural Composite Materials

(Monographs and Textbooks in Material Science Series, Volume 6)

by SABODH GARG, *Systems, Science and Software Company, La Jolla, California,* VYTAS SVALBONAS, *Grumman Aerospace Corporation, Bethpage, New York,* and GERRY GURTMAN, *Systems, Science and Software Company, La Jolla, California*

552 pages, illustrated. 1973

Compares various theories for the static and dynamic analysis of structural composite materials. Deals with the elastic properties of laminated composites and particulate and unidirectional fiber reinforced composites, composite strength, and stress wave propagation. May be used as a textbook for a graduate course in composites and is of interest to researchers and analysts in any industry that uses composites.

CONTENTS: Why composites? • Simple analytic models • Elasticity analyses • Bounds on elastic properties by energy methods • Multilayer laminates • Non-statistical models of composite strength • Statistical tensile strength of fiber and fiber bundles • Composite tensile-strength models • Cumulative weakening model including stress concentrations • Compressive strength of composites • Theory of breaking kinetics • Introduction to elastic wave propagation in composites • Approximate analysis techniques for stress wave propagation in composites • Application of continuum mixture theories to the study of elastic wave propagation in composite materials • Shock waves in composite materials.

HAM *Vinyl Polymerization*
In 2 Parts

(Kinetics and Mechanisms of Polymerization Series, Volume 1)

edited by GEORGE E. HAM, *Geigy Chemical Corporation, Ardsley, New York*

Part I 560 pages, illustrated. 1967

Part II 432 pages, illustrated. 1969

"The book is a good introduction to the series. It has provided a sound basis for subsequent volumes and should serve as an important reference text for students and researchers in polymer chemistry."—B. D. Gesner, Bell Telephone Labs., *SPE Journal*

"The book is highly recommended."—Arthur Tobolsky, *The American Scientist*

CONTENTS:

Part I: General aspects of free-radical polymerization, *G. E. Ham.* The mechanism of cyclopolymerization of nonconjugated diolefins, *W. E. Gibbs and J. M. Barton.* Styrene, *M. H. George.* Mechanism of vinyl acetate polymerization, *M. K. Lindemann.* Polymerization of vinyl chloride and vinylidene chloride, *G. Talamini and E. Peggion.* Occlusion phenomena in the polymerization of acrylonitrile and other monomers, *A. D. Jenkins.* Polymerization of acrolein, *R. C. Schulz.* Heats of polymerization and their structural and mechanistic implications, *R. M. Joshi and B. J. Zwolinski.*

Part II: Mechanism of emulsion polymerization, *J. W. Vanderhoff.* Elucidation of emulsion polymerization mechanism based upon copolymer studies, *W. F. Fowler, Jr.* Mechanism of the emulsion polymerization of ethylene, *H. K. Stryker, G. J. Mantell, and A. F. Helin.* Mechanism of stereospecific polymerization of propylene, *W. "E" Smith.* Anionic polymerization,

(continued)

HAM *(continued)*

M. Morton. Mechanisms of cationic polymerization, *Z. Zlámal.* Radiation-induced polymerization, *Y. Tabata.*

HENCH and DOVE *Physics of Electronic Ceramics*

In 2 Parts

(Ceramics and Glass: Science and Technology Series, Volume 2)

edited by LARRY L. HENCH and DEREK B. DOVE, *College of Engineering, University of Florida, Gainesville.*

Part A 584 pages, illustrated. 1971

Part B 576 pages, illustrated. 1972

A highly useful treatise which deals with the physical basis for the behavior of electronic ceramics.

Fundamental physical theories describing each type of electronic ceramics are presented, with discussions included that relate the theories to the applications of the materials. Of special value to graduate students who have had a course in modern physics and also of interest to materials scientists and engineers in electronics, communications, and ceramics.

CONTENTS:

Part A: Quantum mechanics and ceramics, *J. C. Slater.* Band structure and electronic properties of ceramic crystals, *D. Adler.* Electrical conduction in low mobility materials, *I. Bransky and N. M. Tallan.* Defect structure and electronic properties of ceramics, *R. W. Vest.* Conduction domains in solid mixed conductors and electrolytic domain of calcia stabilized zirconia, *J. Patterson.* Semiconducting glasses, *J. D. Mackenzie.* Electronic processes in amorphous semiconductors, *E. A. Davis.* Heterogeneous semiconducting glasses, *H. F. Schaake.* The determination of local order in amorphous semiconducting films, *D. B. Dove.* Negative capacitance effects in amorphous semiconductors, *M. Allen, P. Walsh, and W. Doremus.* Some conduction phenomena in amorphous materials, *K. L. Chopra.* Applications of thin film dielectrics in microelectronics, *N. N. Axelrod.* Substructure and electrical conduction in amorphous thin films, *N. Fuschillo and A. D. McMaster.* Structure of surface defects, *D. L. Stoltz and J. J. Hren.* Electronic surface properties, *P. Mark.* Electron spin resonance and defects in solids, *W. S. Brey, R. B. Gammage, and Y. P. Virmani.* Theory of linear dielectrics, *A. D. Franklin.* Polycrystalline insulators, *H. C. Graham and N. M. Tallan.* Electrical conduction in glass and glass-ceramics, *D. L. Kinser.* Dielectric breakdown of ceramics, *G. C. Walther and L. L. Hench.*

Part B: Some structural mechanisms in ferroelectricity, *R. Pepinsky.* Thermodynamic phenomenology of ferroelectricity in single crystal

and ceramic systems, *L. E. Cross.* Dynamical effects in solid state phase transformations, *J. D. Axe.* Theory of antiferromagnetism and ferrimagnetism, *J. B. Goodenough.* Microstructure and processing of ferrites, *F. J. Schnettler.* Microwave garnet compounds, *G. R. Harrison and L. R. Hodges, Jr.* The optical absorption of glasses, *N. J. Kreidl.* Light scattering from glass, *J. J. Hammel.* Electro-optical and magneto-optical effects, *Y. R. Shen.* The influence of the composition of the gain of Nd-doped glasses, *C. F. Rapp.* Solid state reactions in the preparation of zircon stains, *R. A. Eppler.* Computer color control for ceramic wall tile, *W. K. Culbreth, Jr.* Ceramics and glasses — some uses in the communications industry, *D. G. Thomas.*

HENCH and GOULD *Characterization of Ceramics*

(Ceramics and Glass: Science and Technology Series, Volume 3)

edited by LARRY L. HENCH and ROBERT W. GOULD, *University of Florida, Gainesville*

672 pages, illustrated. 1971

Focuses on the two major directions which comprise the distinct discipline of ceramic characterization: the exploration of the factors that control the properties of the final product and the rapid development of high resolution analytical technisues used for ceramic materials. A particularly timely textbook for an advanced undergraduate or graduate materials science curriculum. Valuable for all materials scientists and ceramic and materials engineers working on the development of an effective and economical materials characterization program.

CONTENTS: Introduction to the characterization of ceramics, *L. L. Hench.* **Part 1: Chemical Analysis:** General analytical chemistry, *P. Rankin.* X-ray spectroscopy, *R. W. Gould.* Atomic absorption flame spectrometry, *J. D. Winefordner.* **Part 2: Phase State and Structure:** X-ray diffraction, *R. W. Gould.* Transmission electron microscopy and electron diffraction, *C. F. Tufts.* Analysis of microstructural defects, *R. W. Newman.* Petrographic analysis, *V. D. Fréchette.* Thermal analysis, *R. K. Ware.* Point defect analysis, *W. J. James and G. Lewis.* **Part 3: Size, Shape, Strain, and Surface of Powders:** Physical characterization, *D. R. Lankard and D. E. Niesz.* Small angle x-ray scattering, *R. W. Gould.* X-ray line profile analysis, *R. W. Gould.* Scanning electron microscopy, *S. R. Bates.* Light scattering, *J. H. Boughton.* Characterization of powder surfaces, *L. L. Hench.* **Part 4: Microstructure:** Electron microprobe, *G. Lewis.* Quantitative stereology, *R. T. DeHoff.* Applied stereology, *S. W. Freiman.* **Part 5: Surfaces:** Characterization of ceramic surfaces, *L. Berrin and R. C. Sundahl.*

KATON *Organic Semiconducting Polymers*

(Monographs in Macromolecular Chemistry Series)

edited by J. E. KATON, *Miami University, Oxford, Ohio*

328 pages, illustrated. 1968

CONTENTS: Basic physics of semiconductors, *D. E. Hill.* Theoretical aspects of the electronic behavior of organic macromolecular solids, *H. A. Pohl.* Recent experimental aspects of the electronic behavior of organic macromolecular solids, *S. Kanda and H. A. Pohl.* Semiconducting organic polymers containing metal groups, *B. A. Bolto.* Semiconducting biological polymers, *D. D. Eley.*

KETLEY *The Stereochemistry of Macromolecules*

In 3 Volumes

edited by A. D. KETLEY, *W. R. Grace & Co., Clarksville, Maryland*

Vol. 1 424 pages, illustrated. 1967

Vol. 2 400 pages, illustrated. 1967

Vol. 3 476 pages, illustrated. 1968

CONTENTS:

Volume 1: Ziegler-Natta polymerization: Catalysts, monomers, and polymerization procedures, *D. O. Jordan.* The mechanism of Ziegler-Natta catalysis. I. Experimental foundations, *D. F. Hoeg.* Mechanism of Ziegler-Natta polymerization. II. Quantum-chemical and crystal-chemical aspects, *P. Cossee.* Copolymerization of olefins by Ziegler-Natta catalysts, *I. Pasquon, A. Valvassori, and G. Sartori.* Polymerization of dienes by Ziegler-Natta catalysts, *W. Marconi.* Manufacture and commercial applications of stereoregular polymers, *M. Compostella.*

Volume 2: Stereospecific polymerization of vinyl-type monomers and dienes by alkali-metal-based catalysts, *D. Braun.* Stereospecific polymerization of vinyl ethers, *A. D. Ketley.* Ionic polymerization of aldehydes, ketones, and ketenes, *G. F. Pregaglia and M. Binaghi.* Stereospecific polymerization of epoxides, *T. Tsuruta.* Stereochemistry of free-radical polymerizations, *W. Cooper.* Conformational effects induced in polymers by rigid matrices, *N. Marans.* Simple stereoregular polymers in biological systems, *J. N. Baptist.*

Volume 3: Chain conformation and crystallinity, *P. Corradini.* High-resolution nuclear magnetic resonance of synthetic polymers, *J. C. Woodbrey.* Vibrational analyses of the infrared spectra of stereoregular polypropylene, *T. Miyazawa.* Optically active stereoregular polymers, *M. Farina and G. Bressan.* Physical properties of stereoregular polymers in solid state, *J. F. Johnson and R. S. Porter.* Properties of synthetic linear stereoregular polymers in solution, *V. Crescenzi.* Macromolecules as information storage systems, *A. M. Liquori.* Automata theo-

ries of hereditary tactic copolymerization, *H. H. Pattee.* Effect of microtacticity on reactions of polymers, *M. M. van Beylen.* Degradation of stereoregular polymers, *H. H. G. Jellinek.*

KURYLA and PAPA *Flame Retardancy of Polymeric Materials*

a series edited by WILLIAM C. KURYLA and ANTHONY J. PAPA, *Union Carbide Corporation, South Charleston, West Virginia*

Vol. 1 352 pages, illustrated. 1973

Vol. 2 296 pages, illustrated. 1973

A series concerned with the various modes of rendering polymeric materials fire resistant, which emphasizes specific reagents and techniques in use today. Treats each class of polymer separately to aid the fabricator in gaining an understanding of the specific problems associated with its flammability characteristics, and to review the science and practical solution to its flame retardancy. Of special benefit to industrial polymer chemists, plastics engineers, and fabricators of polymeric materials.

CONTENTS:

Volume 1: Available flame retardants, *W. Kuryla.* Inorganic flame retardants and their mode of action, *J. Pitts.* Fire retardation of polyvinyl chloride and related polymers, *M. O'Mara, W. Ward, D. Knechtges, and R. Meyer.* Fire retardation of wool, nylon, and other natural and synthetic polyamides, *G. Crawshaw, A. Delman, and P. Mehta.*

Volume 2: Fire retardation of polystyrene and related thermoplastics, *R. Lindemann.* Fire retardation of polyethylene and polypropylene, *R. Schwarz.* Flame retardation of natural and synthetic rubbers, *H. Fabris and J. Sommer.* Flame retardancy of phenolic resins and urea- and melamine-formaldehyde resins, *N. Sunshine.*

LEFEVER *Aspects of Crystal Growth*

(Preparation and Properties of Solid State Materials Series, Volume 1)

edited by R. A. LEFEVER, *Sandia Laboratories, Albuquerque, New Mexico*

Vol. 1 296 pages, illustrated. 1971

Concerns certain aspects of the growth and properties of single crystals. Directed to both beginning and experienced crystal growers, material scientists, and solid state physicists.

CONTENTS: A review of the preparation of single crystals by fused melt electrolysis and some general properties, *W. Kunnmann.* The role of mass transfer in crystallization processes, *W. Wilcox.* Exploratory flux crystal growth, *A. Chase.*

LOWRY *Markov Chains and Monte Carlo Calculations in Polymer Sciences*

(Monographs in Macromolecular Chemistry Series)

edited by George G. Lowry, *Western Michigan University, Kalamazoo*

344 pages, illustrated. 1970

Written for the polymer chemist who, although not primarily concerned with mathematical theories, desires a working knowledge of the topics treated. Begins with an introduction to the principles involved and later exemplifies some significant applications of Markov chain theory and Monte Carlo methods.

CONTENTS: Introduction: Deterministic and stochastic approaches, *G. G. Lowry.* Markov chains, *J. Myhre.* Monte Carlo methods, *M. Fluendy.* Polymer conformation as a Markov chain problem, *J. Kinsinger.* Polymer conformation and the excluded-volume problem, *S. Windwer.* Higher order Markov chains and statistical thermodynamics of linear polymers, *J. Mazur.* Copolymer composition and tacticity, *F. P. Price.* Molecular-weight distributions, *G. G. Lowry.*

McCULLOUGH *Concepts of Fiber-Resin Composites*

(Monographs and Textbooks in Material Science Series, Volume 2)

by R. L. McCullough, *Boeing Scientific Research Laboratories, Seattle, Washington*

128 pages, illustrated. 1971

Presents basic concepts of composite material systems. Introduces the study of composite systems by discussing where and how composite materials are used and how their components are selected. A valuable reference for materials scientists, and students and research management engaged in the exploration of composite materials.

CONTENTS: Materials • Composite structures • Composite properties • The interphase region • Synopsis.

MAY and TANAKA *Epoxy Resins: Chemistry and Technology*

edited by Clayton May, *Shell Development Company, Emeryville, California* and Yoshio Tanaka, *Research Institute for Polymers and Textiles, Yokohama, Japan*

704 pages, illustrated. 1973

Brings together the contributions of a number of outstanding researchers in the field of epoxy resins. Not only emphasizes the chemistry and technology of epoxy resins, but also deals with many industrial applications. Of great value for polymer chemists and technicians, and a wide variety of engineers.

CONTENTS: Introduction to epoxy resins, *C. May.* Synthesis and characteristics of epoxides, *Y. Tanaka, A. Okada, and I. Tomizuka.* Epoxide-curing reactions, *Y. Tanaka and T. Mika.* Curing agents and modifiers, *T. Mika.* Properties of cured resins, *D. Kaelble.* Epoxy-resin adhesives, *A. Lewis and R. Saxon.* Epoxy-resin coatings, *G. Somerville and I. Smith.* Electrical and electronic applications, *A. Breslau.* Epoxy laminates, *J. DelMonte.* Polymer stabilizers and plasticizers, *W. Port.* Analysis of epoxides and epoxy resins, *H. Jahn and P. Goetzky.* Toxicity, hazards, and safe handling, *H. Borgstedt and C. Hine.*

MILLICH and CARREHER *Interfacial Synthesis*

edited by Frank Millich, *Department of Chemistry, University of Missouri, Kansas City,* and Charles E. Carreher, Jr., *Department of Chemistry, University of South Dakota, Vermillion*

in preparation. 1973

Summarizes the accomplishments in interfacial synthesis to date. Speculates on mechanism, discusses the complex matter of synthetic control, and points out the beneficial aspects of interfacial synthesis in comparison to alternative methods of synthesis. Of fundamental concern to graduate students and teachers of organic chemistry, physical chemists, macromolecular biochemists, and polymer chemists who engage in interfacial synthesis.

CONTENTS: Stirring in organic chemical synthesis, *J. Rushton.* High-speed stirring in interfacial synthesis, *J. Rushton.* Problems and solutions in kinetics and mechanisms, *J. Bradbury and P. Crawford.* Interface effects on chemical reaction rate, *F. MacRitchie.* Liquid-vapor interfacial polycondensations, *L. Sokolov.* Copolycondensation and macroscopic kinetics, *L. Sokolov and V. Nikonov.* The role of the particle-water interface in polymerization, *J. Gardon.* Biochemical reactions at an interface, *R. Baier and D. Cadenhead.* Commercial application of interfacial synthesis, *E. Oliver and Y. Yen.* Polycarboxylic esters, *S. Temin.* Polycarbonates, *H. Vernaleken.* Polycondensations with carbon suboxide, *I. Daniewska.* Polyamides, *V. Nikonov and V. Savinov.* Polyesteramides, *I. Panayotov.* Polyurethanes, *T. Tanaka and T. Yokoyama.* Polyureas, *K. Stueben and A. Barnabeo.* Polyphosphonates, polyphosphates, and polyphosphites, *F. Millich, J. Teague, L. Lambing, and D. Hackathorn.* Other phosphorus containing polymers, *C. Carreher, Jr.* Organometallic polymers, *C. Carreher, Jr.* Modification of natural polymers by interfacial methods, *M, Horio.* Interfacial modifications of poly(vinyl alcohol) and related polymers, *M. Tsuda.* High temperature resistant polymers made by interfacial polymerization, *H. Mark and S. Atlas.*

MYERS and LONG *Characterization of Coatings: Physical Techniques*

In 2 Parts

(Treatise on Coatings Series, Volume 2)

edited by RAYMOND R. MYERS, *Paint Research Institute, Kent State University, Ohio,* and J. S. LONG, *Department of Chemistry, University of Southern Mississippi, Hattiesburg*

Part I   696 pages, illustrated. 1969

Part II in preparation. 1973

Explores the scientific frontier that has developed since the appearance of Mattiello's treatise on coatings. Emphasizes the urgent need of the paint industry to master new technological concepts and instrumental techniques to match the rapid pace of development of its products. Written for the working paint scientist, the laboratory assistant, technician, and superintendent. Also a valuable reference for the formulator and personnel engaged in the production of raw materials for the paint industry.

CONTENTS:

Part I: The intrinsic properties of polymers, *A. Tawn.* Surface areas, *D. Gans.* Adhesion of coatings, *A. Lewis and L. Forrestal.* Mechanical properties of coatings, *P. Pierce.* The ultimate tensile properties of paint films, *R. Evans.* Gas chromatography, *J. Haken.* Thermoanalytical techniques, *P. Garn.* Microscopy in coatings and coating ingredients, *W. Lind.* Radioactive isotopes, *G. Coe.* Infrared spectroscopy *C. Smith.* Ultraviolet and visible spectroscopy, *F. Spagnolo and E. Scheffer.* Color of polymers and pigmented systems, *G. Ingle.* Photoelastic coatings, *A. Blumstein.*

Part II: Dielectric properties, *S. Negami.* Gel permeation chromatography, *K. Boni.* Infrared Fourier transform spectroscopy, *M. Low.* Interfacial energetics, *D. Gans.* Nuclear magnetic resonance, *M. Levy.* Particle sizing, *B. DeWitt.* Scanning electron microscopy, *L. Princen.* Solubility, *J. Gordon and J. Teas.* Transport properties, *G. Park.* Viscometry, *K. Oesterle.* X-ray analysis, diffraction, and emission, *R. Scott.*

MYERS and LONG   *Film-Forming Compositions*

In 3 Parts

(Treatise on Coatings Series, Volume 1)

edited by RAYMOND R. MYERS, *Paint Research Institute, Kent State University, Ohio,* and J. S. LONG, *Department of Chemistry, University of Southern Mississippi, Hattiesburg*

Part I   584 pages, illustrated. 1967

Part II 448 pages, illustrated. 1968

Part III 608 pages, illustrated. 1972

Devoted to materials which form, or aid the formation of, continuous films. Discusses vehicles and resins, placing considerable emphasis on procedures for developing a suitable vehicle for conveying dissolved or suspended solids to a substrate and imparting to the surface those protective and decorative properties for which the coating was designed. Of inestimable value to chemists, formulators, laboratory assistants, technicians, and production superintendents of the coatings industry. Also valuable for laboratory personnel of raw material suppliers, chemists in the plastics and similar industries, and as an excellent reference treatise for libraries.

Part I: Acrylic ester emulsions and water-soluble resins, *G. Allyn.* Acrylic ester resins, *G. Allyn.* Alkyd resins, *W. M. Kraft, E. G. Janusz, and D. Sughrue.* Asphalt and asphalt coatings, *S. H. Alexander.* Chlorinated rubber, *H. E. Parker.* Driers, *W. J. Stewart.* Epoxy resin coatings, *G. R. Somerville.* Hydrocarbon resins and polymers, *D. F. Koenecke.* Hydrocarbon solvents, *W. W. Reynolds.* Natural resins, *C. L. Mantell.* Polyethers and polyesters, *A. C. Filson.* Urethane coatings, *A. Damusis and K. C. Frisch.* Vehicle manufacturing equipment, *A. F. Steioff.*

Part II: Styrene-butadiene latexes in protective and decorative coatings, *F. A. Miller.* Starch polymers and their use in paper coating, *T. F. Protzman and R. M. Powers.* Cellulose esters and ethers, *J. B. G. Lewin.* Drying oils — modifications and use, *A. E. Rheineck and R. O. Austin.* Paint and painting in art, *S. Rees Jones.* Rosin and modified rosins and resins, *C. L. Mantell.* Urea and melamine resins, *H. P. Wohnsiedler and W. L. Hensley.* Vinyl resins *W. H. McKnight and G. S. Peacock.* Vinyl emulsions, *H. D. Cogan and A. L. Mantz.*

Part III: Dimer acids in surface coatings, *J. Boylan.* Emulsion technology, *L. Princen.* Phenolic resins for coatings, *S. Richardson and W. Wertz.* Plasticization of coatings, *F. Ball.* Fatty polyamides and their applications in protective coatings, *D. Wheller and D. Peerman.* Polycarbonate resins, *D. Fox and K. Goldblum.* Reactive polyesters, *F. Ball.* Varnishes, *L. Montague.* Reactive silanes as adhesion promoters to hydrophilic surfaces, *E. Plueddemann.* Surface-active agents, *T. Ginsberg.* Shellac, *J. Martin.* Tall oil in surface coatings, *R. Perez.* Silicones in protective coatings, *L. Brown.*

MYERS and LONG *Formulations*

(Treatise on Coatings Series, Volume 4)

edited by RAYMOND R. MYERS, *Department of Chemistry, Kent State University, Ohio*, and J. S. LONG, *University of Southern Mississippi, Hattiesburg*

Part I in preparation. 1973

NEUSE and ROSENBERG
Metallocene Polymers

(Reviews in Macromolecular Chemistry Series, Volume 5, Part I)

by EBERHARD NEUSE, *F. J. Weck Company, City of Industry, California*, and HAROLD ROSENBERG, *Air Force Materials Laboratory, Wright-Patterson Air Force Base, Ohio*

170 pages, illustrated. 1970

Presents a comprehensive and critical account of the progress made in the synthesis and characterization of metallocene polymers.

CONTENTS: Introduction • Macromolecular compounds with pendent metallocenyl groups • Macromolecular compounds with intrachain metallocenylene groups • Conclusions.

O'CONNOR *Instrumental Analysis of Cotton Cellulose and Modified Cotton Cellulose*

(Fiber Science Series, Volume 3)

edited by ROBERT T. O'CONNOR, *Agricultural Research Service, U.S.D.A., New Orleans, Louisiana*

512 pages, illustrated. 1972

Describes the applications of instrumental procedures specifically developed by the textile chemist to meet today's demands. Particularly geared to textile chemists and others in the textile industry, and to paper and wood manufacturers. Also of interest to polymer chemists, analytical chemists, and other researchers involved with the processes by which fibers are blended and modified.

CONTENTS: Elemental analysis: Detection, identification, and quantitative determination of metals and nonmetallic elements, *R. T. O'Connor*. Infrared spectroscopy and physical properties of cellulose, *C. Y. Liang*. Light microscopy in the study of cellulose, *M. L. Rollins and I. V. de Gruy*. Electron microscopy of cellulose and cellulose derivatives, *M. L. Rollins, A. M. Cannizzaro, and W. R. Goynes*. Instrumental methods in the study of oxidation, degradation, and pyrolysis of cellulose, *P. K. Chatterjee and R. F. Schwenker, Jr.* X-ray diffraction, *V. W. Tripp and C. M. Conrad*. Wide-line nuclear magnetic resonance spectroscopy, *R. A. Pittman and V. W. Tripp*. The infrared spectra of chemically modified cotton cellulose, *R. T. O'Connor*.

PEARL *The Chemistry of Lignin*

by IRWIN A. PEARL, *The Institute of Paper Chemistry, Appleton, Wisconsin*

360 pages, illustrated. 1967

CONTENTS: Nebulous concept of lignin • Isolation of lignin • Chemical structure of lignin • Biosynthesis and formation of lignin • Reactions of lignin in major pulping and bleaching processes • Chemical reactions of lignin • Physical properties of lignin and its preparations • Biological decomposition of lignin • Thermal decomposition of lignin • Linkage of lignin in the plant • Utilization of lignin and its preparations.

PETERLIN *Plastic Deformation of Polymers*

edited by A. PETERLIN, *Research Triangle Institute, Research Triangle Park, North Carolina*

318 pages, illustrated. 1971

Explores the actual mechanism of deformation that occurs during the formation of fibers and films. Focuses primarily on three topics: what happens on a molecular, crystalline, and supercrystalline level; how the original structure influences the deformation; and what determines the useful mechanical properties of fibers and films. A valuable tool for the staff of research and development laboratories working with plastics, rubbers, fibers, and films, and for students and faculty involved in material science and polymer chemistry.

CONTENTS: Infrared studies of the role of monoclinic structure in the deformation of polyethylene, *Y. Kikuchi and S. Krimm*. Infrared studies of drawn polyethylene. Part I. Changes in orientation and conformation of highly drawn linear polyethylene, *W. Glenz and A. Peterlin*. Structure of oriented polyacrylonitrile films, *J. L. Koenig, L. E. Wolfram, and J. G. Grasselli*. Morphology and deformation behavior of "row-nucleated" polyoxymethylene film, *C. A. Garber and E. S. Clark*. Plastic deformation of polypropylene. VI. Mechanism and properties, *F. J. Baltá-Calleja and A. Peterlin*. Polyethylene crystallized under the orientation and pressure of a pressure capillary viscometer. Part I., *J. H. Southern and R. S. Porter*. Heat relaxation of drawn polyoxymethylene, *A. Siegmann and P. H. Geil*.

Retraction of cold-drawn polyethylene and polypropylene, *D. Hansen, W. F. Kracke, and J. R. Falender.* Electron spin resonance studies of free radicals in mechanically loaded nylon 66, *G. S. P. Verma and A. Peterlin.* Transition from linear to nonlinear viscoelastic behavior. Part I. Creep of polycarbonate, *I. V. Yannas and A. C. Lunn.* Yielding behavior of glassy polymers. III. Relative influences of free volume and kinetic energy, *K. C. Rusch and R. H. Beck, Jr.* Yielding of quenched and annealed polymethyl methacrylate, *D. H. Ender.* Yield phenomenon in oriented polyethylene terephthalate, *M. Parrish and N. Brown.* Electron paramagnetic resonance investigation of molecular bond rupture due to ozone in deformed rubber, *K. I. DeVries, E. R. Simonson, and M. L. Williams.* Factors affecting the depth of draw in a cold-forming operation, *H. L. Li, P. J. Koch, D. C. Prevorsek, and H. J. Oswald.* Quantitative structural characterization of the mechanical properties of isotactic polypropylene, *R. J. Samuels.*

RAVVE   Organic Chemistry of Macromolecules: *An Introductory Textbook*

by A. RAVVE, *Continental Can Company, Chicago, Illinois*

512 pages, illustrated. 1967

CONTENTS: **Part I: Introduction:** Historical introduction and definitions • Physical properties of macromolecules • Molecular weights of polymers • **Part II: Polymerization Reactions-Mechanisms:** Addition polymerization: Mechanism of free-radical polymerization • Ionic polymerization • Polymerization with the aid of complex catalysts • Stereospecific polymerization • Bulk, solution, suspension, and emulsion polymerization. **Part III: Common Addition Polymers:** Macroalkanes • Polymers and copolymers from dienes and polyenes • Styrene and styrene-like polymers and polyacrylics • Halogen-bearing addition polymers and vinyl esters and ethers. **Part IV: Condensation Polymers:** Mechanism of polycondensation reactions • polyesters • Polyamides • Polycarbamates, polyureas, and polycarbodiimides • Phenoplasts • Aminoplasts • Ladder and semiladder polymers. **Part V: Naturally Occurring Polymers:** Polysaccharides • Proteins • Polynucleotides. **Part VI: Reactions of Polymers:** Graft and block copolymers • Reactions of polymers • Degradation of polymers.

REICH and STIVALA   Autoxidation of Hydrocarbons and Polyolefins: *Kinetics and Mechanisms*

by LEO REICH, *Picatinny Arsenal, Dover, New Jersey,* and SALVATORE S. STIVALA, *Department of Chemistry, Stevens Institute of Technology, Hoboken, New Jersey*

544 pages, illustrated. 1969

CONTENTS: Introduction • Oxidation of simple hydrocarbons in absence of inhibitors and accelerators • Oxidation of simple hydrocarbons in presence of antioxidants • Oxidation of simple hydrocarbons in presence of metal catalysts • Weak chemiluminescence during hydrocarbon autoxidation • Qualitative aspects of autoxidation of saturated polyolefins • Quantitative aspects of autoxidation of saturated polyolefins • Investigation of polyolefin oxidation by various techniques.

REMBAUM and SHEN   *Biomedical Polymers*

edited by ALAN REMBAUM, *California Institute of Technology, Pasadena* and MITCHEL SHEN, *University of California, Berkeley*

304 pages, illustrated. 1971

A collection of papers given at the Symposium on Biomedical Polymers held in Pasadena in 1969. Of interest to scientists in polymer chemistry, polymer physics, biochemistry, bioengineering, materials science, pharmacology, and surgery.

CONTENTS: Problems in blood-tissue reactions to polymeric materials, *B. Zweifach.* Past, present, and future of artificial kidney treatment, *B. Barbour.* The chemistry and properties of the medical-grade silicones, *S. Braley.* Correlation of the surface charge characteristics of polymers with their antithrombogenic characteristics, *S. Srinivasan and P. Sawyer.* Persistent polarization in polymers and blood compatibility, *P. Murphy, A. Lacroix, S. Merchant, and W. Bernhard.* Selection, characterization, and biodegradation of surgical epoxies, *A. Cupples and R. Schubert.* Foreign body reactions to plastic implants, *D. Ocumpaugh and H. Lee.* Rapid *in vitro* screening of polymers for bio-compatibility, *C. Homsy, K. Ansevin, W. O'Bannon, S. Thompson, R. Hodge, and M. Estrella.* Improved membranes for hemodialysis, *F. Martin, H. Shuey, and C. Saltonstall, Jr.* Control of polymer morphology for biomedical applications, 1. Hydrophilic polycarbonate membranes for dialysis, *R. Kesting.* Surgical adhesives in ophthalmology, *M. Refojo.* Medical uses for polyelectrolyte complexes, *M. Vogel, R. Cross, H. Bixler, and R. Guzman.* Potentialities of a new class of anticlotting and antihemorrhagic polymers, *T. Yen, M. Daver, and A. Rembaum.* Synthesis and properties of a new class of potential biomedical polymers, *A. Rembaum, S. Yen, R. Landel, and M. Shen.* Recognition polymers, *D. Bradley.* The challenge for high polymers in medicine, surgery, and artificial internal organs, *H. Lee and K. Neville.*

RICHARDSON   Optical Microscopy for the Materials Sciences

(Monographs and Textbooks in Material Science Series, Volume 3)

by JAMES H. RICHARDSON, *Aerospace Corporation, El Segundo, California*

(continued)

RICHARDSON *(continued)*
704 pages, illustrated. 1971

Provides in one volume a comprehensive survey of the techniques for preparation and optical examination of specimens in the broad area of materials sciences. A highly useful text for the university or vocational school student studying the microstructure of materials or metallography and also a valuable practical reference for all researchers using microscopy, as well as for engineers and industrial metallographers.

CONTENTS: The Brightfield optical microscope • The microscopy of phase structures • Photomicrography • Photomacrography • Specimen preparation • Reagents and techniques for specimen preparation • Laboratory safety • Examination of the specimen • Accessories • Laboratory design.

ROGERS *Permselective Membranes*

edited by CHARLES E. ROGERS, *Case Western Reserve University, Cleveland, Ohio*
224 pages, illustrated. 1971

Encompasses a broad range of topics pertaining to the expanding field of permselective membranes, including theoretical aspects of transport behavior, new and unusual methods for the preparation or modification of membrane materials, and the effects of experimental conditions on the permselectivity of membranes to both ionic and nonionic penetrants. Of value to biophysicists, polymer chemists, physicists, biochemists, and chemical engineers, as well as biologists and physiologists.

CONTENTS: Transport of dissolved oxygen through silicone rubber membrane, *S. Hwang, T. Tang, and K. Kammermeyer.* Gas transport in segmented block copolymers, *K. Ziegel.* Transport of noble gases in poly(methyl acrylate), *W. Burgess, H. Hopfenberg, and V. Stannett.* Permeation of gases at high pressures, *S. Stern, S. Fang, and R. Jobbins.* Permeation of gases through modified polymer films III. Gas permeability and separation characteristics of gamma–irradiated Teflon FEP copolymer films, *R. Huang and P. Kanitz.* Theoretical interpretation of the effect of mixture composition on separation of liquids in polymers, *M. Fels and R. Huang.* Permselectivity of solutes in homogeneous water–swollen polymer membranes, *H. Yasuda and C. Lamaze.* Ion–exchange selectivity coefficients in the exchange of calcium, strontium, cobalt, nickel, zinc, and cadmium ions with hydrogen ion in variously cross-linked polystyrene sulfonate cation exchangers at 25°C, *M. Reddy and J. Marinsky.* Ion and water transport through permselective membranes, *N. Lakshminarayanaiah.* Permeability of cellulose acetate membranes to selected solutes, *H. Lonsdale, B. Cross, F. Graber, and C. Milstead.* Transport through permselective membranes, *C. Rogers and S. Sternberg.*

SCHEY *Metal Deformation Processes: Friction and Lubrication*
(Monographs and Textbooks in Material Science Series, Volume 1)

edited by JOHN A. SCHEY, *University of Illinois at Chicago Circle*
822 pages, illustrated. 1970

A comprehensive treatment of all aspects of friction and lubrication in metal deformation processes. Of aid to metallurgists, mechanical engineers, chemists, and physicists.

CONTENTS: Background and system of approach, *J. Schey.* Friction effects in metalworking processes, *J. Schey.* Friction, lubrication, and wear mechanisms, *C. Riesz.* Lubricants, *C. Riesz.* Lubricant properties and their measurements, *J. Schey.* Rolling lubrication, *J. Schey.* Wire drawing lubrication, *J. Newnham.* Hot extrusion lubrication, *S. Kalpakjian.* Forging lubrication, *S. Kalpakjian.* Cold forging and cold extrusion lubrication, *J. Newnham.* Sheet metal working lubrication, *J. Newnham.*

SEGAL *High-Temperature Polymers*

edited by CHARLES L. SEGAL, *North American Aviation, Inc., Canoga Park, California*
208 pages, illustrated. 1967

CONTENTS: Introduction, *C. L. Segal.* Thermally stable polymers with aromatic recurring units, *C. S. Marvel.* Inorganic polymer chemistry, *J. R. Van Wazer.* Kinetics and gaseous products of thermal decomposition of polymers, *H. L. Friedman.* Studies of stability of condensation polymers in oxygen-containing atmospheres, *R. T. Conley.* Thermal degradation of polymers. III: Mass spectrometric thermal analysis of condensation polymers, *G. P. Shulman.* Viscoelastic relaxation mechanism of inorganic polymers. V: Counterion effects in bulk polyelectrolytes, *A. Eisenberg, S. Saito, and T. Sasada.* Thermal stability of carborane-containing polymers, *J. Green and N. Mayes.* Synthesis and thermal stability of structurally related aromatic Schiff bases and acid amides, *A. D. Delman, A. A. Stein, and B. B. Simms.* New high-temperature polymers. II: Ordered aromatic copolyamides containing fused and multiple ring systems, *F. Dobinson and J. Preston.* Synthesis of fusible branched polyphenylenes, *N. Bilow and L. J. Miller.*

SEGAL, SHEN, and KELLEY
Polymers in Space Research

edited by CHARLES L. SEGAL, *Whittaker Corporation, San Diego,* MITCHEL SHEN, *University of California, Berkeley,* and FRANK N. KELLEY, *Air Force Rocket Propulsion Laboratory, Edwards, California.* Associate Editors: GEORGE F. PEZDIRTZ, *NASA Langley Research Center, Hampton, Virginia,* and W. DAVID ENGLISH, *Astropower Laboratory, McDonnell Douglas Aeronautics Company, Huntington Beach, California*

480 pages, illustrated. 1970

CONTENTS:

Part I: Recent Developments In the Synthesis, Characterization, and Evaluation of Thermally Stable Polymers

Introduction, *C. Segal and G. Pezdirtz.* Aromatic polymers: Single- and double-stranded chains, *J. Stille.* Thermally stable spiropolymers, *J. Hodgkin and J. Heller.* Isomeric and substituent effects in some dibenzoylbenzene-diamine polymers, *A. Volpe, L. Kaufman, and R. Dondero.* Arylsulfimide polymers. III. The syntheses of some monomeric aryl-1,2-disulfonic acids and derivatives, *G. D'Alelio, Y. Giza, and D. Feigl.* Properties of heterocyclic condensation polymers, *G. Berry and T. Fox.* Relative thermophysical properties of some polyimidazopyrrolones, *R. Jewell.* Thermal decomposition of polyimides in vacuum, *T. Johnston and C. Gaulin.* Thermomechanical behavior of an aromatic polysulfone, *J. Gillham, G. Pezdirtz, and L. Epps.* Panel discussion on thermally stable polymers, *C. Segal, J. Stille, G. Pezdirtz, G. D'Alelio, H. Levine, and W. Gibbs.*

Part II: Properties of Polymers at Low Temperatures

Introduction, *M. Shen and W. English.* Relaxation behavior of polymers at low temperatures, *J. Sauer and R. Saba.* Thermal properties of polymers at low temperatures, *W. Reese.* Multiple transitions in polyvinyl alkyl ethers at low temperatures, *W. Schell, R. Simha, and J. Aklonis.* Internal friction study of diluent effect on polymers at cryogenic temperatures, *M. Shen, J. Strong, and H. Schlein.* Stress-strain behavior of adhesives in a lap joint configuration at ambient and cryogenic temperatures, *G. Tiezzi and H. Doyle.* Some properties of nitroso rubbers in fluorine at ambient and cryogenic temperatures, *S. Toy, W. English, W. Crane, and M. Toy.* Cryogenic properties of a polyurethane adhesive, *R. Robbins.* Some effects of structure on a polymer's performance as a cryogenic adhesive, *R. Gosnall and H. Levine.*

Part III: Solid Propellents

Introduction, *F. Kelley.* Recent developments in solid-propellent binders, *H. Marsh, Jr.* Saturated hydrocarbon polymers for solid rocket propellents, *A. Di Milo and D. Johnson.* Preparation and curing of poly (perfluoroalkylene oxides), *J. Zollinger, J. Throckmorton, S. Ting, R. Mitsch, and D. Elrick.* Functionality and functionality distribution measurements of binder prepolymers, *A. Muenker and R. Hudson, Jr.*

SERAFINI and KOENIG
Cryogenic Properties of Polymers

edited by TITO T. SERAFINI, *NASA-Lewis Research Center, Cleveland,* and JACK L. KOENIG, *Case Western Reserve University, Cleveland, Ohio*

312 pages, illustrated. 1968

CONTENTS: Cryogenic positive expulsion bladders, *R. F. Lark.* Adhesives for cryogenic applications, *L. M. Roseland.* Glass-, boron-, and graphite-filament-wound resin composites and liners for cryogenic pressure vessels, *M. P. Hanson.* Mechanical behavior of poly(ethylene terephthalate), *I. M. Ward.* Effect of film processing on cryogenic properties of poly(ethylene terephthalate), *R. E. Eckert and T. T. Serafini.* Mechanical properties of epoxy resins and glass/epoxy composites at cryogenic temperatures, *L. M. Soffer and R. Molho.* Mechanical relaxation of poly-4-methyl-pentene-1 at cryogenic temperatures, *M. Takayanagi and N. Kawasaki.* Transitions in glasses at low temperatures, *R. A. Haldon, W. J. Schell, and R. Simha.* Mechanical behavior of poly(ethylene terephthalate) at cryogenic temperatures, *C. D. Armeniades, I. Kuriyama, J. M. Roe, and E. Baer.* Infrared studies of chain folding in polymers II. Poly(ethylene terephthalate), *J. L. Koenig and M. J. Hannon.* Crystallization of poly(ethylene terephthalate) from the glassy amorphous state, *G. S. Y. Yeh and P. H. Geil.* Strain-induced crystallization of poly(ethylene terephthalate), *G. S. Y. Yeh and P. H. Geil.* Molecular motion in polytetrafluoroethylene at cryogenic temperatures, *E. S. Clark.* Synthesis of ultrahigh molecular weight poly(ethylene terephthalate), *L.-C. Hsu.* Development of vulcanizable elastomers suitable for use in contact with liquid oxygen, *P. D. Schuman, E. C. Stump, and G. Westmoreland.* Synthesis of fluorinated polyurethanes, *R. Gosnell and J. Hollander.*

SKEIST *Reviews in Polymer Technology*

edited by IRVING SKEIST, *Skeist Laboratories, Inc., Livingston, New Jersey*

260 pages, illustrated. 1972

Consists of intensive, up-to-date reviews on various aspects of polymer technology. Directed to chemists, engineers, technicians, commercial planners, and others working in the polymers and plastics fields.

CONTENTS: Coupling agents as adhesion promoters, *P. Cassidy and W. Yager.* Processing powdered polyethylene, *A. Zimmerman.* Plastics and other polymers in building, *I. Skeist and J. Miron.* Fire retardance of polymeric ma-

(continued)

SKEIST *(continued)*

terials, *I. Einhorn*. Recent advances in photo-cross-linkable polymers, *G. Delzenne*. Organic colorants for polymers, *T. Reeve*.

SOLOMON *Step-Growth Polymerizations*

(Kinetics and Mechanisms of Polymerization Series, Volume 3)

edited by DAVID H. SOLOMON, *C.S.I.R.O., Melbourne, Australia*

416 pages, illustrated. 1972

Presents a critical and constructive assessment of developments in step-growth polymerization and considers the application of theoretical concepts to commercial systems. Highly recommended to students of polymer science and researchers working in the area of step-growth polymerization, including polymer chemists and other scientists in the paint, coatings, and plastics industries.

CONTENTS: Polyesterification, *D. H. Solomon*. Polyamides, *D. C. Jones and T. R. White*. Polyurethanes: The chemistry of the diisocyanate-diol reaction, *D. J. Lyman*. Cyclopolycondensation, *P. M. Hergenrother*. The reactions of formaldehyde with phenols, melamine, aniline, and urea, *M. F. Drum and J. R. LeBlanc*. Diels-Alder polymerization, *W. J. Bailey*. Inorganic polymers, *J. R. MacCallum*.

STEWART *Infrared Spectroscopy: Experimental Methods and Techniques*

by JAMES E. STEWART, *Durrum Instrument Corporation, Palo Alto, California*

656 pages, illustrated. 1970

A guide to instrumentation and experimental methods and techniques for the infrared spectroscopist. Primarily for those involved with spectroscopy research and for graduate students in chemistry and physics interested in spectroscopy.

CONTENTS: Infrared spectroscopy • The infrared spectrophotometer • Elements of geometric optics • Elements of physical optics • Optical components of infrared spectrophotometers • Optical systems of infrared spectrophotometers • Slit functions and spectral modulation transfer functions of monochromators • Interference spectroscopy • Mechanics of infrared spectrophotometers • Elements of electronics • Infrared detectors • Electronic systems of infrared spectrophotometers • Electromechanical transfer functions of infrared spectrophotometers • Photometric accuracy of infrared spectrophotometers • Experimental methods of infrared transmission spectroscopy • Experimental methods of infrared reflection spectroscopy • Experimental methods of infrared emission spectroscopy • Advanced methods of infrared spectroscopy.

SZEKELY *Blast Furnace Technology: Science and Practice*

edited by JULIAN SZEKELY, *State University of New York, Buffalo*

414 pages, illustrated. 1972

Represents the efforts of academic and industrial researchers, metallurgists, plant operators, and designers concerned with the most up-to-date aspects of ironmaking technology. Valuable reading for all production engineers, plant operators, and designers concerned with ironmaking technology, and also of importance to metallurgical engineers, research scientists—chemists, physicists, engineers—and students in this field.

CONTENTS: Single particle studies applied to direct reduction and blast furnace operations, *R. Bleifuss*. Structural effects in gas—solid reactions, *J. Szekely and J. Evans*. The use of catalysts to enhance the rate of Boudouard's reaction in direct reduction metallurgical processes, *Y. Rao and B. Jalan*. Thirty psi high top—gas pressure operation at NSC Nagoya works, *T. Yatsuzuka, Y. Yamada, and A. Tayama*. Practical application of mathematical models in ironmaking, *D. Christie, C. Kearton, and R. Thomas*. A mass—transport model of erosion of the carbon hearth of the iron blast furnace, *J. Elliott and J. Popper*. Contribution to the study of the reaction mechanism occurring in high temperature zone of the blast furnace, *R. Vidal and A. Poos*. The place of direct reduction in a modern blast furnace—BOF plant, *J. Peart and D. George*. The blast furnace control problem, *J. A. Laslo*. Modern blast furnace design in Germany, *F. Lenger*. The blast furnace—a transition, *F. Berczynski*. Projected performance of a blast furnace with prereduced burdens, *J. Agarwal*.

SZEKELY *The Steel Industry and the Environment*

edited by JULIAN SZEKELY, *Center for Process Metallurgy, State University of New York, Buffalo*

312 pages, illustrated. 1973

Contains the proceedings of the Second C. C. Furnas Memorial Conference on The Steel Industry and The Environment held at the State University of New York at Buffalo in November, 1971. Brings together authors representing different viewpoints on the questions raised in examining the interaction of the steel industry and the environment. Of utmost importance to plant operators, designers, metallurgists, environmental scientists, and all others concerned with the impact of the steel industry on the environment.

CONTENTS: The role of the federal government in environmental pollution control, *K. Johnson.* Control of air pollution in the British iron and steel industry, *F. Ireland.* Health and the steel industry environment, *K. Spring.* The economic impact of the installation and operation of pollution abatement devices, *J. Barker.* Desulfurization of coke oven gas: Technology, economics, and regulatory activity, *R. Dunlap, W. Gorr, and M. Massey.* Treatment of cold-mill wastewaters by ultrahigh-rate filtration, *C. Symons.* Experience with pollution abatement, *C. Black and W. Sebesta.* The interaction of the socioeconomic and ecological environment in American steel mill towns, *L. Thaxton and R. Genton.* Plant availability versus clean air: An economic dilemma that can be solved, *R. Heller.* A survey of wastewater treatment techniques for steel mill effluents, *T. Centi.* Emission of sulfurous gases from blast-furnace slags, *R. Kaplan and G. Ringstorff.* Treatment of waste gases from the basic oxygen furnace in West Germany, *E. Weber.* On the oxidation of cyanides in the stack region of the blast furnace, *H. Sohn and J. Szekely.* Reclaimed scrap and solid metallics for steelmaking, *J. Elliott.*

TALLAN *Electrical Conductivity in Ceramics and Glass*

(Ceramics and Glass: Science and Technology Series, Volume 4)

edited by NORMAN M. TALLAN, *Aerospace Research Laboratories, Wright-Patterson Air Force Base, Ohio*

in preparation. 1973

A text which thoroughly describes several aspects of the electrical conductivity of ceramic materials, and discusses their conductivity, physical dependence on their electronic and ionic defect structures, and the transport mechanisms by which charge and mass move through ceramic materials. Additionally stressed is the use of conductivity measurements to characterize the defect structure and transport properties of ceramics. A great aid to advanced students in materials science, ceramics, and glass.

CONTENTS: General concepts of electrical transport, *D. Adler.* Experimental techniques, *R. Blumenthal and M. Seitz.* Defect structure of ceramic materials, *R. Brook.* Electronic conduction mechanisms, *I. Bransky and J. Wimmer.* Controlled valency effects in electronic conductors, *J. Wagner.* Highly conducting ceramics and the conductor-insulator transition, *J. Honig and R. Vest.* Ionic conductivity and electrochemistry of crystalline ceramics, *J. Patterson.* Conductivity of glass and other amorphous materials, *J. Mackenzie.* Microstructural and polyphase effects, *J. Wimmer and H. Graham.*

TSURUTA and O'DRISCOLL *Structure and Mechanism in Vinyl Polymerization*

edited by TEIJI TSURUTA, *Department of Synthetic Chemistry, Faculty of Engineering, University of Tokyo,* and KENNETH F. O'DRISCOLL, *Department of Chemical Engineering, State University of New York, Buffalo*

552 pages, illustrated. 1969

Presents a general survey of studies on this subject in terms of physical organic chemistry. Topics are organized to focus on the most important chemical features of vinyl compounds and their response to variations in chemical and physical circumstances.

CONTENTS: Historical development of the theory of the reactivity of vinyl monomers, *M. Imoto.* Structure and reactivity of vinyl monomers, *T. Tsuruta.* Initiation in free radical polymerization, *K. F. O'Driscoll and P. Ghosh.* Termination mechanism in radical polymerization, *A. M. North and D. Postlethwaite.* Organometallic compounds as radical-type initiators for vinyl polymerization, *S. Inoue.* Heterogeneous metal peroxides, *T. Otsu.* Polymerization of α, β-disubstituted olefins, *Y. Minoura.* Polymerization of α, β-unsaturated carbonyl compounds, *D. M. Wiles.* Cationic polymerization of vinyl monomers by metal alkyl catalysts, *T. Saegusa.* Rate constants of elementary reactions in cationic polymerization, *T. Higashimura.* Elementary steps in anionic vinyl polymerization, *J. Smid.* Molecular rearrangements in polymerization of vinyl monomers, *A. D. Ketley and L. P. Fisher.*

VOGL *Polyaldehydes*

edited by OTTO VOGL, *Central Research Division, E. I. du Pont de Nemours & Company, Wilmington, Delaware*

152 pages, illustrated. 1967

CONTENTS: Preface, *O. Vogl.* Polyaldehydes: introduction and brief history, *O. Vogl.* Polymerization of formaldehyde, *N. Brown.* Polymerization and copolymerization of trioxane, *M. B. Price and F. B. McAndrew.* Polymerization of aliphatic aldehydes, *O. Vogl.* Polymers of haloaldehydes, *I. Rosen.* NMR studies of polyaldehydes, *E. G. Brame, Jr. and O. Vogl.* Polymerization of fluorothiocarbonyl compounds, *W. H. Sharkey.* Crystal structure of polyaldehydes, *P. Corradini.* Morphology of polyoxymethylene, *P. H. Geil.*

VOGL and FURUKAWA *Polymerization of Heterocyclics*

edited by OTTO VOGL, *Department of Polymer Science and Engineering, Univer-*

(continued)

VOGL and FURUKAWA *(continued)*

sity of Massachusetts, Amherst, and
JUNJI FURUKAWA, *Kyoto University,
Japan*

216 pages, illustrated. 1973

Reviews the polymerization of cyclic
ethers and thio ethers, lactones, and lactams.
Also covers preparation, polymerization, and
properties of perfluoro epoxides, kinetics of
cyclic ether polymerization, and the influence of ring strain on the rate of polymerization and living polymers based on cationic ring opening polymerization. Valuable
to scientists interested in polymer science,
heterocyclic chemistry, polymer engineering, and materials science.

CONTENTS: Introduction, *J. Furukawa.* Polymerization of cyclic ethers, *T. Saegusa.* Polymerization of perfluoro epoxides, *H. Eleuterio.*
Specific nature of the polymerization of hetero-
cyclics, *N. Enikolpoyan.* New trioxane copolymers, *H. Cherdron.* Alkylene sulfide polymerizations, *F. Lautenschlaeger.* Lactone polymerization and polymer properties, *G. Brode and J.
Koleske.* Lactam polymerization, *J. Sebenda.*

WALKER and THROWER *Chemistry
and Physics of Carbon:* **A Series
of Advances**

a series edited by PHILIP L. WALKER and
PETER A. THROWER, *Department of Material Sciences, Pennsylvania State University, University Park*

Vol. 1 400 pages, illustrated. 1965
Vol. 2 400 pages, illustrated. 1966
Vol. 3 464 pages, illustrated. 1968
Vol. 4 416 pages, illustrated. 1968
Vol. 5 400 pages, illustrated. 1969
Vol. 6 368 pages, illustrated. 1970
Vol. 7 424 pages, illustrated. 1970
Vol. 8 480 pages, illustrated. 1973
Vol. 9 272 pages, illustrated. 1973
Vol. 10 288 pages, illustrated. 1973
Vol. 11 in preparation. 1974

CONTENTS:

Volume 1: Dislocations and stacking faults in
graphite, *S. Amelinckx, P. Delavignette, and
M. Heerschap.* Gaseous mass transport within
graphite, *G. F. Hewitt.* Microscopic studies of
graphite oxidation, *J. M. Thomas.* Reactions of
carbon with carbon dioxide and steam, *S. Ergun
and M. Mentser.* Formation of carbon from
gases, *H. B. Palmer and C. F. Cullis.* Oxygen
chemisorption effects on graphite thermoelectric
power, *P. L. Walker, Jr., L. G. Austin, and
J. J. Tietjen.*

Volume 2: Electron microscopy of reactivity
changes near lattice defects in graphite, *G. R.
Hennig.* Porous structure and adsorption properties of active carbons, *M. M. Dubinin.* Radiation damage in graphite, *W. N. Reynolds.*
Adsorption from solution by graphite surfaces,
A. C. Zettlemoyer and K. S. Narayan. Electronic transport in pyrolytic graphite and boron
alloys of pyrolytic graphite, *C. A. Klein.* Acti-
vated diffusion of gases in molecular-sieve
materials, *P. L. Walker, Jr., L. G. Austin and
S. P. Nandi.*

Volume 3: Nonbasal dislocations in graphite, *J.
M. Thomas and C. Roscoe.* Optical studies of
carbon, *S. Ergun.* Action of oxygen and carbon
dioxide above 100 millibars on "pure" carbon,
F. M. Lang and P. Magnier. X-ray studies of
carbon, *S. Ergun.* Carbon transport studies for
helium-cooled high-temperature nuclear reactors, *M. R. Everett, D. V. Kinsey, and E.
Römberg.*

Volume 4: X-ray diffraction studies on carbon
and graphite, *W. Ruland.* Vaporization of carbon, *H. B. Palmer and M. Shelef.* Growth of
graphite crystals from solution, *S. B. Auster-
man.* Internal friction studies on graphite, *T.
Tsuzuku and M. H. Saito.* Formation of some
graphitizing carbons, *J. D. Brooks and G. H.
Taylor.* Catalysis of carbon gasification, *P. L.
Walker, Jr., M. Shelef, and R. A. Anderson.*

Volume 5: Deposition, structure and properties
of pyrolytic carbon, *J. C. Bokros.* The thermal
conductivity of graphite, *B. T. Kelly.* The study
of defects in graphite by transmission electron
microscopy, *P. A. Thrower.* Intercalation isotherms on natural and pyrolytic graphite, *J. G.
Hooley.*

Volume 6: Physical adsorption of gases and
vapors of graphitized carbon blacks, *N. N.
Avgul and A. V. Kiseleyv.* Graphitization of soft
carbons, *J. Maire and J. Méring.* Surface complexes on carbons, *B. R. Puri.* Effects of reactor
irradiation on the dynamic mechanical behavior
of graphites and carbons, *R. E. Taylor and D.
E. Kline.*

Volume 7: The kinetics and mechanism of
graphitization, *D. B. Fischbach.* The kinetics of
graphitization, *A. Pacault.* Electronic properties of doped carbons, *A. M. Marchand.* Positive and negative magnetoresistances in carbons,
P. Delhaes. The chemistry of the pyrolytic conversion of organic compounds to carbon, *E.
Fitzer, K. Mueller and W. Schaefer.*

Volume 8: The electronic properties of graphite,
I. Spain. Surface properties of carbon fibers,
D. McKee and V. Mimeault. The behavior of
fission products captured in graphite by nuclear
recoil, *S. Yajima.*

Volume 9: Carbon fibers from rayon presursors,
R. Bacon. Control of structure of carbon for
use in bioengineering, *J. Bokros, L. LaGrange,
and F. Schoen.* Deposition of pyrolytic carbon
in porous solids, *W. Kotlensky.*

Volume 10: The thermal properties of graphite,
B. Kelly and R. Taylor. Lamellar reactions in
graphitizable carbons, *M. Robert, M. Oberlin,
and J. Mering.* Methods and mechanisms of
growth of synthetic diamond, *F. Bundy, H.
Strong, and R. Wentorff, Jr.*

† *Volume edited by Philip L. Walker*

Volume 11: Structure and physical properties of carbon fibers, *W. Reynolds*. Highly oriented pyrolytic graphite, *A. Moore*. Evaporated carbon films, *I. McLintock and J. Orr*. Deformation mechanisms in carbons, *G. Jenkins*.

WARD   *Chemical Modification of Papermaking Fibers*

(Fiber Science Series, Volume 4)
by KYLE WARD, JR., *Institute of Paper Chemistry, Appleton, Wisconsin*
256 pages, illustrated. 1973

Bridges the gap between research and industrial applications in the field of chemical modification of papermaking fibers. Deals with the chemical changes which produce new or improved properties in paper products. Of particular importance to researchers and technologists in the paper, textile, and related industries, and students of polymer and organic chemistry.

CONTENTS: Introduction • Esterification • Etherification • Oxidation • Crosslinking • Graft polymerization onto cellulose.

WASLEY   *Stress Wave Propagation in Solids: An Introduction*

(Monographs and Textbooks in Material Science Series, Volume 5)
by RICHARD J. WASLEY, *Department of Chemistry, University of California, Livermore*
344 pages, illustrated. 1973

Provides the fundamentals necessary for the study of the propagation of short duration, high–intensity, nonelastic, mechanical stress disturbances in solids. The first part of the book treats some of the dynamic analyses of elastic solid media which obey Hooke's law. The last section discusses some of the theoretical and experimental aspects of one–dimensional stress waves and shock loading and response. For scientists and engineers desiring further knowledge in the field of stress wave propagation. Also of interest to advanced undergraduate and graduate students of engineering and physics.

CONTENTS: Elasticity: Quasistatic and dynamic response • Wave propagation in extended media • Wave propagation in semi-extended media: Reflection and refraction • Wave propagation in circular cylindrical rods • Selected applications of concepts of elasticity • Nonelastic material behavior • One–dimensional stress wave investigations • Nonelastic (shock) one–dimensional strain wave investigations.

WEINBERG   *Tools and Techniques in Physical Metallurgy*

In 2 Volumes
edited by FRED WEINBERG, *University of British Columbia, Vancouver*
Vol. 1   416 pages, illustrated. 1970
Vol. 2   376 pages, illustrated. 1970

Aids the non-specialist in understanding and making use of the new instruments and techniques of physical metallurgy.

CONTENTS:
Volume 1: Temperature measurement, *R. Bedford, T. Dauphinee, and H. Preston-Thomas*. X-ray diffraction, *C. M. Mitchell*. Crystal growth and alloy preparation, *F. Weinberg and J. T. Jubb*. Quantitative metallography, *J. R. Blank and T. Gladman*. Metallography, *H. E. Knechtel, W. F. Kindle, J. L. McCall, and R. D. Buchheit*.
Volume 2: Electron microscopy, *E. Smith*. Scanning electron microscopy, *O. Schaaber*. Field-ion microscopy, *B. Ralph*. Thermionic-emission microscopy, *W. L. Grube and S. R. Rouze*. Electron-probe microanalysis, *L. C. Brown and H. Thresh*. Emission spectrography and atomic absorption spectrophotometry, *G. L. Mason*.

WILSON   *Radiation Chemistry of Monomers-Polymers-Plastics*

by JOSEPH E. WILSON, *Department of Chemistry, Bishop College, Dallas, Texas*
in preparation. 1974

Provides an up-to-date survey of the radiation chemistry of monomers, polymers, and plastics. Gives essential information on radiation properties, measurement, and detection, and the primary chemical results of the interaction of radiation with matter. Of particular interest to polymer and radiation chemists.

CONTENTS: (tentative): Types and sources of radiation • Fundamental effects of the irradiation of matter • Short-term chemical effects of radiation absorption • Radiation chemistry of small molecules • Radiolytic polymerization in homogeneous systems • Radiolytic polymerization in the solid state • Radiation-induced polymerization in thermosetting, polyester, and emulsion systems • Irradiation of polymers: Crosslinking versus scission • Radiolytic grafting of monomers on polymeric films • Radiolytic grafting on fibers.

YOCUM and NYQUIST   *Functional Monomers: Their Preparation, Polymerization, and Applications*

In 2 Volumes
edited by RONALD H. YOCUM, *The Dow Chemical Company, Freeport, Texas,* and

(continued)

YOCUM and NYQUIST *(continued)*

EDWIN B. NYQUIST, *The Dow Chemical Company, Midland, Michigan*

Vol. 1 712 pages, illustrated. 1973
Vol. 2 321 pages, illustrated. 1973

A practical reference work which deals with functional monomers. Presents a broad technical background on the preparation and polymerization of individual functional monomers and their applications to various areas of industry. Of special interest to both academic and industrial chemists, particularly those working in the paint, coatings, and textile industry.

CONTENTS:

Volume 1: Acrylamide and other alpha, beta, and unsaturated acids, *D. C. MacWilliams.* Reactive halogenated monomers, *C. F. Raley and R. J. Dólinski.* Hydroxy monomers, *E. B. Nyquist.* Sulfonic acids and sulfonate monomers, *D. A. Kangas.*

Volume 2: Reactive heterocyclic monomers, *D. Tomalia.* Acidic monomers, *L. Luskin.* Basic monomers, vinylpyridines and aminoalkyl (METH) acrylates, *L. Luskin.*

ZIEF *Purification of Inorganic and Organic Materials: Techniques of Fractional Solidification*

edited by MORRIS ZIEF, *J. T. Baker Chemical Co., Phillipsburg, New Jersey*

340 pages, illustrated. 1969

Of interest to the chemist, chemical engineer, and metallurgist.

CONTENTS: Analysis of ultrapure materials, *C. L. Grant.* Optical-emission spectrochemical analysis—arc, spark, and flame, *C. L. Grant.* Spark-source mass spectrography, *P. R. Kennicott.* Atomic-absorption spectroscopy, *J. W. Robinson.* Infrared spectrophotometry, *K. E. Stine and W. F. Ulrich.* Gas-liquid chromatography, *R. A. Keller.* Differential thermal analysis and differential scanning calorimetry, *E. M. Barrall, II, and J. F. Johnson.* Electrical resistance-ratio measurement, *G. T. Murray.* Reduction of cyclohexane content of benzene under steady flow conditions, *J. D. Henry, Jr., M. D. Danyi, and J. E. Powers.* Purification of aromatic amines, *B. Pouyet.* The freezing staircase method, *C. P. Saylor.* Purification of aluminum, *J. L. Dewey.* Concentration of humic acids in natural waters, *J. Shapiro.* Fractionation of polystyrene, *J. D. Loconti.* Purification and growth of large anthracene crystals, *J. N. Sherwood.* Purification of indium antimonide, *A. R. Murray.* Purification of alkaline iodides (KI, RbI, CsI), *D. Ecklin.* Zone melting of metal chelate systems, *K. Ueno, H. Kobayashi, and H. Kaneko.* Purification of dienes, *R. Kieffer.* Purification of kilogram quantities of an organic compound, *J. C. Maire and M. Delmas.* Rapid purification of organic substances, *M. J. van Essen, P. F. J. van der Most, and W. M. Smit.* Investigation of zone-melting purification of gallium trichloride by a radiotracer method, *W. Kern.* Purification of potassium chloride by radio-frequency heating, *R. Warren.* Purification of a metal by electron-beam heating, *R. E. Reed and J. C. Wilson.* Heating by hollow-cathode gas discharge, *W. Class.* Continuous zone refining of benzoic acid, *J. K. Kennedy and G. H. Moates.* Purification of naphthalene in a centrifugal field, *E. L. Anderson.* Zone-melting chromatography of organic mixtures, *H. Plancher, T. E. Cogswell and D. R. Latham.* The concentration of flavors at low temperature, *M. T. Huckle.* Containers for pure substances, *E. C. Kuehner and D. H. Freeman.*

ZIEF and SPEIGHTS *Ultrapurity: Methods and Techniques*

edited by MORRIS ZIEF, *J. T. Baker Chemical Co., Philipsburg, New Jersey,* and ROBERT M. SPEIGHTS, *American Metal Climax, Inc., Golden, Colorado*

720 pages, illustrated. 1972

Brings together for the first time the four essential and interrelated parameters of ultrapurity: preparation, handling, containment, and analysis. Reflects the continuing progress in the preparation of ultrapure chemicals, the explosive growth in developments pertaining to the handling and containment of these materials, as well as the necessity for complete analysis.

Directed to all those working in research, development, or analysis of ultrapure products.

CONTENTS: Purification of alkali halides, *F. Rosenberger.* Purification of organic solvents by frontal-analysis chromatography, *H. Engelhardt.* The preparation of pure sodium and potassium, *R. L. McKisson.* Sublimation of phosphorus pentoxide, *R. D. Mounts.* Purification of proteins by membrane ultrafiltration, *G. J. Fallick.* The purification of p-xylene by partial freezing, *J. R. Gruden and M. Zief.* Purification of isopropylbenzene by preparative gas-liquid chromatography, *J. R. Gruden and M. Zief.* The preparation of ultrapure chemicals by fractional distillation, *H. Plancher and W. E. Haines.* Purification by dry-column chromatography, *F. M. Rabel.* Preparation of ultrapure water, *V. C. Smith.* Preparation and characterization of cholesterol, *I. L. Shapiro.* Contamination problems in trace-element analysis and ultrapurification, *D. E. Robertson.* Airborne contamination, *J. A. Paulhamus.* Glass containers for ultrapure solutions, *P. B. Adams.* Vitreous silica, *G. Hetherington and L. W. Bell.* Ceramics, *C. Garnsworthy.* High-purity chemicals—a challenge to practical analysis, *A. J. Barnard, Jr.* Emission spectroscopy, *E. C. Snooks.* Flame spectrophotometric trace analysis, *D. C. Burrell.* Neutron-activation analysis, *J. J. Kelly.* Visible spectrophotometry, *R. H.*

Weiss. Coulometric titration, *G. W. Higgins.* Information sources for ultrapurification and characterization, *T. E. Connolly.*

ZIEF and WILCOX *Fractional Solidification*

edited by MORRIS ZIEF, *J. T. Baker Chemical Company, Phillipsburg, New Jersey,* and WILLIAM R. WILCOX, *Aerospace Corporation, Los Angeles*

736 pages, illustrated. 1967

CONTENTS: Introduction, *W. R. Wilcox.* **Part I: Basic Principles:** Phase diagrams, *G. M. Wolten and W. R. Wilcox.* Mass transfer in fractional solidification, *W. R. Wilcox.* Constitutional supercooling and microsegregation, *G. A. Chadwick.* Polyphase solidification, *G. A. Chadwick.* Heat transfer in fractional solidification, *W. R. Wilcox.* **Part II: Laboratory Scale Apparatus:** Laboratory scale apparatus, *E. A. Wynne and M. Zief.* Batch zone melting, *E. A. Wynne.* Progressive freezing, *D. Richman, E. A. Wynne, and F. D. Rosi.* Continuous-zone melting, *J. K. Kennedy and G. H. Moates.* Column crystallization, *R. Albertins, W. C. Gates, and J. E. Powers.* Zone precipitation and allied techniques, *I. A. Eldib.* **Part III: Industrial Scale Equipment:** Proabd refiner, *J. G. D. Molinari.* Newton Chambers' process, *J. G. D. Molinari.* Rotary-drum techniques, *J. C. Chaty.* Phillips fractional-solidification process, *D. L. McKay.* Desalination by freezing, *J. C. Orcutt.* **Part IV: Applications:** Ultrapurification, *P. Jannke, J. K. Kennedy, and G. H. Moates.* Ultrapurity in pharamaceuticals, *P. Jannke.* Ultrapurity in electronic materials, *J. K. Kennedy and G. H. Moates.* Ultrapurity in materials research, *J. K. Kennedy, G. H. Moates, and W. R. Wilcox.* Ultrapurity in crystal growth, *G. H. Moates and J. K. Kennedy.* Bulk purification, *J. D. Loconti.* Analytical applications of fractional solidification, *A. S. Yue.* Materials preparation, *D. Richman and F. D. Rosi.* **Part V: Economics:** Economics of fractional solidification, *J. C. Chaty and W. R. Wilcox.* **Part VI: Appendix:** Introduction, *M. Zief and C. E. Shoemaker.* Survey of inorganic materials, *C. E. Shoemaker and R. L. Smith.* Survey of organic materials, *M. Zief.*

———— OTHER BOOKS OF INTEREST————

CUTLER and DAVIS *Detergency: Theory and Test Methods*

In 2 Parts

(Surfactant Science Series, Volume 5)

edited by W. G. CUTLER, and R. DAVIS, *Whirlpool Corporation, Benton Harbor, Michigan*

Part 1 464 pages, illustrated. 1972
Part 2 in preparation. 1973

JUNGERMANN *Cationic Surfactants*

(Surfactant Science Series, Volume 4)

edited by ERIC JUNGERMANN, *Armour-Dial, Inc., Chicago*

672 pages, illustrated. 1970

LINFIELD *Anionic Surfactants*

(Surfactant Science Series)

edited by WARNER M. LINFIELD, *U.S. Department of Agriculture, Philadelphia, Pennsylvania*

in preparation. 1974

MATTSON and MARK *Activated Carbon: Surface Chemistry and Adsorption from Solution*

by JAMES S. MATTSON, *Rosenstiel School of Marine and Atmospheric Sciences, University of Miami, Florida,* and HARRY B. MARK, JR., *Department of Chemistry, University of Cincinnati, Ohio*

248 pages, illustrated. 1971

PATRICK *Treatise on Adhesion and Adhesives*

edited by ROBERT L. PATRICN, *Alpha Research and Development, Inc., Blue Island, Illinois*

Vol. 1 *Theory*
496 pages, illustrated. 1967
Vol. 2 *Materials*
568 pages, illustrated. 1969
Vol. 3 *Special Topics*
264 pages, illustrated. 1973

SCHICK *Nonionic Surfactants*

(Surfactant Science Series, Volume 1)

edited by MARTIN J. SCHICK, *Central Research Laboratories, Interchemical Corporation, Clifton, New Jersey*

1,120 pages, illustrated. 1967

SHINODA *Solvent Properties of Surfactant Solutions*

(Surfactant Science Series. Volume 2)

edited by KOZO SHINODA, *Department of Chemistry, Yokohama National University, Japan*

376 pages, illustrated. 1967

SLADE and JENKINS
Thermal Analysis

(Techniques and Methods of Polymer Evaluation Series, Volume 1)

edited by PHILIP E. SLADE, JR., *Monsanto Company, Pensacola, Florida,* and LLOYD T. JENKINS, *Chemstrand Research Center, Durham, North Carolina*

264 pages, illustrated. 1966

SLADE and JENKINS
Thermal Characterization Techniques

(Techniques and Methods of Polymer Evaluation Series, Volume 2)

edited by PHILIP E. SLADE, JR., *Monsanto Company, Pensacola, Florida* and LLOYD T. JENKINS, *Chemstrand Research Center, Durham, North Carolina*

384 pages, illustrated. 1970

STEVENS *Characterization and Analysis of Polymers by Gas Chromatography*

(Techniques and Methods of Polymer Evaluation Series, Volume 3)

by MALCOLM P. STEVENS, *American University of Beirut, Lebanon*

216 pages, illustrated. 1969

SWISHER *Surfactant Biodegradation*

(Surfactant Science Series, Volume 3)

by R. D. SWISHER, *Monsanto Company, St. Louis, Missouri*

520 pages, illustrated. 1970

WALTON *Radome Engineering Handbook: Design and Principles*

(Ceramics and Glass: Science and Technology Series, Volume 1)

edited by JESSE D. WALTON, JR., *Georgia Institute of Technology, Atlanta*

616 pages, illustrated. 1970

JOURNALS OF INTEREST

BIOMATERIALS, MEDICAL DEVICES, AND ARTIFICIAL ORGANS
An International Journal

editor: T. F. YEN, *University of Southern California, Los Angeles*

The aim of this new international journal is to bridge the gap between the theoretical aspects and practical applications of artificial organs and other medical devices, and implantation materials. The basic principles responsible for the success of artificial organs are stressed in order to encourage new research in this field.

4 issues per volume

JOURNAL OF MACROMOLECULAR SCIENCE—Chemistry

editor: GEORGE E. HAM, *White Plains, New York*

This international journal provides scientists with a cross-section of the outstanding contributions from laboratories around the world—published four and one-half months from publisher's receipt of last manuscript. The fields covered include anionic, cationic, and free-radical addition polymerization and copolymerization, the manifold forms of condensation polymerization, polymer reactions, molecular weight studies, temperature-dependent properties, rheology, effects of radiation of all forms, polymer degradation, and many others.

8 issues per volume

JOURNAL OF MACROMOLECULAR SCIENCE—*Physics*

editor: PHILLIP H. GEIL, *Case Western Reserve University*

A periodical devoted to the publication of significant fundamental contributions concerning the physics of macromolecular solids and liquids. Papers deal with research in transition mechanisms and structure property relationships, the physics of polymer solutions and melts, glassy and rubbery amorphous solids, and individual polymer molecules and natural polymers, as well as all the areas generally contained in polymer state physics.

4 issues per volume

JOURNAL OF MACROMOLECULAR SCIENCE—*Reviews in* Macromolecular Chemistry

editors: GEORGE B. BUTLER, *University of Florida, Gainesville,* and KENNETH F. O'DRISCOLL, *State University of New York, Buffalo,* and MITCHELL SHEN, *University of California, Berkeley*

Topics in this journal are reviews of certain recent chronological periods, and also have the advantage of reflecting the authors' knowledge, interpretation, and concise summary of the state of knowledge in the given area. Because of the nature of the journal, and the short time between completion of a manuscript and its publication, reviews of this nature more closely approximate a current review than could otherwise be accomplished.

2 issues per volume

POLYMER-PLASTICS TECHNOLOGY AND ENGINEERING

editor: LOUIS NATURMAN, *Stamford, Connecticut*

This journal reflects the increasing importance that polymer applications, processing developments, and mass production of new polymer products will have in the coming years. The emphasis of the articles that comprise the journal will also consider plastics technology and engineering as an important new feature of this publication.

2 issues per volume

Examination On-Approval Policy

Our policy allows instructors to examine a particular book for a period of two months without charge. In the event that the book is definitely adopted for a course as a class text, the instructor may retain the copy as *his desk copy,* provided he advises us of the adoption and the number of students enrolled in the class. If, however, the book will not be used as a class text, the instructor may return it or send us his remittance, less the educational discount.

JOURNAL SUBSCRIPTION INFORMATION

Subscriptions are entered on a calendar year basis. When a subscription is entered, it entitles the subscriber to all issues in the particular volume.

All journal subscriptions are processed after payment has been received. Only prepaid orders will receive service.

Cancellations requested for other than publisher's error will be accepted only prior to the publication of the first issue of each current volume, and will be subject to a handling charge.

Please add **foreign postage** for delivery to all countries outside the U.S. and Canada.

Air mail postage is available upon request.

Indexes to journals are bound into the last issue of each volume with the exception of review journals which are not indexed.

Back volume information and prices are available upon request.

A complimentary copy of any journal is available upon request.

Date________________________

MARCEL DEKKER, INC.
95 Madison Avenue
New York, New York 10016

Please send me a complimentary copy of the following journal(s):

____

____

____

Name____

Position____

Company____

Address____

City_______________________State_____________Zip__________